Innovations in Construction Industry towards Sustainable Future: Implementation, Assessment and Opportunities

Innovations in Construction Industry towards Sustainable Future: Implementation, Assessment and Opportunities

Guest Editors

Malindu Sandanayake
Zora Vrcelj

MDPI

Basel • Beijing • Wuhan • Barcelona • Belgrade • Novi Sad • Cluj • Manchester

Guest Editors

Malindu Sandanayake
Built Environment and Engineering
Victoria University
Melbourne
Australia

Zora Vrcelj
Built Environment and Engineering
Victoria University
Melbourne
Australia

Editorial Office
MDPI AG
Grosspeteranlage 5
4052 Basel, Switzerland

This is a reprint of the Special Issue, published open access by the journal *Sustainability* (ISSN 2071-1050), freely accessible at: www.mdpi.com/journal/sustainability/special_issues/QFXFB037HD.

For citation purposes, cite each article independently as indicated on the article page online and as indicated below:

Lastname, A.A.; Lastname, B.B. Article Title. *Journal Name* **Year**, *Volume Number*, Page Range.

ISBN 978-3-7258-3836-3 (Hbk)
ISBN 978-3-7258-3835-6 (PDF)
https://doi.org/10.3390/books978-3-7258-3835-6

Contents

About the Editors

Malindu Sandanayake

Malindu is a Chartered Professional Engineer (CPEng) of the Institute of Engineers Australia. Prior to joining Victoria University, he worked as a post-doctoral research fellow at RMIT University. He is also currently an Honorary Associate Professor at the School of Civil and Infrastructure Engineering at RMIT University. Malindu has been awarded with several Sustainability Victoria grants both as the Lead Investigator and also as partner investigator. He has also been a chief investigator in several projects, including a project funded through the Victorian Higher Education Strategic Investment Fund to investigate "Digital transformation of construction industry for improved quality and safety" and "Social Value Creation for Transport and Infrastructure" worth over $2.5 million.

He has research expertise in sustainable materials, green and sustainable construction, life cycle assessment, Building Information Modelling (BIM), and smart technologies in construction and carbon-neutral construction. Following his research contributions on sustainability in construction, he was a finalist of the RMIT award for research impact and was a recipient of the 2022 Vice Chancellor's Award in the "Sustainability" Category.

A/Prof Sandanayake has published numerous high-quality journal papers, book chapters, conference proceedings, and technical papers. He is also an experienced research supervisor and has supervision interests in building information modelling, green and sustainable construction, smart technologies in construction, the circular economy, sustainable management in construction, and waste-to-value research in construction. He has also delivered several keynote speeches, led research workshops, and presented in conference proceedings.

Zora Vrcelj

Professor Zora Vrcelj is the founder and head of Victoria University's Built Environment program.

Before joining Victoria University in 2012, she served as an academic at UNSW's School of Civil and Environmental Engineering for 12 years. She works across the built environment, engineering, and construction sectors to achieve impactful outcomes.

Her research has focused on developing technologies for improving built environment and engineering education for students and practitioners, improving the functionality of high-performance structures (smart buildings and civil infrastructure), and digitally transforming construction quality control processes and health and safety practices.

She has initiated and led numerous research undertakings, including smart buildings and infrastructure, sustainable construction materials, the digitalisation of construction for improved quality, control, and safety, and multimedia educational tools.

Zora has supervised research projects totalling AUD 3.0M in funding, including large applications (ARC DP, CRC, Australian Teaching & Learning Council, State Victoria Funding, Sustainability Victoria).

Her capacity to manage collaborative research relationships effectively with external partners, including industry, universities, and research institutions, is demonstrated by ongoing collaborations with industry partners and researchers from many prestigious universities and institutes throughout the world.

Since 2010, Zora has held a Visiting Professor position in Civil Engineering at the University of Novi Sad, Serbia. She has published close to 120 refereed technical publications.

sustainability

Article

Potential of and Current Challenges in Reusing Recycled Carbon Fibres in Concrete Construction Applications

Vanessa Overhage * and Thomas Gries

Institut für Textiltechnik, RWTH Aachen University, 52074 Aachen, Germany
* Correspondence: vanessa.overhage@ita.rwth-aachen.de

Abstract: The non-corrosive properties of carbon fibres allow for slimmer concrete components, which may reduce CO_2 emissions during production. Given that cement production contributes approximately 8% of global CO_2 emissions, finding alternatives is crucial. Textile-reinforced concrete (TRC) employs technical textiles instead of steel reinforcements and has been extensively studied for its mechanical properties. Carbon's high tensile strength allows for significantly reduced mass compared to steel while eliminating additional cover requirements. Although producing recycled carbon fibres (rCFs) is energy-intensive, it offers significant energy and raw material savings and can lower global warming risks compared to virgin fibres. This study investigates the potential of rCFs in various forms as concrete reinforcement, highlighting both opportunities and challenges based on experimental results and existing studies. The investigations demonstrated that rCFs, whether used as nonwoven or yarn reinforcement, enhance both the tensile and yield strength of concrete. Furthermore, in many instances, a gradual failure mode rather than an abrupt one is observed. Consequently, the use of rCF textiles as reinforcement in concrete presents significant potential for promoting sustainability within the construction industry. The integration of rCF into carbon concrete presents a promising pathway to enhance the sustainability of construction materials.

Keywords: recycled carbon fibre; construction industry; fibre-reinforced concrete; textile-reinforced concrete; non-metallic-reinforced concrete

Citation: Overhage, V.; Gries, T. Potential of and Current Challenges in Reusing Recycled Carbon Fibres in Concrete Construction Applications. *Sustainability* **2025**, *17*, 2779. https://doi.org/10.3390/su17062779

Academic Editor: Chunlu Liu

Received: 31 October 2024
Revised: 11 March 2025
Accepted: 19 March 2025
Published: 20 March 2025

Citation: Overhage, V.; Gries, T. Potential of and Current Challenges in Reusing Recycled Carbon Fibres in Concrete Construction Applications. *Sustainability* **2025**, *17*, 2779. https://doi.org/10.3390/su17062779

1. Introduction

Natural resources are a prerequisite for sustaining current and future life on our planet. However, many natural resources are only available in limited quantities [1]. Therefore, the protection of natural resources is also of existential importance for future generations. The term sustainability has now become a part of most areas of life, including the construction industry. The inclusion of sustainability is imperative because the construction sector is a prominent contributor to global CO_2 emissions [2]. Some of the reasons for this are the use of fossil raw materials, long delivery routes, outdated construction methods and a waste system whose terminal destination is a landfill. Modern construction means that planners must be more oriented towards the users and the environment, respond to new needs, and build accordingly with the environment and the future in mind [2].

Concrete is the predominant building material and, worldwide, the second most utilised material after water [3]. Concrete primarily consists of cement, sand and water, with cement production being a significant source of CO_2 emissions due to its energy-intensive process, contributing to 8% of worldwide CO_2 emissions [4]. It boasts high compressive strength but has low tensile strength, which necessitates reinforcement with materials capable of withstanding tensile forces.

Steel is the most prevalent reinforcement material, providing high tensile strength to concrete components but with a major drawback: its susceptibility to corrosion. A thick concrete cover creates an alkaline environment to protect steel from corrosion, which increases concrete usage. The construction industry, including building maintenance and the lifecycle of construction materials, contributes to 38% of worldwide CO_2 emissions [5] Reducing these emissions is critical to achieving the Intergovernmental Panel on Climate Change's objective of limiting global warming to a maximum of 1.5 °C [6]. The building sector, with its substantial contribution to greenhouse gas emissions, plays a key role in this effort. Given global population growth and urbanisation trends, the demand for concrete is expected to rise [4].

There are several strategies to achieve this goal at the building component level, including minimising resource consumption, using recycled materials and increasing service life. An approach that addresses all three is employing recycled carbon fibre (rCF) as a reinforcement alternative.

Textile-reinforced concrete (TRC) combines a concrete matrix with advanced reinforcement derived from technical textiles instead of steel reinforcements as a composite material. The viability of carbon when used as a reinforcement material has been extensively studied [7,8]. Carbon's mechanical properties make it particularly suitable for reinforcement. The exceptional tensile strength combined with the low density of carbon fibre (CF) means that, compared to steel, only one-twentieth of the mass is needed to achieve the same tensile strength. Carbon is corrosion-resistant, eliminating the need for an additional concrete cover and potentially saving up to 80% of the concrete [9–11]. In addition, the technical service life of carbon-reinforced concrete components is predicted to be 200 years, more than double that of traditional materials [12,13].

TRC transforms the construction landscape by enabling the creation of thin-walled concrete elements with exceptional mechanical performance, which enhances resource efficiency and supports automated production, as shown by Scheurer et al. [8]. This innovative approach facilitates the effective retrofitting of existing structures, eliminating the need for complete demolition, while also allowing for double-curved architectural designs that traditional steel-reinforced concrete cannot achieve due to the flexibility of textiles. The integration of sensors and other functionalities into TRC textiles adds significant value, enhancing usability through the real-time monitoring of structural integrity and facilitating targeted retrofitting efforts. While many advanced applications have been demonstrated in research and pilot projects, the full market adoption of TRC technology is still underway [8].

Traditional disposal methods of carbon fibre-reinforced plastic (CFRP), such as landfilling and incineration, pose risks by generating harmful substances that can damage the environment. Therefore, utilising CFRP in construction offers a beneficial solution that addresses both economic and environmental concerns [14].

Producing CF is energy-intensive, requiring 198–595 MJ/kg. However, using rCF can reduce the global warming risk compared to virgin fibres. The energy consumption is less than 2.05 MJ/k for mechanical recycling, 3–30 MJ/k for thermal recycling and around 19.2 MJ/k for chemical recycling [15]. The quality of rCFs is lower than that of virgin fibres as they are no longer continuous fibres, but the mechanical properties are similar, and they can be processed into semi-finished textile products [16]. Producing yarns and nonwovens from rCFs shows significant promise [16].

This study aims to explore the use of rCF nonwovens and yarns as reinforcement in concrete. rCF consumes less energy compared to producing conventional steel reinforcement or carbon filaments from primary raw materials [16]. Consequently, rCF reinforcement offers potential energy and raw material savings. The non-corrosive nature of carbon fibres allows for slimmer concrete components compared to those reinforced with steel, potentially reducing CO_2 emissions during production [7,13]. The aim of this study is to investigate the use of rCFs in different forms of delivery as reinforcement in concrete. The potential and current challenges are explained based on our own test results and existing studies.

2. Carbon Fibre Recycling

Manufacturing CF is an energy-intensive process, which means that CFs have high recycling capability from environmental and economic perspectives [17]. The energy consumption of virgin CF (vCF) production is 198–595 MJ/kg. Compared to virgin fibres, rCF has a lower energy consumption and lower costs [14,15]. To reduce the energy required, it is reasonable to use CF after the first life cycle again. Therefore, a recycling process for the recovery of used fibres is required. In Table 1, the advantages, disadvantages and characteristics of different recycling processes are summarised.

Table 1. Advantages, disadvantages and characteristics of various fibre recycling methods according to [14,15,18–20].

Recycling Method	Advantages	Disadvantages	Characteristics
Mechanical recycling	• Efficient and high capacity. • Scalable on an industrial scale • No air pollution from gas emissions or water pollution from chemicals. • Low-cost equipment and no skilled labour required. • Processing costs are less than incineration or landfilling costs in Europe [14].	• Health and safety concerns due to the risk of ignition during the shredding process. • Low-value recyclates that are barely competitive with virgin material. • 35% of the tensile strength of rCF compared to vCF [19].	• No recovery of individual fibres. • Produces inferior products; only used for GRP. • Requires special plants with a closed area to limit dust emissions. • Recycling energy consumption: 0.27–2.03 MJ/k [15].
Thermal Recycling	• The by-products (gas and oil) can be used as an energy source. • Easily scalable. • Already used on a commercial scale for the recycling of CFRP. • Pyrolysis has the highest technology readiness level for a discarded CF recycling process and is possible on an industrial scale [14].	• Oxidation residues or carbonisation can be on recovered fibres. • Loss of strength of 5–20% of the fibres due to high temperatures [19]. • Not economically viable. • CFRP wastes are shredded to typically between 6 and 20 mm before using fluidised bed [15].	• Possible leakage of gases from waste treatment chambers. • Recycling energy consumption: 3–30 MJ/k [15]—less than 16% of energy consumption of vCF [14]. • Fluidised bed, microwave-assisted pyrolysis and the addition of superheated water steam into pyrolysis are, in addition to pyrolysis, novel or adapted thermal recycling processes [14].
Chemical Recycling	• Recovery of clean, full-length fibres. • rCF tensile strengths of 80–100% compared to vCF [14,19]. • Recovery of resin that can be reused. • Low-risk solvents are used, such as alcohols, glycols and supercritical water.	• Low efficiency and high costs. • High energy consumption due to high temperature and high pressure compared to other recycling methods, but still a fraction of those for vCF. • Large quantities of solvent required.	• Effects on human health due to greenhouse gases. • Recycling energy consumption: 19.2 MJ/k [15]. • Variants are using supercritical fluids and solvolysis or, as a more sustainable and cost-effective novel method, electrochemicals [14].

2.1. Mechanical Recycling

As stated in [20], slow-running cutting mills break down end-of-life components into pieces measuring between 50 mm and 100 mm. In contrast, fast-running mills generate fragments with sizes ranging from 50 μm to 10 mm for homogeneous components [20,21]. The first investigations of mechanical laminae separations of laminated composites were made by Imbert et al. to obtain longer recycled CFRP pieces [22].

Mechanical shredding is often preceded by thermal or chemical recycling processes [21]. Therefore, mechanical recycling processes are categorised as pre- and post-treatment processes for waste containing carbon fibres in the recycling sequence [21]. The process involves shredding, sorting, compacting and pelletising waste containing carbon fibres. This process step must be applied to both end-of-life waste and prepreg materials made from carbon concrete. Prior to shredding, the end-of-life products containing carbon fibres, such as car body components, rotor blades or textile concrete, must be shredded into small, manageable pieces. Production waste is processed directly or shredded, depending on its size. CF waste is then cut into short fibres in granulators or reduced to particles in various mills, such as hammer mills, jaw crushers or impact mills [23].

Hammer mills crush materials with several hammers arranged along a grinding path of the rotor axis, allowing control over particle size through screen size. Hammer mills are frequently utilised for coarse and medium size reduction, while jaw crushers involve feeding the materials to be crushed between two plates and crushing them through compressive stress [24,25]. The jaw crusher is suitable for initial coarse to medium crushing On the other hand, in the impact mill, the material to be ground is fed in from above through a feed hopper and comminution is performed. This process is used for medium size reduction [24,25].

To separate and classify waste into powdery and fibrous components, screens, laser diffraction spectrometry or image analysis are commonly used. Dry waste is typically shredded using grinding units with cutting and shearing functions. When treating wet waste containing carbon fibres, mechanical recycling requires sorting to occur first. Mixing different polymers during the recycling process makes it more difficult to release fibres as they are treated with different process parameters such as temperature and pressure. This results in a reduction in quality [23].

It is not possible to completely separate the resin from the CF with mechanical stress; therefore, the mechanical recycling of wet CF waste is a process in which no individual fibres are recovered. During the recycling process, only the dry and wet waste containing carbon fibre is shredded, and the resulting components are separated based on particle size. The shape and size of the shredded components vary depending on the mechanical shredding method used, which can significantly impact subsequent carbon fibre treatment processes. This process is known as the degradation of carbon fibre.

Isa and Nosbi et al. [26] addressed the degradation of CF during mechanical recycling. The authors emphasised the necessity of degradation studies to assess the mechanical properties of waste containing CF. In the context of this paper, degradation studies refer to the process of the degradation of mechanical properties. Fine-grained waste has the potential to be reused as a reinforcing material in the original matrix application. In contrast, the reinforcing properties of the coarse-grained recyclate are lower compared to vCF [26].

Studies have shown that fibre shortening and damage to the fibre structure during the recycling process result in brittle rCF compared to vCF. Additionally, mechanical rCFs generally have a lower tensile strength and modulus of elasticity than vCF. This reduction in mechanical properties is caused by the recycling process, which leads to damage to the fibre structure. Mechanical shredding causes fibre shortening, which makes it difficult to use rCF [17,18,27]. As Li et al. summarised in [14], mechanical recycling is suitable for glass fibre-reinforced polymers, rather than CFRP. For CFRP recycling, methods are preferred in which the fibres are separated from the matrix and then obtained in fibre form.

2.2. Thermal Recycling

Thermal recycling processes are used to separate CF from thermoplastic matrices in particular [14]. Two procedural options for thermal recycling are the fluidised-bed process and pyrolysis [20]. Both of these processes are described in more detail in the following.

Pyrolysis is a thermal recycling process that separates fibre and matrix through the thermal decomposition of the fibre-binding polymers. The process is carried out at temperatures ranging from 300 to 800 °C under various atmospheric conditions, such as inert gas. The temperature resistance differences between the CF and matrix materials are utilised for the decomposition of the matrix material, which produces oily, solid and gaseous products. The gases in question are primarily composed of carbon monoxide, CO_2 and other hydrocarbons. These gases are recycled in a condenser and used as furnace fuel or separated into solid and liquid hydrocarbons. The recirculation of combustible gases causes solid coke residues to adhere to the furnace surfaces, which negatively impacts further processing options. The incomplete exposure and damage of the rCF due to the unclear preparation of the composite material poses a problem when using pyrolysis [28].

Pyrolysis recycling is the only technology currently in use at an industrial scale [14]. Applying pyrolysis on an industrial level presents a challenge as the CF to be recycled often consists of a variety of starting materials, each requiring different temperatures for effective fibre recovery. However, the process is flexible and can be adapted to various fibre–matrix combinations by optimising the process parameters [28,29]. Classic pyrolysis is a process carried out under inert gas, such as nitrogen. The process temperatures depend on the matrix. For polyester resins, low temperatures (approx. 450 °C) are used, while epoxies or thermoplastics are treated at higher temperatures between 550 and 650 °C. Thermal treatments below 450 °C are beneficial as they cause minimal damage to the fibre surfaces and have little effect on the mechanical properties. However, lower temperatures result in longer process times, which can make economic implementation more challenging.

The pyrolysis process presents a challenge due to the formation of soot on the rCF, rendering resulting fibres unsuitable for further processing into yarns or new reinforcement structures. Furthermore, the interfacial bonding is negatively affected [14]. Finding a compromise between the resulting mechanical properties and the remaining matrix residue is necessary as complete soot removal is not possible afterwards. Chemical post-treatment or post-heating of the rCF contributes to a certain reduction of the carbon black [14,27].

Onwudli et al. and Kim et al. investigated the reduction and complete avoidance of soot formation using pyrolysis techniques under different atmospheric conditions. Significant improvements in the composite material were achieved through chemical treatment with additives and light oxidation post-treatment of rCF [14,20,30]. The addition of $ZnCl_2$ as a catalyst in the pyrolysis process reduced the degradation temperature to below 400 °C while retaining 95% of the tensile strength compared to vCF [31].

Over the past two decades, the potential of microwave-assisted pyrolysis has been investigated. It is purported to be a more expeditious, energy-efficient and efficacious method of producing rCF that additionally has a cleaner fibre surface [14]. Another way to produce cleaner fibre surfaces was investigated recently by the use of super-heated water streams in addition to the pyrolysis process [14].

The fluidised bed process is an effective thermal recycling method for recovering CF from polymer composites. This technique involves suspending particulate matter, typically the composite material, in a rising stream of gas, usually air, which creates a fluid-like state and a silica bed. Within this system, the organic matrix is decomposed in the combustion chamber at temperatures generally ranging from 450 °C to 550 °C. The elevated temperatures cause the polymer matrix to decompose, while the carbon fibres are preserved due to their high thermal stability [14,15].

One of the main advantages of the fluidised bed process is its ability to handle mixed and contaminated waste streams, providing a versatile solution for composite recycling The resulting carbon fibres often retain a significant portion of their original mechanical properties, allowing them to be used in a range of secondary applications [15]. Compared to the pyrolysis process, rCFs from the fluidised bed recycling process have less fibre strength due to the presence of sand [14].

2.3. Chemical Recycling

Solvolysis is a chemical recycling process that dissolves the matrix from the CF using solvents. In comparison to pyrolysed rCF, the rCF surfaces are clean after the recycling process and the rCF mechanical properties are comparable to those of vCF [14]. To increase the surface area for the chemical recycling process, CFRP components are typically mechanically ground before chemical recycling. The processes are divided into low-temperature solvolysis and solvolysis with supercritical fluids [20].

Low-temperature solvolysis is a process that recycles CF under normal pressure at temperatures below 200 °C. Acids or solvents, such as water, alcohol, ammonia, nitric acid or sulphuric acid, are used to separate the fibre from the matrix. Catalysts and additives must be used due to the low process temperatures. The mechanical properties of the recovered fibres and resin are almost identical to those of the original materials. Unlike the pyrolysis process, the matrix resins are also recovered in addition to the fibres during decomposition. The resins that have been recovered can be reused either as fuel or as a chemical raw material [18]. The solvolysis process has the capability of obtaining long fibres but it presents challenges due to the use of harmful substances and the long process time [18,32,33].

Supercritical fluids are used to separate CF from the matrix in CFRP. For example, supercritical water is used with a pressure of 221 bar and temperatures over 374 °C. Besides supercritical water, other solvents, such as different alcohols (methanol, ethanol, propanol), are used with lower pressures and temperatures for the separation of CF and different matrices. But, recycling with supercritical fluids requires reactors that withstand high temperatures as well as high process pressure. Therefore, it is considered a more expensive recycling method due to the required equipment and the necessary processing energy [14].

Recently, electrochemical recycling was investigated for extracting rCF from CFRP. To retain the rCF properties, the parameters of current, temperature and concentration have to be elected properly. This recycling method is considered more sustainable and cost-effective [14].

2.4. Properties of Recycled Carbon Fibre

Nevertheless, the extent of this reduction of the material properties of rCF and vCF is marginal. Table 2 presents a material properties comparison and evaluates them as a percentage. The CFs that are commonly used in the construction industry were used as representatives of vCF.

Table 2. Quantification of the remaining material properties of CF after recycling [34–36].

Fibre Type	Tensile Strength [N/mm^2]	E-Modulus [kN/mm^2]	Length	Diameter [µm]
Carbon fibre	4300	240–250	Endless or cut to a defined length	7
Recycled carbon fibre	>3500	230	Milled–short fibre	6 (±1)
Maintaining quality after recycling [%]	>81%	92–96%	-	98.5

Currently, there are no known recycling methods that can be used on an industrial scale for CF that allow the fibre to be recovered without shortening its length. Therefore, rCF is typically sold in ground form or as short fibres of varying lengths ranging from less than 100 µm to several centimetres. During the pyrolysis process, any residues from the original composite matrix, impregnation or sizing are typically removed from the rCF. Even in this state, rCF retains 80% of the tensile strength of vCF [23,37]. Accordingly, the main quality losses are not caused by pyrolysis but by the significant shortening of the carbon fibre during recycling.

Due to their mechanical properties, CFs are ideal for use in composite materials in various forms of semi-finished textile products (woven fabrics, scrims, braids, etc.) [38]. The processability in different processes is influenced and partly limited by the fibre lengths. This means that for use in injection moulding or as an additive, maximum fibre lengths limit use. In contrast, for yarn or surface production processes, the processability is limited by a minimum fibre length. The term 'short' defines different length ranges depending on the fibre application. In textile and man-made fibre technology, fibres with a length of <40 mm are referred to as short staples, whereas, in plastics technology, fibres with a length of <6 mm are defined as short fibres [38]. The definition of fibre length varies depending on the application. In Table 3, a list of the terms used in this paper is given.

Table 3. Fibre types sorted by length and application in the construction industry [10].

Fibre Type	Length [mm]	Application
Milled	<0.5	Filler
Short-fibre	<40	Wet laid tape, FRC
Long-fibre	>60	Yarn, dry-laid nonwoven

Milled CFs are produced by mechanical mills and chopped fibres are produced by mechanical separation processes. Ground fibres are in powder form and have a length of less than 0.5 mm [38]. Chopped fibres have a fibre length of more than 0.5 mm. Depending on the parameter settings, the fibre length varies during the cutting process. Milled and chopped fibres are mixed into plastic to, for example, improve the mechanical properties and conductivity of the composite material.

For example, long fibres can be used in the dry-laying process to produce CF nonwovens. A CF nonwoven is defined as a flat, semi-finished product consisting of nonwoven made from CF. In the dry-laid process, the CF is laid on a substrate and then impregnated with a binder or resin [38,39].

Composite materials have been used in the construction industry for centuries. The principle of combining different materials has been continuously developed, from simple building materials like clay and straw. Textile reinforcement elements made of non-metallic fibre materials represent a new generation of reinforcement for concrete components [Sch07]. In addition to alkali-resistant glass fibres (AR-GF), CFs are particularly suitable for use in concrete due to their chemical resistance.

3. Materials and Methods

In this work, different rCFs were tested as textile reinforcements in concrete. The results are presented in Section 4.2. Two nonwovens with different manufacturing processes were tested. Firstly, the nonwovens were pre-impregnated with the matrix in order to increase the textile–matrix bond. In the second manufacturing process, another production variant was used: the two nonwovens were embedded in the concrete using the lamination process, but, as an addition, before the second layer of concrete was added, the nonwoven was manually pressed into the first layer of concrete using a roller. In the third variant the nonwoven was cut into stripes beforehand and was therefore not embedded over the entire concrete but only as reinforcement strips in the laminating process. Reinforcement with unreinforced slivers in the concrete was also investigated. The specimens were also produced by the laminating process. The slivers were an intermediate product in the production of rCF yarns. Two different rCF yarns, friction and wrapping yarn, were also tested as part of the investigations. The test results of the nonwoven and yarn variants have been published by the authors previously in [40] but are analysed further in this paper. Finally, the use of yarn fragments after recycling was tested as a possible 3rd rCF life cycle.

In this work, the four-point bending test of the specimen tested at Institut für Textiltechnik of RWTH Aachen University (ITA), Aachen, Germany adhered to DIN EN 1170-5 [41] for measuring bending tensile strength (shown in Figure 1).

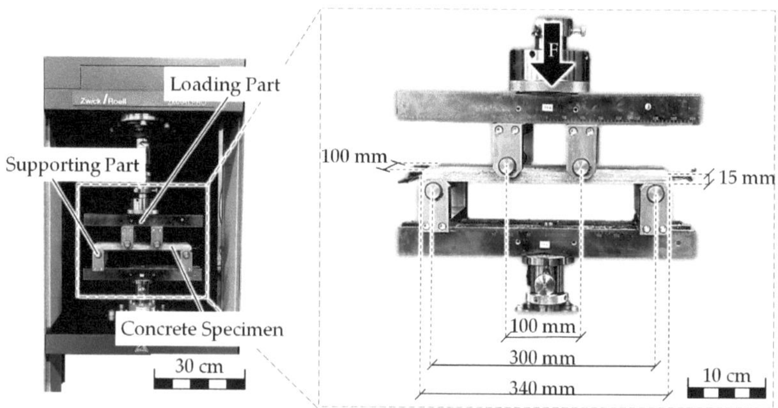

Figure 1. Four-point bending test according to DIN EN 1170-5.

Elongation was measured during the flexural test by reference to the displacement of the loading member during the four-point bending test. Samples with a size of $34 \times 10 \times 15$ cm^3 were produced for each series of tests. The fine concrete mix used was formulated based on an investigation carried out by Brockmann in sub-project C1 of the Special Research Area (SFB) 532 "Textile Reinforced Concrete" at RWTH Aachen University [42]. Within the scope of this paper, this composite is referred to as TRC in the following, though it is also occasionally designated as textile-reinforced mortar or fabric-reinforced cementitious matrices [7]. The quantities and volume proportions of all solid and liquid components are given in Table 4.

The samples produced were stored in the mould for 24 h before being demoulded. Afterwards, they were stored in water for approximately 6 days and a further 21 days in room climate before being tested at the age of 28 days.

Table 4. Fine concrete mixture.

	Component	Quantity [kg/m^3]	Quantity [Vol.%]
solid	Cement CEM I 42.5 R	490	22
	Fly ash	175	8
	Silica fume	35	3
	Quartz powder	500	23
	Sand (aggregate size 0.2–0.6 mm)	713	32
liquid	Water	280	11
	Superplasticiser	7	0.3

4. Application of Recycled Carbon Fibre in the Construction Industry

Table 5 gives an overview of different textile versions of rCFs in the construction industry.

Table 5. Recycled carbon fibre textile application variants: (a) fibre reinforcement; (b) nonwoven reinforcement; (c) textile reinforcement.

Short Fibre	Nonwoven	Wrapping Yarn
Fibre reinforcement	Nonwoven as stripes as textile reinforcement	Yarn as 1D as textile reinforcement
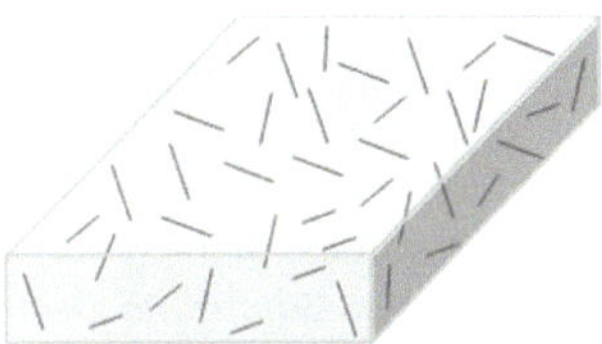		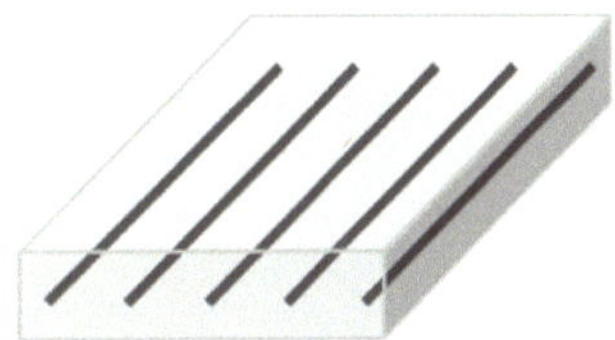
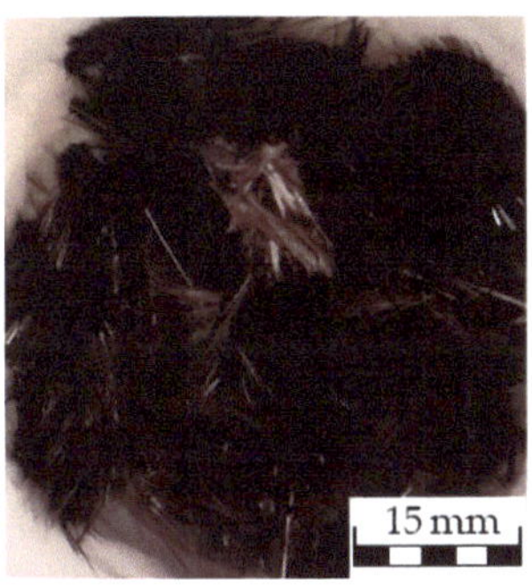	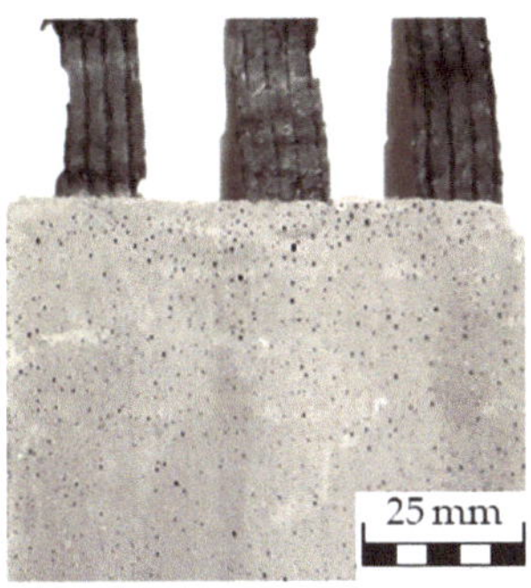	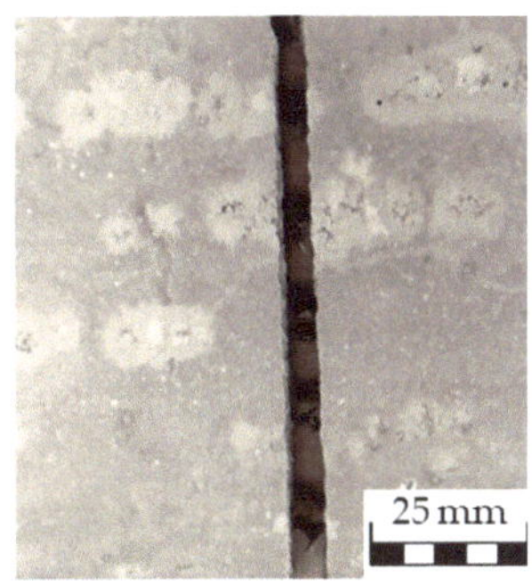

4.1. Fibre Reinforcement

Fibre-reinforced concretes (FRCs) belong to the class of building materials known as fibre-reinforced composites. The production process involves adding short fibres ranging from 10 to 40 mm in length to fresh concrete and evenly distributing them in the cement matrix during mixing [43]. The resulting FRC components can be cast or shotcrete and are shown schematically in Table 5. Both standard installation methods create an undirected, three-dimensional reinforcement structure. The random and stochastic distribution of the CF in the concrete results in isotropic material properties in the component [10,44].

The fibre reinforcement typically does not have a structurally load-bearing function in the component, but it does improve both the load-bearing behaviour and the concrete properties of the composite material. This results in the following changes in the material [10,45]:

- Increase in tensile and compressive strength

- Increase in impact resistance
- Increase in ductility
- Avoidance of shrinkage and shrinkage cracks during setting
- Avoidance of cracking in use/reduction of crack widths
- Increased weather-tightness
- Increased fire safety

FRC can be used for almost all types of construction in building construction and civil engineering, both as in situ concrete and in prefabricated construction. The shotcrete construction method is mostly used in specialised civil engineering or repair work [45,46]

There are several problems with the use of CF compared to other macro fibre types that are mainly used. The low density of CF can cause the fibres to float, resulting in varying material properties across the cross-section of the overall component. Additionally, using high fibre volume contents can lead to a change in the consistency of the fresh concrete The use of rCF in concrete reduces the workability of the concrete [14].

However, the technical advantages of CF in carbon fibre-reinforced concrete (CFRC) lie in its high tensile strength, which allows for the achievement of high strengths in the composite material. Furthermore, the chemical resistance of this material makes it suitable for use in environments that are humid and contaminated with chloride [10].

Despite these design advantages, the use of CF as CFRC is not yet established on the market due to its high cost [46]. Additionally, shortening the endless vCF does not make technical sense. Currently, research is being conducted on the alternative use of rCF for application in FRC [10].

The production of recycled CFRC (rCFRC) is the simplest method of using rCF as an aggregate in FRC. This method only requires the fibres to be recovered from the original matrix material and possibly resized, without the need for processing them into semi-finished textile products.

CFRC is produced by mixing short fibres into fresh concrete, which affects both its fresh and hardened properties. Kimm [10] identified fibre volume content and length as relevant parameters for the resulting properties of rCFRC. As part of the dissertation, the influence of fibre length variation and material on rCF was investigated [10]. Further investigations evaluated the effect of surface modification on rCF adhesion to the concrete matrix and fibre distribution in fresh concrete [10,47].

The properties of FRC are influenced by parameters such as fibre volume content, fibre length and surface modification. A non-linear relationship exists between fibre volume content and flexural tensile strength, with a critical threshold beyond which strength decreases, ranging from 0.5% to 1.0% by volume, depending on fibre length. Longer carbon fibres (>20 mm) improve flexural strength across all tested volumes (0.25–2.0%), achieving increases of 1.3–2.5 MPa compared to shorter fibres (<20 mm). The maximum recorded strengths were 8.47 MPa for short recycled carbon fibres (rCFs) at a 1% volume and 10.32 MPa for long rCFs at 0.5%, exceeding conventional FRC strengths by 40%. Analyses of fibre length variation showed no significant impact on flexural tensile strength when standard deviations were below or above 10 mm; however, extreme fibre contents resulted in over 10% standard deviations in strength. Within the optimal range of 0.5–1.0% by volume, variations in rCF length did not affect the flexural tensile strength of the specimens [10].

Li et al. compared various studies and found that the inclusion of pyrolysed carbon fibre, although it reduces workability, enhances both the flexural and compressive strength of cement-based materials at room and elevated temperatures, with 1% being the optimal dosage. This addition also benefits the strength development in alkali-activated materials. Furthermore, chemically rCFs have also shown an increase in strengthening cementitious matrices. Nevertheless, further research is required to optimise the chemical recycling of rCF for practical use in the construction industry [14].

The objective of a subsequent study was to analyse the fibre–matrix adhesion of rCF in concrete [48]. The utilisation of pyrolysed unsized or acrylate-sized fibres facilitated an enhanced fibre–matrix adhesion, exceeding the adhesion observed for vCF. In general, these pyrolysed fibres demonstrated superior absolute and average shear strengths. Notably, an increase in apparent interfacial shear strength between 150 and 250% was observed in comparison to vCF. In comparison, Li et al. stated that the inferior bonding behaviour of rCF compared to vCF exhibits less adhesive behaviour with cement [14]. It appears that the addition of sizing has a beneficial effect on the homogeneity of the fibres, as evidenced by the consistency observed in the apparent interfacial shear strength values for sized fibres, without the presence of outliers. The fibre distribution affects the resulting strength values. In the event of fibre agglomeration, the bond between fibres facilitates their extraction rather than direct fracture at the fracture edge. The distribution of fibres within cells results in a strong bond between the fibre and matrix within composite components, which can cause fibres to break at the breaking edge [48].

The findings demonstrated that the sizing type, in addition to the resulting fibre–fibre adhesion, influences the test outcomes. When fibres undergo resizing subsequent to pyrolysis, they demonstrate a reduced capacity to adhere to one another in comparison to vCF fibres treated with sizing during the production process and were subsequently cut to short fibres [48].

Li et al. showed that acidic and alkaline treatments enhance the bonding of mechanically recycled carbon fibres with mineral matrices, thereby improving composite properties. Surface modification is of particular importance for inert pyrolysed fibres, with variations based on recycling methods influencing the selection of an appropriate treatment. For pyrolysed fibres, electrochemical modification and oxygen plasma treatment were compared. Chemical recycling could achieve fibre recovery and modification simultaneously. But, regardless, fibre surface modification was crucial for hydrophobic rCF. Li et al. stated that further research is required to improve fibre–matrix stress transfer efficiency in cement-based composites [14].

In terms of the durability of rCF in FRC, further investigation is needed. Regarding durability, there is still a lack of publications on the use of rCF in cementitious matrices as well as the use of rCF in alkali-activated composites [14].

4.2. Textile Reinforcement

'Textiles', according to the standard [49], is the collective term for textile fibres, semi-finished products and finished products made of textiles. This study examined textiles based on recycled carbon fibres with untreated and treated rCF at the fibre level for use in fibre-reinforced concrete. At the semi-finished product level, linear and flat structures were distinguished. Staple fibre yarns were examined at the linear level. Nonwovens were flat, semi-finished products that were examined at the textile level.

Different rCF textiles were used for the reinforcement of concrete. Nonwovens were used as whole, flat textiles and also cut into stripes and yarns as one-dimensional reinforcements. As part of this research, other rather unusual variants for the use of rCF as reinforcement in concrete were investigated. Variants that had fewer process steps in textile production were tested. The approach aimed to make better use of the fibre orientation and position of the fibres in relation to the mechanical properties than with isotopically distributed fibres in the FRC. Slivers from a pre-production process in the nonwoven production were used. Furthermore, recycled rCF yarn pieces after the second life cycle were tested as reinforcement in concrete. Here, the pieces were lain in a given orientation in one layer and were not distributed isotopically.

Nonwoven 1 was composed of 100% rCF with a grammage of 100 g/m^2 and was not subject to any additional bonding after needling. In contrast, nonwoven 2 was bonded using the Maliwatt process. Consequently, the composition of the material was 96% rCF and 4% polyethylene terephthalate (PET) via the binder thread. The grammage remained consistent at 100 g/m^2. In Figure 2, the variants of nonwoven 1 are shown in blue, the nonwoven 2 in black and the plain concrete in grey. Pre-impregnated nonwovens are visualised with a dashed line. The samples in which the nonwoven was additionally rolled in the laminating production process are shown by the solid line. The nonwoven that was cut into strips beforehand and not applied over the entire surface is shown by the dotted line. A representative curve is shown for each variant. All detailed curves can be found in [40]. Furthermore, the mean values and standard deviations of nonwoven reinforced concrete variants are listed in Table 6.

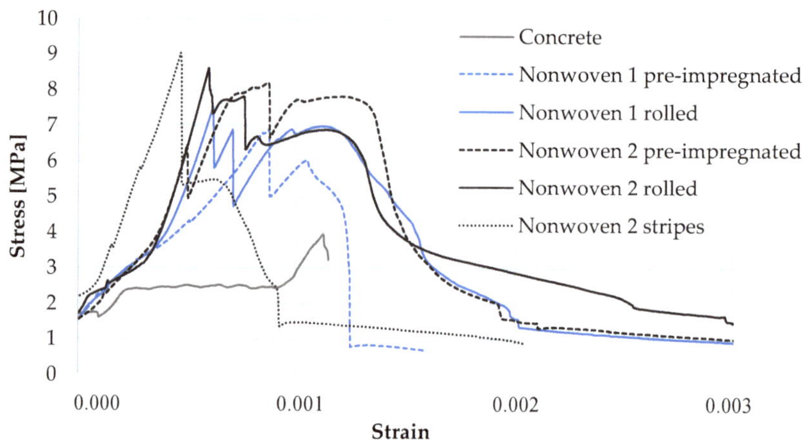

Figure 2. Comparison of flexural stress–strain curves of nonwoven reinforced concrete in accordance with [40].

Table 6. Mean value, standard deviation and fracture energy of nonwoven reinforced concrete variants.

	Nonwoven 1		Nonwoven 2		
	Pre-Impregnated	Rolled	Pre-Impregnated	Rolled	Stripes
Mean value [MPa]	5.97	7.95	8.10	8.79	9.59
Standard deviation [MPa]	0.87	0.50	0.44	1.21	0.45
Fracture energy [J/m^2]	1299	941	1484	874	1023

The findings for nonwoven sample 1 demonstrated a progressive failure mechanism. In the rolled variation, three distinct peaks were observed in the range of 7–8 MPa, with this variant achieving a tensile strength mean value of 7.95 MPa compared to 5.97 MPa for pre-impregnated nonwoven 1. When comparing this with nonwoven 2, it was noted that the pre-impregnated version exhibited the greatest plastic deformation zone and exhibited a yield strength greater than its tensile strength. Conversely, after reaching the tensile strength, other variations exhibited a faster decline in strength. Therefore, the yield strength was also represented by the tensile strength. The mean value of the tensile strength of both rolled nonwoven 2 with 8.79 MPa and striped nonwoven 2 with 9.59 MPa slightly surpassed that of pre-impregnated nonwoven 2 with 8.10 MPa. But the standard deviation was three times as high for rolled nonwoven 2. This was probably due to the greater variation that occurred when rolling by hand. Remarkably, the highest tensile strength was reached by striped nonwoven 2 while utilising only half the amount of nonwoven material compared to other variants. This observation suggests that the bonding between the concrete layers was significantly enhanced in this instance, preventing the striped nonwoven from functioning as a separating layer and thereby facilitating its superior tensile strength performance [40].

Slivers are similar to nonwoven without the needling or Maliwatt bonding processes. Initially, the sliver from the rCF yarn pre-production was tested as a reinforcement. The challenge here was the instability of the textile because the fibres were not consolidated with each other. Therefore, handling was very difficult and even the smallest movements had an influence on the fibre orientation. Figure 3 shows the sliver in the mould (left), while the slivers during concreting are shown in the picture on the right. The bottom concrete layer was in the mould and the sliver lay on top of it. This picture was taken during the concrete lamination process. The upper concrete layer was added on top of the sliver following the step in the illustration.

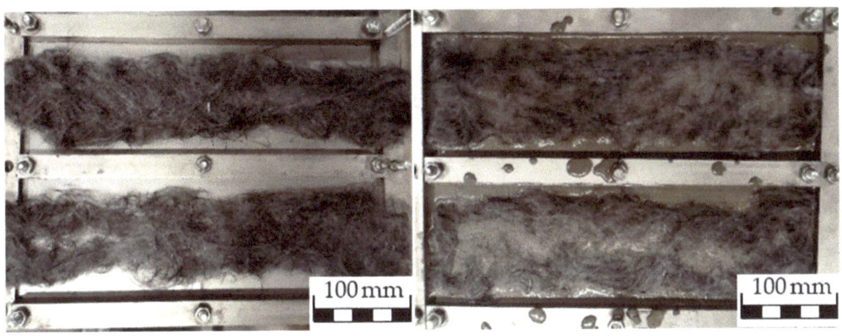

Figure 3. Tape from the rCF slivers without concrete (**left**) and on top of one layer of concrete during the laminating process (**right**).

The test results for the flexural stress elongation curve are shown in Figure 4.

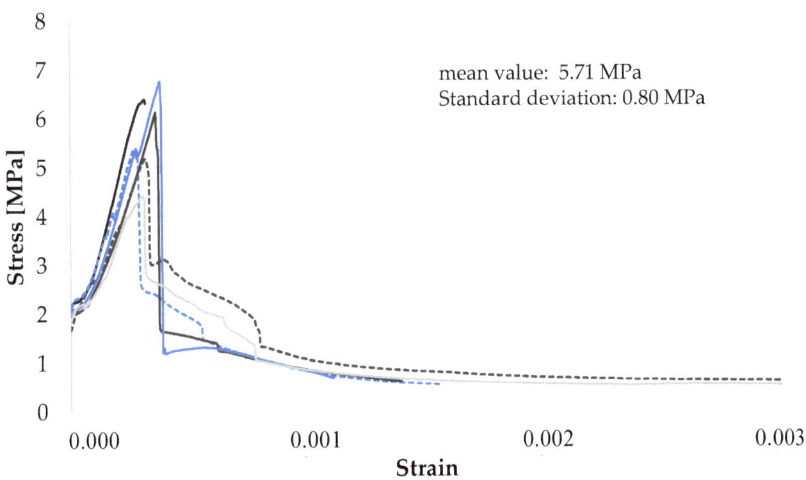

Figure 4. Flexural stress–strain curves of rCF slivers as reinforcement in concrete.

The first test results of sliver-reinforced concrete showed a positive effect. The height of the peaks varied between 4.5 and almost 7 MPa. Therefore, the mean value was 5.71 MPa with a standard deviation of 0.80 MPa. However, the strength values achieved were lower than those of nonwoven reinforcements. The fracture energy varied between 232 J/m^2 and 461 J/m^2.

rCF yarn-reinforced concrete specimens (shown in Table 5 (c)) exhibited an increase in flexural strength across all tested variants. Figure 5 illustrates the flexural stress–elongation curves. Yarn 1, shown with the black curves, was a friction-spun yarn composed of a blend of rCF and polyamide 6 (PA6). In contrast, yarn 2 was produced as a wrapping yarn whereby a filament encased a core material; this yarn maintained the same material composition. The production of the yarns was conducted within the framework of the CarboYarn project. Prior to their integration into the concrete matrix with the lamination process, both yarns received an epoxy resin coating [40].

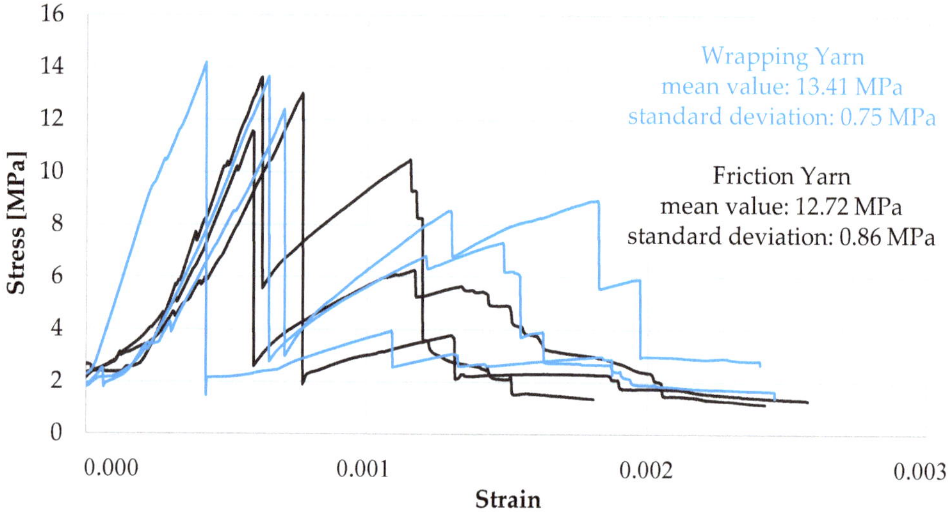

Figure 5. Flexural stress–strain curves of wrapping yarn (blue curves) and friction yarn (black curves) as reinforcement in concrete according to [40].

Comparing both results for yarn reinforcement, the flexural stress–elongation curves exhibited a comparable gradient, and a gradual failure was discernible in all samples. The tensile strength was equivalent to the yield strength for both yarn reinforcements. The mean value for the tensile strength of the wrapping yarn reinforcement was 13.41 MPa, slightly higher than the tensile strength of the friction yarn reinforcement with 12.72 MPa. The standard deviations were comparable, with 0.75 MPa and 0.86 MPa. The fracture energy was around 1205 J/m^2 for the wrapping yarn and 996 J/m^2 for the friction yarn.

However, in contrast to the yarn reinforcement, the impregnated yarn was recycled as short rCFs following the comminution process. The short rCF pieces were not isotopically distributed within the concrete matrix; rather, they were oriented mostly in a single direction within the tensile zone, with the majority of their length aligned in the same direction, as shown in Figure 6. The aim was to achieve the greatest possible reinforcement influence.

Recycled rCF yarn pieces after the second recycling process were also tested as reinforcement. There was a possibility of using this material without the whole recovery of the fibres after the second life cycle with the rCF yarn pieces. The used material was the output of the recycling of TRC in the study by Bayram et al. [50]. The corresponding flexural stress-strain curve is shown in Figure 7.

The first test results for the recycled rCF pieces as reinforcement in concrete showed a positive effect. The height of the peaks varied between 5 and almost 8 MPa with a mean value of 6.38 MPa and a standard deviation of 0.87 MPa. The fracture energy varied between 311 J/m^2 and 786 J/m^2. Therefore, the strength values and the fracture energy achieved were higher maximum forces than those of the sliver reinforcements. The energy absorption capacity was not examined in this study and could be considered in future studies in order to ensure a comprehensive overview.

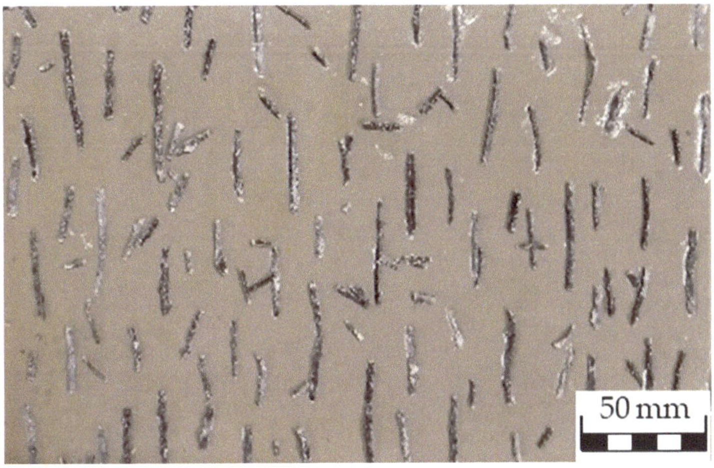

Figure 6. Recycled rCF yarn pieces as reinforcement in concrete.

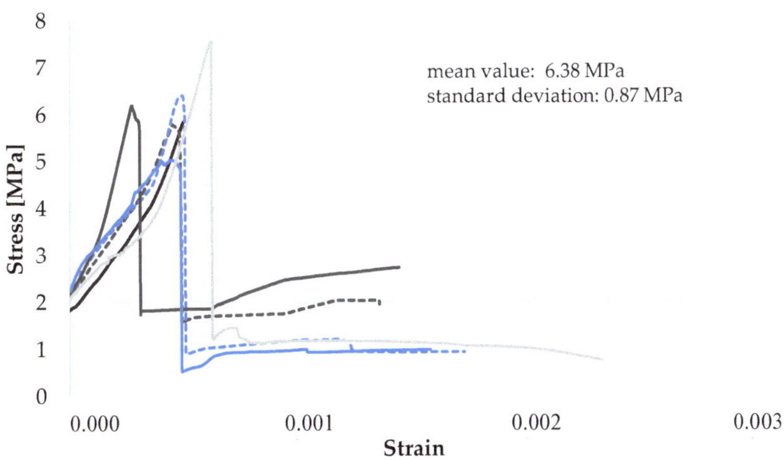

Figure 7. Flexural stress–strain curves of recycled rCF yarn pieces as reinforcement in concrete.

5. End of Life of CFRC

In the framework of the circular economy, it is essential to consider the manufacturing of components, encompassing material extraction and production and their end-of-life phase. This encompasses the potential for recycling and the utilisation of recycled materials in lieu of new materials.

In order to investigate the possibility of separating textile-reinforced concrete at the end of its first life cycle, Kimm et al. [37,51] and Kortmann [37] conducted studies on separability. It was demonstrated that the separation of coated textiles from concrete is a viable process. The separability of materials is significantly influenced by factors such as material composition, construction methods and comminution techniques. Appropriate coatings were found to have a positive effect on separability, with epoxy coatings giving the highest recovery rates. In addition, increasing the cross-sectional area of the roving improved the efficiency of the recycling process. The best recovery rates for carbon reinforcements were achieved using a hammer mill, resulting in a textile recovery rate of over 90% and a residual organic fibre content in the minerals of less than 0.2% [51].

As previously mentioned, while these recycled fibre materials can be used as short fibres in FRC, at present, there is currently no immediate reuse of the textile structure due to the resulting strength losses and cutting.

The potential of rCF over two or more life cycles was considered by Bayram et al. [50]. In this study, the end-of-life separability of rCF yarns in concrete was investigated for the second life cycle [50]. rCFs were employed as wrapping yarn for concrete reinforcement. Subsequently, mechanical processing was utilised to determine the potential for separation of the rCF from the concrete. The results demonstrated that the separation efficiency of rCF yarns varied between 69% and 97%, depending on the machine settings employed. It is important to note that the proportion of the fine fraction (less than 2 mm) also increased with increasing separation efficiency. The separation of TRC made from rCF could only be achieved when the yarn material was coated, in contrast to unimpregnated rCF yarns, which did not allow for effective separation. Additionally, the study conducted a comparative LCA between the utilisation of recycled and virgin CF, establishing that rCFs are environmentally beneficial when a substitution coefficient of over 10% can be achieved [50]. In the study by Bayram et al. [50], the same yarn material and mechanical results were used.

Within the framework of the FaBeR project (funded by the German Federal Ministry of Education and Research (BMBF)), the recycling of CFRC into a recycled CFRC industrial floor was demonstrated.

The project goal was the utilisation of mostly all fractions of recycled concrete components. The fine fraction was investigated for further processing in the cement industry and the coarse fraction was investigated as an alternative to natural aggregates in recycled concrete. The fibre residues were also reused. Floor slabs were produced as part of the project, which were analysed by Luthin et al. as part of a circular life cycle sustainability assessment [52]. The study showed that the recycled CFRC industrial floor demonstrated significant circularity, primarily due to the use of recycled materials and their potential for future recycling. In terms of environmental impact, the global warming risk was lower, whereas the human toxicity potential was higher compared to similar products [52].

Additionally, production costs were notably higher than those of comparable products, likely due to the laboratory-scale production process during the development phase [52]. But, as soon as a higher technology readiness level is reached and production shifts to an industrial scale, the costs would change accordingly.

6. Discussion

FRCs are composites formed by adding short fibres, typically 10–40 mm in length, to fresh concrete, creating a three-dimensional reinforcement structure that enhances the material's properties. While fibre reinforcement is not primarily load-bearing, it significantly improves tensile and compressive strength, impact resistance, ductility, crack prevention, weather resistance, and fire safety. FRC can be utilised in various construction applications but faces challenges such as fibre floatation due to low density and changes in concrete consistency with high fibre volumes.

CFRC benefits from high tensile strength and chemical resistance, making it suitable for harsh environments. However, its high cost limits market adoption. Research is ongoing into using rCF, which simplifies production by recovering fibres without extensive processing. The properties of rCFRC are influenced by factors like fibre volume content and length; longer fibres improve flexural strength significantly compared to shorter ones.

Further studies have shown that surface modifications enhance the adhesion of rCF to a concrete matrix. Pyrolysed or sized rCFs demonstrate improved interfacial shear strengths compared to virgin fibres. Fibre distribution also affects strength outcomes; proper distribution leads to stronger bonds within the composite material. Overall, optimising these parameters can enhance the performance of FRC using recycled materials.

Investigations have demonstrated that rCFs, whether used as nonwoven or yarn reinforcement, enhance both the tensile and yield strength of concrete. Furthermore, in many instances, a gradual failure mode rather than an abrupt one is observed. Consequently, the use of rCF textiles as reinforcement in concrete presents significant potential for promoting sustainability within the construction industry.

For the nonwoven variants, the highest tensile strength with a standard deviation of 9.59 MPa was reached by nonwoven 2, which was cut into stripes and inserted into the laminating process using the entire surface. Although the amount of rCF nonwoven was smaller in this variant, the highest tensile strength was reached because the nonwoven in this case did not act as a separating layer and the concrete matrix bonded together. Compared to the nonwoven, the sliver reinforcement had no progressive failure and a lower highest tensile strength with a mean value of 5.71 MPa. The highest tensile strength was reached with the wrapping rCF yarn variants, with results of 13.41 MPa, closely followed by the friction yarn, with a tensile strength mean value of 12.72 MPa. Both yarns showed a gradual and direct failure. When the rCF yarn specimens were recycled with the hammer mill and the resulting rCF yarn pieces were used as a short fibre reinforcement, they still reached a tensile strength mean value of 6.38 MPa. Therefore, in this investigation, the different variants of rCF were examined and the first results were analysed.

Based on the research carried out so far, further investigations can now be made into the durability, long-term use, weathering resistance and chemical erosion resistance of the materials. Based on this, suitable use cases can then be evaluated for the variants and investigations can be carried out considering the entire life cycle.

In the context of the circular economy, it is essential to evaluate the entire life cycle of components, including material extraction, production and end-of-life recycling. Studies by Kimm et al. [51] and Kortmann [32] demonstrated that separating TRC at the end of its life cycle is feasible, with factors such as material type and construction method significantly influencing separability. An epoxy coating was found to enhance recovery rates, with hammer mills achieving over 90% recovery for carbon reinforcements.

Although recycled fibre materials can be utilised as short fibres in FRC, the immediate reuse of the textile structure is constrained by associated strength losses. However, the mineral content derived from recycled TRC can be repurposed as aggregates for a range of construction applications. Bayram et al. [50] explored the potential for recycling rCF yarns from concrete in a second life cycle. The use of the recycled rCF yarn pieces after the second life cycle was also tested as reinforcement in concrete for a quasi third life cycle. The mechanical properties showed an increase in the mechanical properties.

Once the mechanical properties, including durability and long-term behaviour, have been investigated in more detail, a further LCA as well as an economic assessment will be carried out.

The FaBeR project demonstrated the successful recycling of CFRC into industrial flooring, aiming to utilise all fractions of recycled components. Although initial production costs were high compared to similar products, advancing technology could lead to cost reductions. This study confirmed the significant circularity and lower global warming risks of recycled CFRC compared to conventional options.

Future research topics include the utilisation of textile surface structures made from rCF, i.e., grid structures such as those used in new TRCs made from recycled yarns. This would enable reinforcement in 2D instead of just 1D. The positioning of the yarn in the test specimen will also be possible more precisely than with 1D yarns due to the textile structure, as these partly float in the centre, and the 2D textiles achieve greater stability especially after impregnation.

Furthermore, the electrical conductivity of the carbon fibres can also be used as a sensor system. There are already several approaches for new carbon rovings. The transferability to recycled fibres due to the difficulty of the short fibre lengths also requires in-depth research and is planned for the future.

7. Conclusions

Carbon fibres' non-corrosive properties enable the creation of slimmer concrete components, potentially reducing the CO_2 emissions associated with cement production, which contributes about 8% of global emissions. While steel is the primary reinforcement material due to its high tensile strength, its corrosion susceptibility requires thicker covers, increasing material use. The construction sector accounts for 38% of global CO_2 emissions, necessitating effective emission reduction strategies. TRC uses technical textiles as reinforcements and has been studied for its mechanical advantages. Although producing rCF is energy-intensive, it offers significant energy savings and can lower global warming potential compared to virgin fibres. This study explored rCF's potential as a concrete reinforcement, addressing both opportunities and challenges.

In summary, the integration of recycled carbon fibres into carbon concrete presents a promising pathway to enhance the sustainability of construction materials. This approach not only reduces the environmental impact associated with fibre production but also improves the mechanical properties of concrete, offering both ecological and economic benefits. The continuous development and optimisation of recycling technologies are essential to fully realise these advantages and promote the adoption of sustainable practices in the construction sector.

Author Contributions: Conceptualisation, V.O. and T.G.; methodology, V.O. and T.G.; validation, V.O.; formal analysis, V.O.; investigation, V.O.; resources, T.G.; data curation, V.O.; writing—original draft preparation, V.O. and T.G.; writing—review and editing, V.O. and T.G.; visualisation, V.O.; supervision, T.G.; project administration, T.G.; funding acquisition, T.G. All authors have read and agreed to the published version of the manuscript.

Funding: This research was funded by the German Federal Ministry of Education and Research (BMBF) in the framework of the FONA Research for Sustainability Resource-efficient circular economy (ReMin) as part of the project FaBeR (Faser-und Beton-Recycling von Carbon-und Textilbeton) (grant no. 033R253A) and the PhD-program "Forschungskolleg Verbund. NRW—Verbunde im Bauwesen—Von der Ressourceneffizienz zur validierten Nachhaltigkeit", supported by the Ministry of Innovation, Science and Research North Rhine-Westphalian The financial support from the BMBF and the Ministry of Innovation, Science and Research North Rhine-Westphalian are gratefully acknowledged.

Institutional Review Board Statement: Not applicable.

Informed Consent Statement: Not applicable.

Data Availability Statement: All relevant data are available in the article.

Conflicts of Interest: The authors declare no conflicts of interest.

References

1. United Nations Environment Programme. *Global Resources Outlook 2019: Natural Resources for the Future We Want*; United Nations: New York, NY, USA, 2020; ISBN 978-92-807-3741-7.
2. United Nations Environment Programme. *Building Materials and the Climate: Constructing a New Future*; United Nations: New York, NY, USA, 2023; ISBN 978-92-807-4064-6.
3. Hensen, F.; Kolbmüller, H.; Stahr, M.; Weber, J.; Wild, U. Beton- und Stahlbeton. In *Bausanierung: Erkennen und Beheben von Bauschäden*; mit 145 Tabellen, 4., vollst. überarb. und aktualisierte Aufl.; Stahr, M., Hensen, F., Eds.; Vieweg + Teubner: Wiesbaden, Germany, 2009; pp. 115–147, ISBN 978-3-8348-0246-0.
4. WWF Deutschland. *Klimaschutz in der Beton- und Zementindustrie Berlin (2019)*; WWF Deutschland: Berlin, Germany, 2019.
5. United Nations Environment Programme. *2020 Global Status Report for Buildings and Construction Nairobi (Kenia) (2020)*; United Nations Environment Programme: Nairobi, Kenya, 2020.
6. *Intergovernmental Panel on Climate Change (IPCC) Global Warming of 1.5 °C.—(UK)*; Cambridge University Press: Cambridge, UK; New York, NY, USA, 2018.
7. Friese, D.; Scheurer, M.; Hahn, L.; Gries, T.; Cherif, C. Textile reinforcement structures for concrete construction applications—A review. *J. Compos. Mater.* **2022**, *56*, 4041–4064. [CrossRef]
8. Scheurer, M.; Friese, D.; Penzel, P.; Dittel, G.; Bhat, S.; Overhage, V.; Hahn, L.; Heins, K.; Cherif, C.; Gries, T. Current and Future Trends in Textiles for Concrete Construction Applications. *Textiles* **2023**, *3*, 408–437. [CrossRef]
9. Porta, N.W. Sustainability and Flexural Behaviour of Textile Reinforced Concrete. Bachelor's Thesis, Department of Civil and Environmental Engineering, Division of Structural Engineering, Concrete Structures, Chalmers University of Technology, Gothenburg, Sweden, 2012.
10. Kimm, M.K. Ressourceneffizientes und Recyclinggerechtes Design von Faserverbundwerkstoffen im Bauwesen. Ph.D. Dissertation, Rheinisch-Westfälische Technische Hochschule Aachen, Aachen, Germany, 2021.
11. Aschebrock, G.C. *Textile Reinforced Concrete: A Review*; Engineering School, Universidade Federal do Rio Grande do Sul (UFRGS): Porto Alegre, Brazil, 2022.
12. Haist, M.; Bergmeister, K.; Curbach, M.; Forman, P.; Gaganelis, G.; Gerlach, J.; Mark, P.; Moffatt, J.; Müller, C.; Müller, H.S.; et al. *Nachhaltigkeit, Digitalisierung, Instandhaltung*; 111. Jahrgang; Ernst & Sohn a Wiley Brand: Berlin, Germany, 2022.
13. Seifert, W.; Lieboldt, M. Ressourcenverbrauch im globalen Stahlbetonbau und Potenziale der Carbonbetonbauweise. *Beton Stahlbetonbau* **2020**, *115*, 469–478. [CrossRef]
14. Li, H.; Yang, J.; Yang, D.; Zhang, N.; Nazar, S.; Wang, L. Fiber-reinforced polymer waste in the construction industry: A review. *Environ. Chem. Lett.* **2024**, *22*, 2777–2844. [CrossRef]
15. Meng, F.; McKechnie, J.; Turner, T.A.; Pickering, S.J. Energy and environmental assessment and reuse of fluidised bed recycled carbon fibres. *Compos. Part A Appl. Sci. Manuf.* **2017**, *100*, 206–214. [CrossRef]
16. Meiners, D.; Eversmann, B. Recycling von Carbonfasern. In *Recycling und Rohstoffe*; TK-Verl.: Neuruppin, Germany, 2014; ISBN 978-3-944310-09-1.
17. Rani, M.; Choudhary, P.; Krishnan, V.; Zafar, S. A review on recycling and reuse methods for carbon fiber/glass fiber composites waste from wind turbine blades. *Compos. Part B Eng.* **2021**, *215*, 108768. [CrossRef]
18. Qureshi, J. A Review of Recycling Methods for Fibre Reinforced Polymer Composites. *Sustainability* **2022**, *14*, 16855. [CrossRef]
19. Gharde, S.; Kandasubramanian, B. Mechanothermal and chemical recycling methodologies for the Fibre Reinforced Plastic (FRP). *Environ. Technol. Innov.* **2019**, *14*, 100311. [CrossRef]
20. Karuppannan Gopalraj, S.; Kärki, T. A review on the recycling of waste carbon fibre/glass fibre-reinforced composites: Fibre recovery, properties and life-cycle analysis. *SN Appl. Sci.* **2020**, *2*, 433. [CrossRef]
21. Baumgaertel, E.; Marx, S. The Recycling of Carbon Components and the Reuse of Carbon Fibers for Concrete Reinforcements. *Appl. Sci.* **2023**, *13*, 7091. [CrossRef]
22. Imbert, M.; Hahn, P.; Jung, M.; Balle, F.; May, M. Mechanical laminae separation at room temperature as a high-quality recycling process for laminated composites. *Mater. Lett.* **2022**, *306*, 130964. [CrossRef]
23. Quicker, P.; Stockschläder, J. *Möglichkeiten und Grenzen der Entsorgung Carbonfaserverstärkter Kunststoffabfälle in Thermischen Prozessen: Abschlussbericht*; Umweltbundesamt: Dessau-Roßlau, Germany, 2021.
24. *Einführung in die Kreislaufwirtschaft: Planung-Recht-Verfahren*; Kranert, M., Ed.; 5. Auflage; Springer Vieweg: Wiesbaden/Heidelberg, Germany, 2017; ISBN 3834818372.
25. Horstmann, V. Vergleichende Untersuchungen zur Mechanischen Aufbereitung von Carbonbeton. Master's Thesis, Bauhaus-Universität Weimar, Weimar, Germany, 2018.
26. Isa, A.; Nosbi, N.; Che Ismail, M.; Md Akil, H.; Wan Ali, W.F.F.; Omar, M.F. A Review on Recycling of Carbon Fibres: Methods to Reinforce and Expected Fibre Composite Degradations. *Materials* **2022**, *15*, 4991. [CrossRef] [PubMed]
27. Rabe, D.; Häntzsche, E.; Cherif, C. Recycling of Carbon Fibres and Subsequent Upcycling for the Production of 3D-CFRP Parts. *Materials* **2022**, *15*, 5052. [CrossRef] [PubMed]

28. Benka, C. Systematische Untersuchung Thermischer Prozesse für das Recycling von Kohlenstofffaserverstärkten Kunststoffen. Bachelor's Thesis, University of Applied Sciences, Augsburg, Germany, 2015.
29. Almushaikeh, A.M.; Alaswad, S.O.; Alsuhybani, M.S.; AlOtaibi, B.M.; Alarifi, I.M.; Alqahtani, N.B.; Aldosari, S.M.; Alsaleh, S.S.; Haidyrah, A.S.; Alolyan, A.A.; et al. Manufacturing of carbon fiber reinforced thermoplastics and its recovery of carbon fiber: A review. *Polym. Test.* **2023**, *122*, 108029. [CrossRef]
30. Onwudili, J.A.; Miskolczi, N.; Nagy, T.; Lipóczi, G. Recovery of glass fibre and carbon fibres from reinforced thermosets by batch pyrolysis and investigation of fibre re-using as reinforcement in LDPE matrix. *Compos. Part B Eng.* **2016**, *91*, 154–161. [CrossRef]
31. Kim, K.-W.; Lee, H.-M.; An, J.-H.; Chung, D.-C.; An, K.-H.; Kim, B.-J. Recycling and characterization of carbon fibers from carbon fiber reinforced epoxy matrix composites by a novel super-heated-steam method. *J. Environ. Manag.* **2017**, *203*, 872–879. [CrossRef]
32. Liu, Y.; Meng, L.; Huang, Y.; Du, J. Recycling of carbon/epoxy composites. *J. Appl. Polym. Sci.* **2004**, *94*, 1912–1916. [CrossRef]
33. Feraboli, P.; Kawakami, H.; Wade, B.; Gasco, F.; DeOto, L.; Masini, A. Recyclability and reutilization of carbon fiber fabric/epoxy composites. *J. Compos. Mater.* **2012**, *46*, 1459–1473. [CrossRef]
34. Mitsubishi Chemical Advanced Materials GmbH. *Produktdatenblatt CarboNXT Chopped NP 5*; Mitsubishi Chemical Advanced Materials GmbH: Lotte, Germany, 2019.
35. Mitsubishi Chemical Advanced Materials GmbH. *Produktdatenblatt CarboNXT Chopped PURE*; Mitsubishi Chemical Advanced Materials GmbH: Lotte, Germany, 2021.
36. Teijin Carbon Europe GmbH. *Produkttadenblatt (EU): Tenax Filament Yarn: Version 1.2*; Teijin Carbon Europe GmbH: Wuppertal, Germany, 2020.
37. Kortmann, J. *Verfahrenstechnische Untersuchungen Zur Recyclingfähigkeit Von Carbonbeton*; Springer Fachmedien Wiesbaden GmbH: Wiesbaden, Germany, 2020; ISBN 9783658301255.
38. Veit, D. *Fasern: Geschichte, Erzeugung, Eigenschaften, Markt*; Springer Vieweg: Berlin/Heidelberg, Germany, 2023, ISBN 978-3-662-64468-3.
39. Gries, T.; Veit, D.; Wulfhorst, B. *Textile Fertigungsverfahren*; Carl Hanser Verlag GmbH & Co. KG: Munich, Germany, 2018.
40. Overhage, V.; Gries, T. Use of Recycled Carbon Fibres in Textile Reinforced Concrete for the Construction Industry. In *Proceedings of the International RILEM Conference on Synergising Expertise Towards Sustainability and Robustness of Cement-Based Materials and Concrete Structures: SynerCrete'23—Volume 1, 1st ed., Milos, Greece, 14–16 June 2023*; Jędrzejewska, A., Kanavaris, F., Azenha, M., Benboudjema, F., Schlicke, D., Eds.; Springer Nature Switzerland, Imprint Springer: Cham, Switzerland, 2023; pp. 540–551, ISBN 978-3-031-33210-4.
41. *DIN EN 1170-5*; Precast Concrete Products—Test Method for Glass-Fibre Reinforced Cement—Part 5: Measuring Bending Strength by the 'Complete Bending Test' Method. Deutsches Institut für Normung: Berlin, Germany, 1997.
42. Brockmann, T. *Mechanical and Fracture Mechanical Properties of Fine Grained Concrete for Textile Reinforced Composites*; Verlag Günter Mainz: Mainz, Germany, 2006; ISBN 3-86130-631-X.
43. Kimm, M.; Bhat, S.; Gries, T. *Recycling von Textilbeton*; Thomé-Kozmiensky Verlag GmbH: Neuruppin, Germany, 2020.
44. Cuntze, R. *Fachbegriffe Für Kompositbauteile—Technical Terms for Composite Parts: Glossar Für Bauwesen und Maschinenbau—Glossary for Civil- and Mechanical Engineering*; Springer Fachmedien Wiesbaden GmbH: Wiesbaden, Germany, 2019; ISBN 9783658256357.
45. Wietek, B. *Beton—Stahlbeton—Faserbeton: Eigenschaften und Unterschiede*; Springer Fachmedien Wiesbaden GmbH: Wiesbaden, Germany, 2019; ISBN 9783658277079.
46. Wietek, B. *Faserbeton: Im Bauwesen*, 3rd ed.; Springer Fachmedien Wiesbaden GmbH: Wiesbaden, Germany, 2020; ISBN 9783658308759.
47. Kimm, M.; Overhage, V.; Gries, T.; Sodagar, M. Recycled carbon fibers with improved surface characteristics as reinforcing material for fiber concrete. In Proceedings of the 2021 Optical Fiber Communications Conference and Exhibition (OFC), San Francisco, CA, USA, 6–11 June 2021.
48. Overhage, V.; Heins, K.; Kimm, M.; Ryu, G.-S.; You, Y.-J.; Kim, H.-Y.; Gries, T. Investigation of pull-out and mechanical performance of fibre reinforced concrete with recycled carbon fibres. *Mater. Struct.* **2024**, *57*, 184. [CrossRef]
49. Textilnorm, Fachnormenausschuß Textil- und Textilmaschinenindustrie im Deutschen Normenausschuß. In *Textilien: Grundbegriffe*; Deutsche Normen: Berlin, Germany, 1969.
50. Bayram, B.; Overhage, V.; Löwen, M.; Terörde, K.; Raulf, K.; Greiff, K.; Gries, T. Recycling potential of carbon fibres in the construction industry: From a technical and ecological perspective. *Waste Manag. Res.* **2024**, *42*, 726–737. [CrossRef]
51. Kimm, M.; Gerstein, N.; Schmitz, P.; Simons, M.; Gries, T. On the separation and recycling behaviour of textile reinforced concrete: An experimental study. *Mater. Struct.* **2018**, *51*, 122. [CrossRef]
52. Luthin, A.; Crawford, R.H.; Traverso, M. Demonstrating circular life cycle sustainability assessment—A case study of recycled carbon concrete. *J. Clean. Prod.* **2023**, *433*, 139853.

sustainability

MDPI

Article

Spatiotemporal Characterization of the Three-Dimensional Morphology of Urban Buildings Based on Moran's I

Tao Shen *, Wenshiqi Zhou, Shuai Yuan and Liang Huo *

School of Geomatics and Urban Spatial Information, Beijing University of Civil Engineering and Architecture, Beijing 100044, China; zhouwenshiqi@163.com (W.Z.); yuanshuai1203@163.com (S.Y.)
* Correspondence: shentao@bucea.edu.cn (T.S.); huoliang@bucea.edu.cn (L.H.)

Abstract: The three-dimensional morphological analysis of urban buildings constitutes a pivotal component of urban planning and sustainable development. Nevertheless, the majority of current research is two-dimensional in nature, which constrains the comprehensive understanding of urban spatial–temporal evolution. The existing body of three-dimensional studies frequently fails to consider the temporal dimension of architectural change and lacks a detailed examination of micro areas such as communities and streets. In order to accurately identify the patterns of spatial–temporal evolution in urban architectural morphology, this study focuses on the Yau Tsim Mong District in Hong Kong, utilizing three-dimensional data. By innovatively integrating temporal factors, constructing a spatial–temporal weight matrix, and applying the spatial–temporal Moran's I, this study conducts an in-depth quantitative analysis of Coverage, Staggeredness, and Duty Cycle at the community scale, neighborhood scale, and urban scale. From 2014 to 2023, the global spatial–temporal Moran's I of key urban morphology indicators in Yau Tsim Mong District has exhibited a marked increase, underscoring the close interrelationship and significant optimization between urban morphology and overall development. The findings illustrate that urban architecture is undergoing a process of agglomeration and high homogeneity, with strategic shifts emphasizing efficient spatial utilization and refined design. The analysis at the neighborhood scale is of particular importance, as its independent and complete spatial structure effectively captures local dynamics, revealing high-value agglomeration and low-value dispersion characteristics. This suggests that buildings in the Yau Tsim Mong District are being constructed in a more compact manner at the neighborhood level, which reflects the precision and efficiency of urban planning and the rationality of spatial planning. These significant findings provide valuable references for the development planning and governance of sustainable cities. They enhance urban governance capabilities and promote the optimization of urban development strategies, ensuring steady progress on the path of efficiency, harmony, and sustainability.

Keywords: spatial weight matrix; spatial–temporal autocorrelation analysis; Moran's I; urban morphology

Citation: Shen, T.; Zhou, W.; Yuan, S.; Huo, L. Spatiotemporal Characterization of the Three-Dimensional Morphology of Urban Buildings Based on Moran's I. *Sustainability* **2024**, *16*, 6540. https://doi.org/10.3390/su16156540

Academic Editors: Malindu Sandanayake and Zora Vrcelj

Received: 5 June 2024
Revised: 19 July 2024
Accepted: 29 July 2024
Published: 31 July 2024

1. Introduction

The rapid acceleration of large-scale urbanization processes worldwide is driving cities to expand vertically as well as horizontally, due to the surge in urban populations and the scarcity of land resources [1,2]. The continuous expansion of urban areas, the rise of megacities, and the swift development of high-rise buildings are collectively shaping a more modern and three-dimensional urban landscape [3]. Consequently, there is an increasing demand from the international community for higher standards to be applied to urban development. This has led to a greater emphasis on the scientific assessment of urban development, the planning of urban structures and the assurance of sustainability and livability [4,5].

The discipline of urban morphology is concerned with the physical form and structure of cities [6]. By examining a range of indicators pertinent to urban morphology, it is possible to express the spatial structure and internal relationships of cities in quantitative terms [7]. An examination of urban morphology from a sustainability perspective, coupled with an evaluation of urban architecture, facilitates the formulation of urban planning strategies and provides robust support for sustainable urban development [8,9].

Since the concept of sustainable development was introduced, it has attracted considerable attention from scholars engaged in research on the built environment and cities [8]. The analysis of changes in urban morphology constitutes a fundamental reference point for the pursuit of sustainable urbanization [1]. Historically, a significant proportion of research on urban morphology has concentrated on the two-dimensional aspect, which is inadequate for meeting the needs of modern cities [10].

In recent years, the rapid advancements in computer technology and data acquisition techniques have facilitated a new phase of academic exploration of urban morphology [11–20]. A considerable number of scholars have engaged in regional analysis of urban three-dimensional morphology, precise measurement of residential space morphology, and in-depth analysis of the spatial–temporal differentiation of urban three-dimensional morphology, thereby significantly enriching the research in this field [21–36]. It is noteworthy that morphological indicators, which are of great importance in urban planning research, are widely employed to elucidate the intrinsic logic and mechanisms of urban spatial development. They provide substantial support for the analysis of urban development changes [36–48].

Concurrently, the scope of urban morphology research has been progressively extended. Scholars have broadened their perspectives beyond the morphology itself to encompass various aspects of urban development [17,22–25]. This has led to a closer link between urban morphology and issues such as traffic congestion, environmental quality and social equity. This has resulted in interdisciplinary and multidimensional comprehensive discussions [37,49–51]. This shift is indicative of the contemporary value and social significance of urban morphology research.

A substantial corpus of accumulated research evidence provides compelling evidence of the close link between urban morphology and the quality of the urban environment and sustainable development patterns [8]. The quantitative analysis of urban morphology indicators enables the scientific assessment of urban development sustainability, thereby providing a robust theoretical foundation for the construction of green, low-carbon, and livable urban environments [14–27,43]. Furthermore, this provides a highly valuable reference point and guidance for optimizing urban planning and layout [4–9]. The findings of this series of research studies not only enhance our comprehension of urban morphology but also indicate potential avenues for fostering more harmonious and sustainable urban development.

Notwithstanding the current emphasis on the evolving nature of urban spatial structures, there remains a notable research gap in the exploration of the specific changes in urban architectural forms over time [7,27–36,38,51–53]. Although some studies have sought to integrate spatial and temporal factors, they frequently adopt a macro-level perspective when examining the evolutionary patterns of cities, thereby overlooking the intricate changes occurring at the micro level, such as those observed in streets and communities [42–47,54]. This limitation constrains our comprehensive understanding of the complexity of urban architectural morphology and also fails to precisely capture and reveal the critical details that shape the unique characteristics of cities. It is therefore imperative that future research should seek to strengthen the detailed analysis of urban architecture over time, while balancing macro and micro perspectives.

In order to address the aforementioned research gap, this study employs the Spatial–temporal Moran's I as an analytical tool in order to conduct a detailed quantitative analysis of urban morphology indicators across multiple scales, including the community scale, the neighborhood scale and the urban scale. This study aims to provide a detailed and comprehensive exploration of the spatial–temporal evolution patterns of urban architectural three-dimensional morphology at different spatial scales. The objective is to depict the dynamic characteristics and complex structures of urban architectural morphology in a more accurate and precise manner. The objective of this research is to enhance urban architectural morphology in a manner that reinvigorates sustainable urban development optimizes resource utilization, enhances the quality of life for residents and contributes to the creation of greener, more livable and harmonious urban environments.

2. Study Area and Data Source

2.1. Study Area

The Yau Tsim Mong District is situated at the dynamic southern extremity of the Kowloon Peninsula and represents one of Hong Kong's most emblematic areas, largely due to its distinctive geographical location and rich cultural heritage (Figure 1). This district is a densely populated area comprising the vibrant commercial districts of Yau Ma Tei, Tsim Sha Tsui, and Mong Kok. Each of these areas has a distinctive historical legacy and a contemporary urban character. The architectural space morphology within Yau Tsim Mong District is characterized by a high degree of complexity and variety. It encompasses a diverse array of building types, including historic traditional structures and contemporary skyscrapers, which collectively contribute to the formation of the district's distinctive urban landscape. An examination of the spatial characteristics of these buildings is essential for a more profound comprehension of the structure, function, and evolution of Hong Kong's urban space.

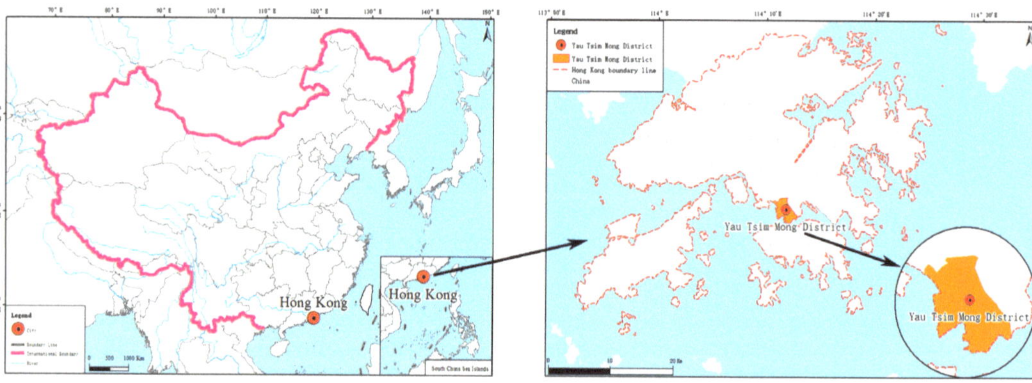

Figure 1. Study Area Yau Tsim Mong District, Hong Kong, China.

Furthermore, the cultural facilities within Yau Tsim Mong District, such as museums, art galleries, and theatres, are fundamental elements in the study of urban space. Such facilities not only enrich the cultural and spiritual lives of residents but also facilitate cultural exchange and dissemination. In the planning of future urban spaces, it is of the utmost importance to give full consideration to the layout and functions of these cultural facilities. This will ensure that they contribute to sustainable urban development while creating a more diverse and higher quality living experience for citizens [55].

2.2. Data Sources

The data employed in this study are principally derived from the Hong Kong Special Administrative Region Planning Department and Google Earth's photorealistic 3D data service. The dataset includes oblique photogrammetric models from 2014 to 2023, which provide detailed three-dimensional representations of the built environment in the Yau Tsim Mong District, illustrating the temporal changes that have occurred (Figure 2). Concurrently, data on urban land use types and spatial information during the same period were collected, thereby providing a scientific basis for understanding the distribution and changes in different land use types in the Yau Tsim Mong District.

(a) (b)

Figure 2. (**a**) The slope model display of Yau Tsim Mong District in Hong Kong; (**b**) the corresponding three-dimensional model data.

3. Methodology

3.1. Urban Morphology Indicators

This study builds upon existing research to further explore the spatiotemporal characteristics of urban architectural 3D morphology in the Yau Tsim Mong District of Hong Kong through a quantitative analysis of urban morphology indicators [1,4–9,22]. In order to meet the research objectives, four relevant urban morphology indicators (Table 1) were selected with great care to provide a comprehensive and accurate representation of the urban architectural morphology characteristics of the area in question. In particular, the "Staggeredness" index, defined as the ratio of Building Height Standard Deviation to Building Height, was introduced to visually illustrate the distribution differences in building heights. Furthermore, the concepts of "Duty Cycle" and "Coverage" offer a novel interpretation of traditional measures such as Building Volume Density and Building Floor Area Ratio [7,29]. Table 1 provides detailed descriptions of each indicator, offering clear and standardized references for subsequent analysis and citation purposes.

Table 1. Urban morphology indicators.

Name	Morphology Indicators	Definition	Significance
Staggeredness	Building Height Standard Deviation	Height of all buildings in the space unit standard deviation	Evaluating the vertical development status of urban areas, a larger Staggeredness value indicates more significant height differences among urban buildings.
	Building Height	The average height of buildings in the area	
Duty Cycle	Building Volume Density	The ratio of building volume to the total volume within a space unit	This metric illustrates the volume disparity among building entities within an urban area; a larger value reflects a higher building density in the region.
Coverage	Building Floor Area Ratio	The ratio of gross floor area to building land area in a spatial unit	This metric measures the proportion of building footprint area relative to the area of each study unit within an urban region.

3.2. Moran's I

The objective of spatial autocorrelation studies is to investigate the correlation between the observation values of a given attribute within disparate spatial regions. This is achieved by focusing on the spatial structural characteristics of variables, with the aim of elucidating the inherent connections between regions [56,57]. The concept is divided into two main categories: global and local autocorrelation. Global autocorrelation typically employs statistics such as Moran's I to quantify the overall spatial correlation, whereas local autocorrelation utilizes metrics like the local Moran's I and Getis-Ord Gi/Gi* to elucidate local spatial heterogeneity [27].

In this study, Moran's I is employed in conjunction with a time factor to construct a spatiotemporal weight matrix, thereby enabling the innovative calculation of global and local spatial–temporal Moran's I. This approach is applied in the analysis of urban architectural morphology, enabling a profound exploration of the spatiotemporal correlation patterns of building forms. It provides an accurate delineation of spatial agglomeration and dispersion characteristics across different scales, offering a unique perspective and valuable reference for studies on temporally sensitive architectural morphology.

3.2.1. Global Moran's I

Global Moran's I statistic provides a precise assessment of the similarity of attributes between adjacent study units, thereby revealing their spatial correlation [58]. Furthermore, it effectively identifies the overall spatial distribution patterns of these attributes. The mathematical expression is concise and intuitive, as demonstrated by Equations (1) and (2):

$$I = \left(\frac{n}{\sum_{i=1}^{n}\sum_{j=1}^{n} W_{ij}}\right)\left(\frac{\sum_{i=1}^{n}\sum_{j=1}^{n} W_{ij}(a_i-\overline{a})(a_j-\overline{a})}{\sum_{i=1}^{n}(a_i-\overline{a})^2}\right) \quad (1)$$

$$\overline{a} = \frac{1}{n}\sum_{i=1}^{n} a_i \quad (2)$$

In the aforementioned equation, n represents the total number of study units, a_i denotes the observed value of the research element within the i-th study unit, $\overline{a}$ is the mean observation value of the research element, and W_{ij} represents the spatial weight matrix indicating adjacency relationships between study units.

Global Moran's I statistic ranges from −1 to 1, with negative values indicating dispersion and positive values indicating clustering. In the context of research, it is essential to select an appropriate spatial weight matrix that takes into account both adjacency and distance-based types [59]. The adjacency matrix is based on the assumption of binary adjacency, whereby spatial interaction is presumed to occur exclusively between adjacent sub-regions. The aforementioned adjacency relationships can be exemplified by the Bishop (point contact), Rook (edge contact), and Queen (edge and point contact) patterns, as illustrated in Figure 3, which provide a precise spatial relationship framework for analysis.

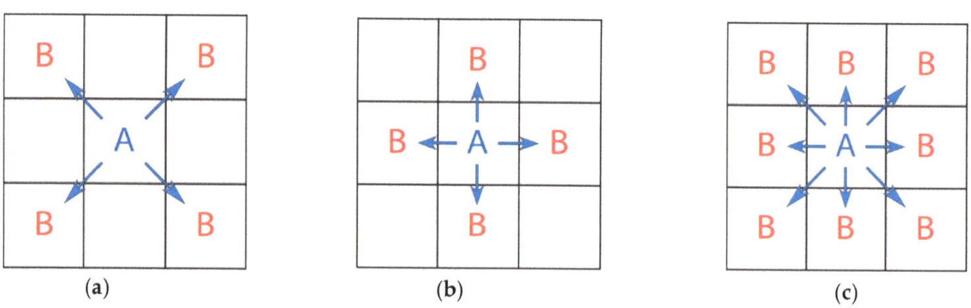

Figure 3. (**a**) Bishop adjacency; (**b**) Rook adjacency; (**c**) Queen adjacency.

Distance-based matrices are instrumental in characterizing the distance relationships between sub-regions, and are therefore of paramount importance in the context of spatial autocorrelation analysis. The most common types are inverse distance weight matrices and binary geographical distance weight matrices. The spatial weight matrix, represented by the symbol W_{ij}, can be expressed by the following Formula (3):

$$\text{Distance} - W_{ij} = \begin{cases} 1, \ d\{ij\} < d \\ 0, \ d\{ij\} \geq d \end{cases} \tag{3}$$

In this context, d_{ij} represents the distance between sub-regions i and j.

3.2.2. Local Moran's I

Local Moran's I provides a more detailed examination of the local relationships and heterogeneity of spatial attribute values, presenting the distribution characteristics of elements at a local scale in a visual format. The results are presented in the form of scatter plots, with specific clustering types listed in Table 2. This comprehensive approach reveals the intricate relationships between spatial units [60]. This analysis not only enhances understanding of spatial relationships but also captures local autocorrelation phenomena that may be missed by Global Moran's I. The principles are detailed in Formulas (4)–(6).

$$I_i = \frac{Z_i}{S^2} \sum_{j \neq 1}^{n} w_{ij} Z_j \tag{4}$$

$$Z_j = y_j - \overline{y} \tag{5}$$

$$s^2 = \frac{1}{n} \sum \left(y_i - \overline{y}\right)^2 \tag{6}$$

Table 2. Types of clustering and their meanings for the local Moran's I.

Cluster Type	Hidden Meaning
High–High Clustering	High-value regions are also surrounded by high-value regions, showing a positive spatial correlation
High–Low Clustering	High-value regions are surrounded by low-value regions, showing a negative spatial correlation
Low–Low Clustering	Low-value regions are also surrounded by low-value regions, showing a positive spatial correlation
Low–High Clustering	Low-value regions are surrounded by high-value regions, showing negative spatial correlation

In the aforementioned equation, the variable I_i represents the local Moran's I for the i-th region, n denotes the total number of study areas, and W_i is the spatial weight for region i. The spatial weight, represented by W_{ij}, is the measure of the relationship between regions i and j. The attribute values for regions i and j, represented by y_i and y_j, respectively, are the observed values. The mean attribute value, represented by $\overline{y}$, is the overall value.

3.3. Spatiotemporal Weighting Matrix

The construction of a spatiotemporal weight matrix requires the integration of spatial and temporal dimensions in order to accurately depict the evolution of geographical phenomena over time and space. The matrix is of size $n \times n$, where n is the number of spatial units, and the elements reflect the spatiotemporal proximity between units.

In this study, Inverse Distance Weighting is employed to quantify spatial weights, whereas temporal weights are derived from the disparities in time series. Adjacent time points are assigned a weight of 1, while non-adjacent points are assigned a weight of 0, thereby ensuring temporal continuity. To illustrate, temporal weights are set to 1 for consecutive years and 0 for non-consecutive years when data from 2014, 2019, and 2023 are employed. By integrating oblique model data from the study area, the spatial weights of urban buildings are precisely defined using the IDW method. This configuration considers both spatial distance and temporal continuity, thereby accurately capturing the dynamic changes and evolution patterns of urban architectural morphology over time and space.

The generation of the spatiotemporal weight matrix, designated as W, necessitates the consideration of the interrelationships between spatial locations and time series. The input data is a two-dimensional array, with rows representing spatial locations and columns representing time points. Consequently, the dimensions of the W matrix are $n \times t \times n \times t$ where n is the number of spatial locations and t is the length of the time series. The following section provides a brief overview of the four-layer loop employed for the construction of the W matrix (Figure 4).

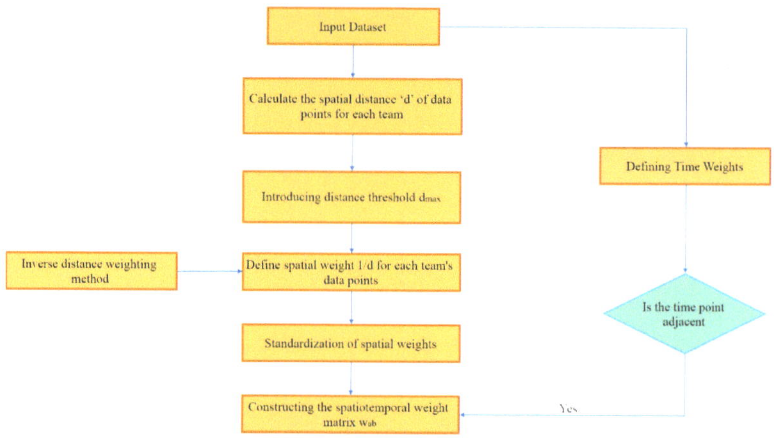

Figure 4. Spatiotemporal weight matrix construction process.

1. The outer loop iterates through all spatial locations, designated as i;
2. The inner loop one traverses all time points, designated as j;
3. The outer loop is then executed again, traversing all spatial locations, designated as k and comparing the location designated as i;
4. The innermost loop traverses all time points, designated as l and compares the time point designated as j;

For each combination of spatial locations (i, k) and time points (j, l) the value of W is calculated based on the spatial distance and time difference. This process comprehensively covers all possible spatiotemporal combinations, thus ensuring that the resulting matrix W, reflects both the spatial adjacency and temporal sequence relationships.

3.4. Spatial–Temporal Moran's I

This study broadens the application of Global Moran's I to encompass the spatiotemporal dimension. This is achieved by defining the spatiotemporal object, denoted as $ST_{(a,i)}$, which incorporates information pertaining to both the spatial location a and the time point i. The construction of the spatiotemporal weight matrix is based on the principle that two spatiotemporal objects are considered to be spatiotemporally adjacent if they are both spatially and temporally adjacent. In such cases, the corresponding weight matrix element, designated as $W_{(a,i)(b,j)}$, is set to a value of 1; otherwise, it is set to 0.

Global spatial–temporal Moran's I is calculated using this weight matrix, with its positive or negative values directly reflecting the overall spatiotemporal evolution characteristics of the three-dimensional morphology of buildings within the study area. The calculation formula for this index is provided in Equation (7), which allows for a comprehensive analysis and quantification of the impact of spatiotemporal adjacency relationships on the dynamic changes in building morphology.

$$I = \frac{nt\sum_{i=1}^{n}\sum_{a=1}^{t}\sum_{j=1}^{n}\sum_{b=1}^{t} W(a,i)(b,j)\left(y_{(a,i)} - \overline{y}\right)\left(y_{(q,j)} - \overline{y}\right)}{\sum_{i=1}^{n}\sum_{a=1}^{t}\sum_{j=1}^{n}\sum_{b=1}^{t} W(a,i)(b,j)\sum_{i=1}^{n}\sum_{a=1}^{t}\left(y_{(a,i)} - \overline{y}\right)^{2}} \tag{7}$$

Local spatial–temporal Moran's I is an extension of the global version that is capable of taking values beyond the range of $[-1,\ 1]$, thereby allowing for a more flexible reflection of spatiotemporal correlations. In the local analysis of the spatiotemporal object, denoted by $ST_{(a,i)}$, the results of this study indicate that positive values signify a positive correlation between the object and its surrounding area, with the strength of this correlation increasing as the value rises. Conversely, negative values indicate a negative correlation, with the negative correlation strength intensifying as the absolute value increases. A value of 0, on the other hand, denotes the absence of a spatiotemporal correlation. This calculation takes into account both local spatial and temporal relationships in a comprehensive manner. The specific formulas are provided in Equations (8)–(10), which offer a robust tool for in-depth analysis of the spatiotemporal evolution of three-dimensional building morphology.

$$I_{(a,i)} = Z_{(a,i)} W_{Z(a,i)} \tag{8}$$

$$Wz_{(a,i)} = \frac{\sum_{a=0}^{n}\sum_{j=0}^{t} w_{(a,i)(b,j)} Z_{(b,j)}}{\sum_{a=0}^{n}\sum_{j=0}^{t} w_{(a,i)(b,j)}} \tag{9}$$

$$Z_{(a,i)} = \frac{\left(y_{(a,i)} - \overline{y}\right)}{s^{2}} \tag{10}$$

4. Results

4.1. Analysis of Multiscale Spatial–Temporal Variations in Urban Buildings Three-Dimensional Morphology

In examining the evolution of urban buildings spatial morphology, Coverage serves as a principal indicator for assessing building density and spatial utilization. Its trajectory reveals significant implications at different spatial scales. As illustrated in Figure 5, the global spatial–temporal Moran's I for Coverage demonstrates a notable upward trajectory from 2014 to 2023, discernible at various spatial scales.

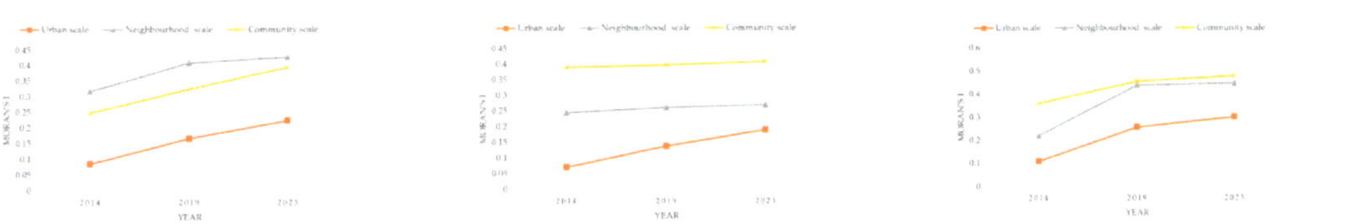

(a) (b) (c)

Figure 5. Sequential trend graphs of the global spatial–temporal Moran's I for Coverage (**a**), Staggeredness (**b**), and Duty Cycle (**c**).

At the community scale, the Coverage index demonstrated a gradual increase from 0.248 to 0.393 (Table 3), indicating a tendency towards greater building density within communities and a shift towards more intensive spatial utilization patterns. As the smallest units of urban space, optimized and compact building layouts within communities provide a solid foundation for enhancing residents' quality of life and the efficient use of urban space.

Table 3. Global spatial–temporal Moran's I for Coverage, Staggeredness, and Duty Cycle at the Community Scale in the years 2014, 2019, and 2023.

Year	Coverage	Staggeredness	Duty Cycle
2014	0.248	0.389	0.359
2019	0.324	0.396	0.455
2023	0.393	0.407	0.479

At the neighborhood scale, the Coverage index also demonstrated an upward trajectory (Table 4), increasing from 0.317 to 0.426. This shift towards balanced and dense building coverage within neighborhoods indicates a trend towards overall functionality and spatial optimization.

Table 4. Global spatial–temporal Moran's I for Coverage, Staggeredness, and Duty Cycle at the Neighborhood Scale in the years 2014, 2019, and 2023.

Year	Coverage	Staggeredness	Duty Cycle
2014	0.317	0.244	0.219
2019	0.408	0.261	0.438
2023	0.426	0.269	0.446

From a macro perspective, the significant increase in the Coverage index (Table 5), from 0.085 to 0.223, directly reflects the accelerated urbanization process at the urban scale. The aggregation and distribution of urban buildings over a larger area not only shape the unique spatial patterns and landscapes of the city but also enhance the optimized allocation and efficient utilization of urban spatial resources, thereby providing a strong impetus for sustainable urban development.

Table 5. Global spatial–temporal Moran's I for Coverage, Staggeredness, and Duty Cycle at the Urban Scale in the years 2014, 2019, and 2023.

Year	Coverage	Staggeredness	Duty Cycle
2014	0.085	0.070	0.108
2019	0.166	0.137	0.256
2023	0.223	0.189	0.300

As a pivotal indicator for gauging disparities in building height and assessing the coherence of spatial layouts, an in-depth examination of Staggeredness is imperative. Figure 5 presents a variation curve of Staggeredness, which illustrates the evolution of building height differences and spatial layout harmony.

At the community scale, Global Moran's I for Staggeredness exhibited a slight increase from 0.389 to 0.407 (Table 3), reflecting a gradual reduction in building height differences and the harmonious unification of spatial layouts within communities. This transformation serves to enhance the overall aesthetic appeal and visual attractiveness of the community.

At the neighborhood scale, the alterations in Staggeredness manifest in a more intricate manner (Table 4). A slight increase in the Staggeredness index results in a more balanced and coordinated height distribution of buildings within neighborhoods, thereby enhancing the spatial hierarchy and visual impact of the neighborhood.

From a macro perspective at the urban scale (Table 5), the rapid growth of the Staggeredness index demonstrates the richness and diversity in the city's skyline height levels. The staggered arrangement of high-rise and low-rise buildings not only creates a distinctive urban skyline but also reflects the meticulous planning and design of urban spaces in terms of height levels, thereby demonstrating the efficacy of such an approach.

As illustrated in Figure 5, alterations in Duty Cycle also manifest notable discrepancies across a range of spatial scales. As a principal indicator for the assessment of building footprint and spatial utilization efficiency, the variations in this index are of critical importance for the evaluation of the optimized allocation of urban spatial resources and sustainable development.

At the community scale, the considerable increase in the Duty Cycle index (Table 3), from 0.359 to 0.479, directly reflects the enhancement in land use efficiency and the optimization of building layouts within communities. This optimization provides residents with a more spacious and comfortable living environment.

At the neighborhood scale, the alterations in the Duty Cycle index (Table 4) provide further evidence of the effective integration and utilization of spatial resources. As the Duty Cycle index increases, the spatial aggregation of building footprints within neighborhoods is enhanced, thereby promoting the coordinated development of neighborhood functions.

From a macro perspective at the urban scale (Table 5), the increase in the Duty Cycle index demonstrates the optimized layout and efficient use of building footprints over a larger spatial extent. The implementation of scientific spatial planning and rational architectural design strategies has enabled the precise matching and efficient integration of building footprints with spatial resources, thereby establishing a robust foundation for the sustainable development of the city.

4.2. Spatial–Temporal Analysis of Urban Buildings 3D Morphological Features at the Neighborhood Scale

In light of the preceding discussion concerning the findings of global spatial–temporal Moran's I analysis, this study concentrates on the neighborhood scale with the objective of elucidating the spatiotemporal transformations in urban edifices within the Yau Tsim Mong District in greater detail. The neighborhood scale, with its relatively independent and complete spatial structure, is more effective in capturing phenomena of agglomeration, dispersion, and heterogeneity, thereby providing a precise perspective for urban planning and management. Figure 6 provides a clear illustration of the dynamic changes in the number and percentage of scatter points for the Coverage, Duty Cycle, and Staggeredness indices at the community, neighborhood, and urban scales from 2014 to 2023 in terms of local Spatial–temporal Moran's I. In this chart, the height of the bar graph represents the number of local Spatial–temporal Moran's I scatter points, with specific values referenced by the primary vertical axis (left). The line graph indicates the proportion of local Spatial–temporal Moran's I scatter points, with specific values referenced by the secondary axis (right). The horizontal charts represent different urban morphology indicators and the vertical charts describe the quadrants where local Spatial–temporal Moran's I scatter points are located.

With regard to the Coverage index, the continuous increase in the number of scatter points in the first quadrant (H-H) from 111 to 131 clearly indicates a significant agglomeration of high-Coverage units at the neighborhood scale. This agglomeration not only reflects efficient land use but also indicates a trend towards concentrated urban development. The reduction in the number of scatter points in the second quadrant (L-H) from 48 to 31 indicates a gradual decline in the prevalence of low-Coverage units situated adjacent to high-Coverage units. This may be attributed to the implementation of urban planning and renewal policies. Although the number of scatter points in the third quadrant (L-L) has increased from 80 to 88, the percentage rise indicates that while the number of low-Coverage units has grown, their relative importance within the entire study area has not significantly increased. The increase in the number of scatter points in the fourth quadrant (H-L) from 27 to 34 illustrates alterations in the spatial relationship between High- and low-Coverage units, offering vital insights to urban planners into regional development disparities.

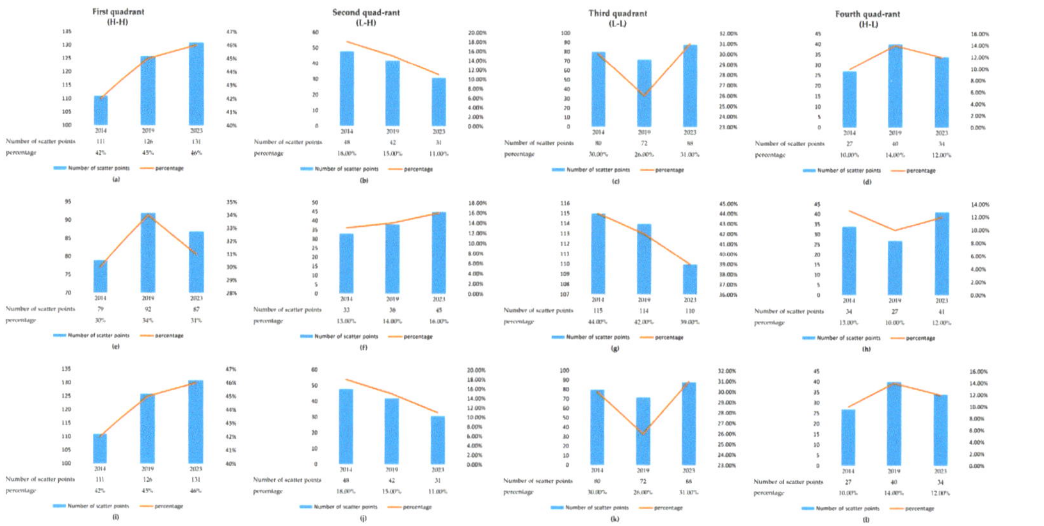

Figure 6. The horizontal charts represent urban morphology indicators: (**a–d**) for Coverage, (**e–h**) for Staggeredness, and (**i–l**) for Duty Cycle. The vertical charts indicate the quadrants of the scatter points, such as (**a**,**e**,**i**) for the first quadrant.

A detailed examination of the Staggeredness-related charts (Figure 6) reveals the presence of distinct trends across the four quadrants. The scatter plot in the first quadrant (H-H) demonstrates an initial increase from 79 in 2014 to 92 in 2019, followed by a decline to 87 by 2023. This pattern indicates that units with comparable height disparities tend to form relatively stable clusters at the neighborhood scale. However, over time, some areas may undergo readjustments or redevelopment, which could result in a slight decrease in clustering intensity.

Concurrently, the second quadrant (L-H) demonstrates an increase from 33 to 45 points, indicating an augmented spatial heterogeneity between low- and high-Staggeredness units. This reflects the diversity and complexity of urban spatial morphology. Conversely, the third quadrant (L-L) demonstrates a decline from 115 to 110 points, indicating a spatial dispersion trend among low-Staggeredness units. The fourth quadrant (H-L) demonstrates an increase from 34 to 41 points, indicating a heightened encirclement of high-Staggeredness units by low-Staggeredness ones.

It is also important to consider Duty Cycle, which is a crucial indicator of building spatial efficiency. The notable rise in scatter points in the initial quadrant (H-H), from 92 to 136, underscores the pronounced clustering of high-Duty Cycle units at the neighborhood scale. This directly indicates urban spatial compactness and efficiency. The second quadrant (L-H) exhibits minor fluctuations from 56 to 52 points, indicating an increase in the percentage of low-Duty Cycle units in proximity to high-Duty Cycle units. This may be attributed to historical urban layout issues or specific planning requirements. The reduction in scatter points in the third quadrant (L-L) from 103 to 93 and the relative stability in the fourth quadrant (H-L) from 28 to 27 provide further insight into the characteristics of Duty Cycle spatial distribution.

In conclusion, local Spatial–temporal Moran's I analysis of building characteristics at the neighborhood scale demonstrates the complexity and dynamism of Staggeredness, Coverage, and Duty Cycle in spatial distribution. The occurrence of phenomena such as high-value clustering and low-value dispersion, enhanced heterogeneity, and changes in spatial relationships collectively illustrate the existence of a diverse urban spatial structure. These findings not only provide valuable data support for urban planners but also offer essential references for optimizing urban development strategies and enhancing urban governance.

5. Discussion

The close connection between urban morphology and sustainable urban development has become a topic of considerable interest and debate in academic circles. The potential of in-depth research into urban morphology to inform the construction of sustainable urban pathways is both profound and complex [8,61–63]. This study builds upon existing research to identify four key urban morphology indicators, including Building Height Standard Deviation, Building Height, Building Volume Density, and Building Floor Area Ratio. It is noteworthy that Staggeredness is defined as the ratio of Building Height Standard Deviation to building height. Similarly, Duty Cycle and Coverage represent Building Volume Density and Building Floor Area Ratio, respectively. These indicators provide a comprehensive assessment of the dynamic changes in urban buildings in both the horizontal and vertical spatial dimensions.

Prior research has amply demonstrated the efficacy of morphology indicators in analyzing a range of urban phenomena, including urban development expansion, urban heat island effects, and environmental impacts [47,54–57,62]. This study further applies these methods to the Yau Tsim Mong District, utilizing Moran's I for quantitative analysis, thereby achieving a deep integration of morphological indicators with spatiotemporal dynamics. The findings demonstrate a notable increase in Coverage, Staggeredness, and Duty Cycle at the urban scale, with Duty Cycle exhibiting a particularly pronounced rise of approximately 178%. This serves to illustrate the extent of the spatial optimization achievements of the Yau Tsim Mong District in the context of urban planning and design. These findings are consistent with those of numerous domestic and international studies, which demonstrate the intimate connection between urban development and spatial utilization [41,46–52].

It is notable that this study identifies an upward trajectory in indicator values at three distinct geographical scales: community, neighborhood and urban. This reflects a growing phenomenon of building agglomeration and a tendency towards reduced differences in building heights in the Yau Tsim Mong District. This change can be attributed to a number of factors, including adjustments in urban development strategies, urban renewal initiatives, and population mobility. Additionally, it aligns with the global trend of high-rise building intensification in urbanization processes [1]. At the neighborhood scale, the high values of local Spatial–temporal Moran's I indicating clustering and low values indicating dispersion serve to further underscore the complexity of the rapid spatial evolution of the three-dimensional urban structure over time.

By incorporating the temporal dimension and employing quantitative analysis of urban spatial morphology indicators based on Spatial–temporal Moran's I, this study represents a methodological innovation, markedly enhancing the precision with which spatiotemporal characteristics of urban buildings dynamics can be captured. Notwithstanding the absence of significant differences in the evaluation results across different scales, they consistently indicate an increase in urban buildings density and convergence in average height, thereby providing new insights into the understanding of urban morphological evolution.

In conclusion, this study contributes to the theoretical framework of the relationship between urban morphology and sustainable development. By employing a refined scale division and spatiotemporal dynamic analysis, this study provides a scientific basis for urban planning and management. These findings are of crucial reference value for government decision-makers, urban planners, and architects, aiding them in more accurately grasping the nuances of urban development and promoting sustainable and harmonious urban growth in future urban planning and construction endeavors [5].

6. Conclusions

This study employs a comprehensive approach utilizing Global and local Spatial-temporal Moran's I to conduct a quantitative analysis of urban morphology indicators (Coverage, Staggeredness and Duty Cycle) in the Yau Tsim Mong District of Hong Kong across the community, neighborhood and urban scales. The research is focused on the examination of spatial morphology and its evolving trends within this geographical area The multilevel spatial–temporal analysis yielded the following principal conclusions:

- Between 2014 and 2023, there were notable enhancements in Coverage, Staggeredness, and Duty Cycle in global spatial–temporal Moran's I, suggesting robust spatial correlations between the examined urban morphology indicators and the overall urban development in Yau Tsim Mong District. This trend reflects a notable optimization of urban buildings forms and spatial utilization throughout the district's urbanization process.
- A comparison of trends across different scales reveals a shift in urban development strategies. A notable trend is the aggregation of buildings and the gradual reduction in height differences, which suggests that urban planning is increasingly focused on overall spatial efficiency and refined architectural design with clear regional functional divisions.
- The application of global spatial–temporal Moran's I reveals that, in comparison to city-wide and community-level scales, the neighborhood scale exhibits a relatively autonomous and comprehensive spatial configuration. This scale is an effective means of capturing clustering, dispersion, or heterogeneity phenomena in local areas. It is therefore reasonable and necessary to conduct a local spatial–temporal analysis at the neighborhood scale, as this provides detailed information on the internal spatial structure of the city and enhances our comprehensive understanding of urban morphological changes.
- At the neighborhood scale, the local Spatial–temporal Moran's I for Coverage, Staggeredness, and Duty Cycle demonstrates a notable clustering of high values and a dispersion of low values. This finding provides further evidence of the trend of building expansion at the neighborhood level in Yau Tsim Mong, whereby buildings are gradually clustering horizontally and converging vertically in height. This phenomenon reflects the district's commitment to rational spatial planning within compact urban environments, thereby demonstrating the efficacy and precision of urban planning.

This study employs a multiscale analytical approach to examine the overall development trends in Yau Tsim Mong District. In addition, it investigates the neighborhood level, providing comprehensive and accurate discussions on the three-dimensional morphology of urban buildings. This multiscale analytical approach facilitates a more comprehensive understanding of the spatiotemporal evolution of urban buildings characteristics, providing a more nuanced perspective and a richer foundation for urban planning and management. The research emphasizes the necessity of incorporating spatial heterogeneity and diversity into urban planning in order to achieve scientifically sound planning layouts.

In conclusion, this study, which employs a multiscale approach and combines global and local spatial–temporal indices analysis, offers valuable insights and theoretical support for understanding the spatiotemporal evolution of urban buildings characteristics in Yau Tsim Mong District, Hong Kong. The findings contribute to the existing body of knowledge on urban morphology, and provide valuable insights that can inform future urban planning, land use, and sustainable urban development.

Author Contributions: Conceptualization, T.S. and L.H.; methodology, W.Z.; software, S.Y.; validation, T.S., W.Z. and S.Y.; formal analysis, T.S.; investigation, W.Z.; resources, S.Y.; data curation, W.Z.; writing—original draft preparation, T.S.; writing—review and editing, T.S.; visualization, W.Z.; supervision, T.S.; project administration, T.S.; funding acquisition, L.H. All authors have read and agreed to the published version of the manuscript.

Funding: This work was supported by the Key R&D Program of Shanxi Province (Project No. 202202010101005).

Institutional Review Board Statement: Not applicable.

Informed Consent Statement: Not applicable.

Data Availability Statement: The raw data supporting the conclusions of this article will be made available by the authors on request.

Conflicts of Interest: The authors declare no conflict of interest.

References

1. Domingo, D.; van Vliet, J.; Hersperger, A.M. Long-term changes in 3D urban form in four Spanish cities. *Landsc. Urban Plan.* **2023**, *230*, 104624. [CrossRef]
2. Wu, W.; Chen, M.; Wang, N.; Chen, D. Research Progress on the Spatial-Temporal Heterogeneity of Urban Land Expansion in China. *Geogr. Geo-Inf. Sci.* **2017**, *33*, 57–63. [CrossRef]
3. Long, Y.; Li, P.; Hou, J. Analysis of urban form types based on three-dimensional block morphology: A case study of major cities in China. *Shanghai Urban Plan. Rev.* **2019**, *3*, 10–15.
4. Liu, Y.; Meng, Q.; Zhang, J.; Zhang, L.; Allam, M.; Hu, X.; Zhan, C. Evaluation of Urban Spatial Structure from the Perspective of Socioeconomic Benefits Based on 3D Urban Landscape Measurements: A Case Study of Beijing, China. *Remote Sens.* **2022**, *14*, 5511. [CrossRef]
5. Liang, F.; Liu, J.; Liu, M.; Zeng, J.; Yang, L.; He, J. Scale-Dependent Impacts of Urban Morphology on Commercial Distribution: A Case Study of Xi'an, China. *Land* **2021**, *10*, 170. [CrossRef]
6. Wang, Y.; Pan, W.; Liao, Z. Impact of Urban Morphology on High-Density Commercial Block Energy Consumption in Severe Cold Regions. *Sustainability* **2024**, *16*, 5795. [CrossRef]
7. Liu, B.; Guo, X.; Jiang, J. How Urban Morphology Relates to the Urban Heat Island Effect: A Multi-Indicator Study. *Sustainability* **2023**, *15*, 10787. [CrossRef]
8. Mobaraki, A.; Oktay Vehbi, B. A Conceptual Model for Assessing the Relationship between Urban Morphology and Sustainable Urban Form. *Sustainability* **2022**, *14*, 2884. [CrossRef]
9. Zhu, S.; Ma, C.; Wu, Z.; Huang, Y.; Liu, X. Exploring the Impact of Urban Morphology on Building Energy Consumption and Outdoor Comfort: A Comparative Study in Hot-Humid Climates. *Buildings* **2024**, *14*, 1381. [CrossRef]
10. Xie, Q.; Chang, Z. Spatial structure and driving factors of public cultural facilities in Jiangsu Province: From quantitative statistics to efficiency evaluation. *J. Nanjing Normal Univ.* **2023**, *46*, 50–59.
11. Xiong, L.; Tang, G.; Yang, X.; Li, F. Geomorphology-oriented digital terrain analysis: Progress and perspectives. *Acta Geogr. Sin.* **2021**, *76*, 595–611. [CrossRef]
12. Duan, Y. Research on the Multi-Scale Expression Method of Feature-Driven Urban 3D Models. Master's Thesis, Beijing University of Civil Engineering and Architecture, Beijing, China, 2022.
13. Liu, W. Study on the Spatiotemporal Evolution of Floor Area Ratio of Urban Residential Communities. Master's Thesis, Liaoning Normal University, Dalian, China, 2021.
14. Hou, F. Study on the Spatiotemporal Differentiation and Limit Evaluation of Land Development Intensity in Xi'an Based on GIS. Master's Thesis, Chang'an University, Xi'an, China, 2019.
15. Yang, Y.; Yu, C.; Fu, A.; Sun, Z.; Zhou, K. Investigation of the relationship between urban three-dimensional landscape patterns and the spatial-temporal distribution of atmospheric pollution: A case study of Changsha. *J. Hunan Normal Univ.* **2024**, *47*, 12–21. [CrossRef]
16. Ranjbar, H.R.; Gharagozlou, A.R.; Nejad, A. 3D Analysis and Investigation of Traffic Noise Impact from Hemmat Highway Located in Tehran on Buildings and Surrounding Areas. *J. Geogr. Inf. Syst.* **2012**, *4*, 322–334. [CrossRef]
17. Zhu, R.; Wong, M.S.; You, L.; Santi, P.; Nichol, J.; Ho, H.C.; Lu, L.; Ratti, C. The effect of urban morphology on the solar capacity of three-dimensional cities. *Renew. Energy* **2020**, *153*, 1111–1126. [CrossRef]
18. He, W. Study on the Three-Dimensional Openness Based on the Urban Digital Elevation Model. Master's Thesis, Nanjing Normal University, Nanjing, China, 2013.
19. Lu, X. Study on the Spatial form and Thermal Environment of the Central Urban Area of Jinan based on UDEM. Master's Thesis, Shandong Jianzhu University, Jinan, China, 2020.
20. Qu, X. Study on the 3D Visual Analysis Method of Urban Space based on UDEM. Master's Thesis, Liaoning Normal University, Dalian, China, 2018.
21. Guo, C. Study on the Spatiotemporal Dynamic Simulation of Urban Spatial Growth based on XGBoost-CA. Master's Thesis, Liaoning Normal University, Dalian, China, 2023.
22. Silva, I.; Santos, R.; Lopes, A.; Araújo, V. Morphological Indices as Urban Planning Tools in Northeastern Brazil. *Sustainability* **2018**, *10*, 4358. [CrossRef]
23. Gao, J.; Liu, C.; He, Q.; Chen, M.; Gan, L. Application and Exploration of 3D Real Scene Technology in Ecological Environment. *Urban Surv. Mapp.* **2024**, *1*, 16–19. [CrossRef]

24. Li, P.; Sun, Y. Research and application of real-scene 3D and multi-source data fusion technology. *Bull. Surv. Mapp.* **2019**, *S1* 133–136.
25. Li, L.; Liu, X.; Ou, J.; Niu, N. Analysis of the spatiotemporal changes and mechanisms of three-dimensional characteristics of urban expansion based on the random forest model. *Geogr. Geo-Inf. Sci.* **2019**, *35*, 53–60.
26. Cheng, Q. Study on the Scale Effect of the Relationship between Urban Sky View Factor and Surface Temperature. Master's Thesis, Beijing University of Civil Engineering and Architecture, Beijing, China, 2023.
27. Jin, A.; Ge, Y.; Zhang, S. Spatial Characteristics of Multidimensional Urban Vitality and Its Impact Mechanisms by the Built Environment. *Land* **2024**, *13*, 991. [CrossRef]
28. Lin, C.; Li, G.; Zhou, Z.; Li, J.; Wang, H.; Liu, Y. Enhancing Urban Land Use Identification Using Urban Morphology. *Land* **2024**, *13*, 761. [CrossRef]
29. Li, J.; Wang, H.; Cai, X.; Liu, S.; Lai, W.; Chang, Y.; Qi, J.; Zhu, G.; Zhang, C.; Liu, Y. Quantifying Urban Spatial Morphology Indicators on the Green Areas Cooling Effect: The Case of Changsha, China, a Subtropical City. *Land* **2024**, *13*, 757. [CrossRef]
30. Ibrahim, S.; Younes, A.; Abdel-Razek, S.A. Impact of Neighborhood Urban Morphologies on Walkability Using Spatial Multi-Criteria Analysis. *Urban Sci.* **2024**, *8*, 70. [CrossRef]
31. Wang, H.; Shan, Y.; Xia, S.; Cao, J. Traditional Village Morphological Characteristics and Driving Mechanism from a Rural Sustainability Perspective: Evidence from Jiangsu Province. *Buildings* **2024**, *14*, 1302. [CrossRef]
32. Du, S.; Wu, Y.; Guo, L.; Fan, D.; Sun, W. How Does the 2D/3D Urban Morphology Affect the Urban Heat Island across Urban Functional Zones? A Case Study of Beijing, China. *ISPRS Int. J. Geo-Inf.* **2024**, *13*, 120. [CrossRef]
33. Zou, C.; Tang, X.; Tan, Q.; Feng, H.; Guo, H.; Mei, J. Constructing Ecological Networks for Mountainous Urban Areas Based on Morphological Spatial Pattern Analysis and Minimum Cumulative Resistance Models: A Case Study of Yongtai County. *Sustainability* **2024**, *16*, 5559. [CrossRef]
34. Shi, T.; Zhao, L.; Liu, F.; Zhang, M.; Li, M.; Peng, C.; Li, H. Conditional Diffusion Model for Urban Morphology Prediction. *Remote Sens.* **2024**, *16*, 1799. [CrossRef]
35. Li, F.; Zhou, T.; Dong, Y.; Zhou, W. Design of the 3D Digital Reconstruction System of an Urban Landscape Spatial Pattern Based on the Internet of Things. *Int. J. Inf. Technol. Syst. Approach* **2023**, *16*, 1–14. [CrossRef]
36. Cheng, G.; Li, Z.; Xia, S.; Gao, M.; Ye, M.; Shi, T. Research on the Spatial Sequence of Building Facades in Huizhou Regional Traditional Villages. *Buildings* **2023**, *13*, 174. [CrossRef]
37. Zhou, R.; Xu, H.; Zhang, H.; Zhang, J.; Liu, M.; He, T.; Gao, J.; Li, C. Quantifying the Relationship between 2D/3D Building Patterns and Land Surface Temperature: Study on the Metropolitan Shanghai. *Remote Sens.* **2022**, *14*, 4098. [CrossRef]
38. Yu, Z.; Chen, L.; Li, L.; Zhang, T.; Yuan, L.; Liu, R.; Wang, Z.; Zang, J.; Shi, S. Spatiotemporal Characterization of the Urban Expansion Patterns in the Yangtze River Delta Region. *Remote Sens.* **2021**, *13*, 4484. [CrossRef]
39. Gapeyenko, M.; Moltchanov, D.; Andreev, S.; Heath, R.W. Line-of-Sight Probability for mmWave-Based UAV Communications in 3D Urban Grid Deployments. *IEEE Trans. Wirel. Commun.* **2021**, *20*, 6566–6579. [CrossRef]
40. Tang, L.; Ying, S.; Li, L.; Biljecki, F.; Zhu, H.; Zhu, Y.; Yang, F.; Su, F. An Application-Driven LOD Modeling Paradigm for 3D Building Models. *ISPRS J. Photogramm. Remote Sens.* **2020**, *161*, 194–207. [CrossRef]
41. Zaki, S.A.; Othman, N.E.; Syahidah, S.W.; Yakub, F.; Muhammad-Sukki, F.; Ardila-Rey, J.A.; Shahidan, M.F.; Mohd Saudi, A.S. Effects of Urban Morphology on Microclimate Parameters in an Urban University Campus. *Sustainability* **2020**, *12*, 2962. [CrossRef]
42. Li, S.; Wu, C.; Lin, Y.; Li, Z.; Du, Q. Urban Morphology Promotes Urban Vibrancy from the Spatiotemporal and Synergetic Perspectives: A Case Study Using Multisource Data in Shenzhen, China. *Sustainability* **2020**, *12*, 4829. [CrossRef]
43. Xu, Y.; Liu, M.; Hu, Y.; Li, C.; Xiong, Z. Analysis of Three-Dimensional Space Expansion Characteristics in Old Industrial Area Renewal Using GIS and Barista: A Case Study of Tiexi District, Shenyang, China. *Sustainability* **2019**, *11*, 1860. [CrossRef]
44. Shukla, A.; Jain, K. Critical Analysis of Spatial-Temporal Morphological Characteristic of Urban Landscape. *Arab. J. Geosci.* **2019**, *12*, 112. [CrossRef]
45. Yang, T. A Study on Spatial Structure and Functional Location Choice of the Beijing City in the Light of Big Data. In Proceedings of the 10th Space Syntax Symposium, London, UK, 13–17 July 2015.
46. Qiao, W.; Wang, Y.; Ji, Q.; Hu, Y.; Ge, D.; Cao, M. Analysis of the Evolution of Urban Three-Dimensional Morphology: The Case of Nanjing City, China. *J. Maps* **2019**, *15*, 30–38. [CrossRef]
47. Lai, P.-C.; Chen, S.; Low, C.-T.; Cerin, E.; Stimson, R.; Wong, P.Y.P. Neighborhood Variation of Sustainable Urban Morphological Characteristics. *Int. J. Environ. Res. Public Health* **2018**, *15*, 465. [CrossRef] [PubMed]
48. Zhang, Q.; Huang, X.; Zhang, G. Urban Area Extraction by Regional and Line Segment Feature Fusion and Urban Morphology Analysis. *Remote Sens.* **2017**, *9*, 663. [CrossRef]
49. Wang, T. Urban Spatial Analysis and Visualization System Construction from the Perspective of Public Transportation. Master's Thesis, Beijing Jiaotong University, Beijing, China, 2023.
50. Liu, Y. Study on the Evaluation of Socioeconomic Benefits of Urban Spatial Structure Based on 2D and 3D Landscape Quantitative Analysis. Master's Thesis, Liaoning Technical University, Fuxin, China, 2023.
51. Yu, X.; Xu, G.; Liu, Y.; Xiao, R. The impact of three-dimensional urban building morphology on the surface thermal environment in the Yangtze River Delta region. *Chin. Environ. Sci.* **2021**, *41*, 5806–5816.

52. Li, B.; Peng, M.; Tan, R. Analysis of the Evolution Characteristics and Driving Factors of Urban 3D Spatial Morphology: A Case Study of the Main Urban Area of Wuhan. *J. Wuhan Univ. (Inf. Sci. Ed.)* **2023**, *48*, 1–13. [CrossRef]
53. Qu, Y.; Chi, Y.; Gao, J.; Zhang, Z.; Liu, Z.; Wang, Y.-P. Measuring the Multi-Scale Landscape Pattern of China's Largest Archipelago from a Dual-3D Perspective Based on Remote Sensing. *Remote Sens.* **2023**, *15*, 5627. [CrossRef]
54. He, Q. Analysis of the Spatial Impact of Urban Vitality on Carbon Emissions Integrating Multi-Source Data. Ph.D. Thesis, China University of Mining and Technology, Beijing, China, 2024.
55. Zhang, L.; Qing, A.; Kakar, S.; Cui, M. 3D Accessibility Evaluation of Public Space in Three-dimensional Urban Design: A Case Study of Taikoo Place, Hong Kong. *New Archit.* **2021**, *4*, 48–54. [CrossRef]
56. Long, L. Research on Spatial Autocorrelation Analysis Methods and Applications. Ph.D. Thesis, Kunming University of Science and Technology, Yunnan, China, 2016.
57. Zhang, X.; Lin, P.; Chen, H.; Yan, R.; Zhang, J.; Yu, Y.; Liu, E.; Yang, Y.; Zhao, W.; Lv, D.; et al. Understanding land use and cover change impacts on run-off and sediment load at flood events on the Loess Plateau, China. *Hydrol. Process.* **2018**, *32*, 576–589. [CrossRef]
58. Zhang, L. Research on the Geographically Weighted Regression Modeling Method Based on Flow Data. Ph.D. Thesis, Wuhan University, Wuhan, China, 2021.
59. Yue, X.; Mingxing, C. Population evolution at the prefecture-level city scale in China: Change patterns and spatial correlations. *J. Geogr. Sci.* **2022**, *32*, 1281–1296.
60. Wang, X. Analysis of Accessibility and Layout Optimization of Urban Medical Institutions. Master's Thesis, Nanjing University of Posts and Telecommunications, Nanjing, China, 2023.
61. Hillier, B.; Hanson, J. *The Social Logic of Space*; Cambridge University Press: Cambridge, UK, 1984.
62. Sheng, Q.; Liu, N. Comparing the Use of Actual Space and Virtual Space: A Case Study on Beijing's Wangfujing Area. In Proceedings of the 10th Space Syntax Symposium, London, UK, 13–17 July 2015.
63. Sheng, Q.; Jiao, J. Understanding the Impact of Street Patterns on Pedestrian Distribution: A Case Study in Tianjin. In Proceedings of the 12th Space Syntax Symposium, Beijing, China, 8–13 July 2019.

sustainability MDPI

Review

Energetic Valorization of the Innovative Building Envelope: An Overview of Electric Production System Optimization

Eléa Oudot †, Kawtar Gholmane, Damien Ali Hamada Fakra *,†,‡ and Riad Benelmir ‡

Sciences and Technologies Faculty of Nancy, University of Lorraine, Campus Aiguillettes, BP 70239, 54506 Vandoeuvre-Les-Nancy, France
* Correspondence: damien.fakra@univ-lorraine.fr; Tel.: +33-7-8111-3532
† These authors contributed equally to this work.
‡ Current address: Laboratory for Studies and Research on Wood Materials (LERMAB), University of Lorraine, 54506 Vandoeuvre-Les-Nancy, France.

Abstract: The world population increased from 1 billion in 1800 to around 8 billion today. The Population Division of the United Nations predicts a global population of approximately 10.4 billion people by the end of the century. That represents over 2 billion more people. Moreover, the global community is currently experiencing a precarious state due to the enduring repercussions of the COVID-19 pandemic across all sectors, including energy. Given the rising global population and the limited availability of primary energy resources, we must reach a balance between the demands of a growing human population and the planet's carrying capacity. The dreadful conflict in Ukraine has precipitated an enormous energy crisis. This crisis has served as a warning to the world population of how much it depends on this resource to survive. In France, the building sectors, specifically residential and tertiary, account for 45% of the total final energy consumption. It is the first energy consumer of the country and one of the most polluting (i.e., about 34% of CO_2 emitted by France). Consequently, we must consider alternative energy resource forms (i.e., substitution energy forms) Harvesting energy from the building envelope may be a viable technique for partially satisfying the electricity demands of building users. In this context, scientific research offers considerable potential for developing more innovative and efficient systems. This article aims to review the state-of-the-art of advances on the subject to orient and further optimize energy production systems, particularly electricity. This work addresses several points of view: it discusses the overall backdrop of the present study and introduces the subject; details the research strategy and procedures used to produce this paper; develops the state-of-the-art on the potential for generating or recovering power from the building envelope; presents the SWOT analysis of the earlier-described systems. Finally, it concludes by offering findings and viewpoints.

Keywords: systems; power; building envelope; innovation; assembly

Citation: Oudot, E.; Gholmane, K.; Fakra, D.A.H.; Benelmir, R. Energetic Valorization of the Innovative Building Envelope: An Overview of Electric Production System Optimization. *Sustainability* **2024**, *16*, 2305. https://doi.org/10.3390/su16062305

Academic Editors: Zora Vrcelj and Malindu Sandanayake

Received: 3 January 2024
Revised: 21 February 2024
Accepted: 5 March 2024
Published: 11 March 2024

1. Introduction

In the face of a rapidly expanding global population and the finite nature of primary energy resources, it is imperative to reconcile burgeoning human demands with the Earth's energy production capacity. Principal concerns arise from the inadequacy of energy supplies to meet the escalating global demand and the accompanying environmental ramifications associated with fossil fuel utilization. The data from the Agence de la Transition Écologique (ADEME) and the Ministère de la Transition Écologique in FRANCE reveal that the building sector, encompassing residential and tertiary structures, singularly accounts for 45% of final energy consumption in France [1,2]. Furthermore, the United Nations Environment Programme (UNEP) reports that this sector contributes 38% of carbon dioxide (CO_2) emissions, positioning it as the primary energy consumer and one of the most environmentally harmful sectors in the country. In light of contemporary environmental expectations and objectives for sustainable development, concrete measures are imperative to curtail energy consumption within the building sector. These measures necessitate a dual approach to reducing energy consumption and incorporating more renewable and sustainable energy sources. Given its prominent position among the most polluting and energy-intensive sectors, the building sector represents a focal point for potential improvements. Addressing the need for enhanced energy efficiency in buildings and the obligation to meet the energy demands of a burgeoning global populace underscores the relevance of bolstering our infrastructures and innovations. Central to this endeavour is integrating energy production systems, particularly those generating electricity, into the building envelope. Within this context, scientific research emerges as a formidable avenue for developing innovative and efficient systems.

The broader context in sustainable building practices involves examining the larger framework and global trends that contribute to understanding and implementing environmentally friendly construction methods. The idea of the broader context of the topic can be enumerated by some aspect illustrations such as:

Environmental Challenges:

- Climate Change: The growing awareness of climate change has led to an increased focus on sustainable practices in various industries, including construction. Rising global temperatures, extreme weather events, and the depletion of natural resources highlight the urgent need for sustainable building solutions.
- Resource Scarcity: The depletion of traditional construction materials and the environmental impact of their extraction has prompted a shift towards more sustainable alternatives. Exploring recycled materials, renewable resources, and innovative construction techniques can respond to this challenge.

Regulatory Framework:

- International Standards: Organizations and governments worldwide are establishing and updating standards for sustainable construction. The Leadership in Energy and Environmental Design (LEED) certification and similar global standards provide guidelines for eco-friendly building practices, influencing the construction industry worldwide.
- Government Policies: Many countries are implementing policies that incentivize or mandate sustainable building practices. For example, some countries include tax incentives for green buildings, stricter environmental regulations, and government-backed initiatives to promote energy-efficient construction.

Technological Advancements:

- Innovative Materials: Advancements in material science have introduced new, sustainable construction materials. These materials, from bamboo and recycled steel to high-performance concrete, offer environmental benefits and improved structural performance.
- Smart Technologies: Integrating innovative technologies in building design and management helps optimize energy use, monitor environmental impact, and enhance overall sustainability. Notably, we have the case of buildings using sensors and tools for automation and data analytics to create more efficient and eco-friendly buildings.

Social Awareness and Responsibility:

- Consumer Demand: Increasing consumer environmental awareness has created a demand for sustainable and eco-friendly buildings. Developers and builders respond to this demand by incorporating green features and certifications.
- Corporate Social Responsibility (CSR): Many corporations are adopting sustainable building practices as part of their CSR initiatives. The concept proposed by each country aligns with societal expectations and contributes to long-term cost savings and a positive brand image.

Global Collaborations:

- Knowledge Sharing: With the global nature of environmental challenges, there is a significant emphasis on international collaboration and knowledge sharing. Research institutions, industry experts, and governments collaborate to exchange ideas, technologies, and best practices in sustainable construction.

Economic Considerations:

- Cost-Effectiveness: Initially, there might be a perception that sustainable building practices are more expensive than conventional construction. However, many sustainable initiatives are cost-effective over the long term due to reduced energy consumption, lower maintenance costs, and potential government incentives.
- Job Creation: The shift towards sustainable building practices also contributes to creating jobs in sectors related to renewable energy, green construction, and the development of eco-friendly technologies.

Understanding this broader context is essential for researchers, policymakers, and industry professionals to make informed decisions and advancements in sustainable building practices that address global challenges while considering economic, social, and environmental dimensions.

Numerous studies have been conducted on the integration of thermal systems into the envelope of a building and their capability to heat or refresh an indoor ambience. For example, Bigot et al. [3] demonstrated that building indoor temperature is considerably influenced by the BAPV. Very few of them deal with the electrical capacity that these walls, roofs, and floors that constitute the building envelope can produce. The scientific issue addressed in this bibliography revolves around identifying the components that form integral parts of the building's architecture and possess the ability to generate electricity directly. The case studies will enable the analysis of the most efficient systems currently available on the market and the scientific challenges that need to be addressed to comprehend and enhance their functioning. Additionally, the evaluation will encompass the financial, technical, and environmental impacts of these integrated components within the building envelope in a socioeconomic, technological, and climatic context.

This article comprehensively reviews state-of-the-art advancements in this domain to guide funders and designers in optimizing electrical production systems integrated into the building envelope. An exhaustive search and selection process was undertaken to conduct this review, encompassing all scientific articles about the study of direct electricity generation systems within buildings and any form of energy potentially convertible into electricity. The inquiry spanned prominent publication platforms, including Elsevier, ResearchGate, Google Scholar, MDPI, and Taylor & Francis. Additionally, a scrutiny of patents filed for relevant technologies was conducted through Google Patents.

2. State-of-the-Art

The literature search was conducted focusing only on the three components forming a building envelope: glazing, walls, and roofing. The ground component was not considered in the bibliography due to the absence of any electricity-generating system utilizing it. The technologies, according to consideration, can generate electricity through either direct or indirect means (e.g., with the conversion of thermal energy to electrical energy). As a result, we exclude technologies that generate energy other than electricity from the bibliography.

2.1. Roof Technologies

2.1.1. Photovoltaic and Thermal Panels Integrated with the Roof

Solar thermal systems (STSs) have significantly improved efficiency compared to their earlier versions. The driving force behind the advancement of STSs lies in the expanding research on alternative energy sources, recognized as an integral component of low-carbon energy systems essential for generating affordable and reliable electricity [4]. This section delves into the latest developments in STS applications, mainly focusing on PVT (i.e., photovoltaic thermal collectors) or "photovoltaic/thermal" systems—currently the most widely employed green energy technology for power production. This hybrid system seamlessly integrates the output of both thermal and electrical energy. The PVT system capitalizes on the photovoltaic (PV) effect, which generates electric energy through solar irradiation [5]. It finds applications in BIPVs (building-integrated photovoltaic), replacing traditional construction materials [5,6]. PVs can be incorporated as BIPV or building-attached photovoltaic (BAPV) systems. Although BAPV systems yield more electricity, BIPV systems excel in overall building performance due to better control over solar gain. The standard definition for available roof space in BIPV deployment is 40% of the ground-level size. Most solar cells are suitable for BIPV roof applications [7]. Beyond photovoltaic (PV) energy, which directly converts solar radiation into electrical energy, thermal energy can also be harnessed for electricity generation. One promising method involves using thermoelectric generators (TEGs) [8]. Utilizing the Seebeck effect, thermoelectric generators (TEGs) demonstrate their capability to convert thermal energy directly into electrical energy. Consequently, combining PV and TE to enhance electricity production becomes a viable option [9]. This hybrid system incorporates thermoelectric generators attached to a solar panel. Notably, the photovoltaic panels absorb heat and store thermal energy during operation. Applying this technique to the opposite face of the thermoelectric generators on solar panels efficiently recovers the underutilized thermal energy in conventional panels [8]. It constitutes a hybrid photovoltaic and thermoelectric (PV-TE) module that concurrently leverages the photovoltaic and Seebeck effects.

2.1.2. Photobioreactor Roofs

In the pursuit of advancing renewable and sustainable energy sources, the cultivation of algae presents intriguing possibilities. Due to their rapid growth compared to most other plants, algae can yield substantial biomass. Two primary facilities for algae cultivation exist: open ponds and photobioreactors. Open ponds, which do not apply to buildings, are excluded from this study. Photobioreactors, though more costly, boast superior yields and consist of transparent closed tanks filled with water. Microalgae within these reactors can thrive in various water sources, including seawater, wastewater, and harsh water. The cultivation process involves harnessing daylight, carbon dioxide, and organic carbon simultaneously for energy production [10]. A pump circulates water by introducing CO_2-enriched air bubbles into the system, and whereas laboratory studies typically enrich the air with CO_2 using gas canisters, real-world applications aim to capture CO_2 from the surrounding air or recover on-site combustion gases, as demonstrated by the BIQ building and its cogenerator [11]. Regular stirring is essential for proper distribution [12]. An automated anaerobic digestion (AD) unit meets nutrient requirements [13]. The resulting microalgae biomass can be valorized as biomass and/or oil. Microalgae strains also hold potential as a source of H_2 energy, as they can split water into H_2 and O_2 using solar energy [14]. In the AD unit, algae biomass is converted into biogas, such as methane, which powers a biogas generator for electricity and heat production [12]. This biomass can alternatively be transformed into pellets, generating power through combustion [10], or processed to extract lipids for biofuel production, subsequently used in a biofuel generator for electricity [10,15]. Building rooftops can be effectively utilized by integrating these photobioreactors. The choice between tubular and flat photobioreactor (i.e., PBR) panels within both horizontally and vertically oriented buildings presents options. Vertical tubular PBRs, due to their geometry, do not require a specific orientation for optimal solar exposure, whereas flat panels slightly outperform vertical tubular PBRs [12]. Innovative designs like I. Berzin's triangular airlift PBR blend bubble column principles with built-in static mixers [16]. Despite the technical viability of such systems, the economic aspect raises concerns. A. Bender's findings suggest that producing electricity from algae biomass on a building's roof may not be economically feasible [12], and whereas the energy production potential from microalgae remains promising, efficiency improvements are essential, given the myriad factors influencing performance [17]. S. Wilkinson and colleagues delve into the various challenges associated with algae-building technology, offering perspectives for enhancement [18,19].

2.1.3. Building-Integrated Wind Turbines

The development of photovoltaics and wind fields has become evident in recent years. Although the feasibility of integrating photovoltaic (PV) panels into building envelopes is well-established, the same cannot be said for wind turbines. Public acceptance of wind turbines is hindered, primarily due to concerns about visual and auditory disturbances they may cause. Unlike rural areas where wind energy systems are commonplace, harnessing wind as an energy source in urban settings is challenging. Studies have revealed that urban wind flows are predominantly characterized by low speeds, particularly in city centres [20]. Nevertheless, specific urban locations, such as rooftops of large buildings less susceptible to turbulence, exhibit significant potential for wind energy production [21]. Integrating wind turbines with the aerodynamic designs typical of rural areas is often impractical or impossible. Two main types of wind turbines exist: classic horizontal axis wind turbines (HAWTs) and vertical axis wind turbines (VAWTs). A study by M. Casini delves into various VAWTs, exploring their advantages, disadvantages, and potential applications in urban building contexts [22]. In the context of building integration, wind turbines can be strategically placed on rooftops, between buildings, within through-building openings, or incorporated into the building skin [23]. Rooftop installations are standard, capitalizing on unused space where wind speeds are often optimal at higher elevations. Installing turbines between two buildings requires careful planning during the design phase, ensuring structure compatibility. Integration within building openings and envelopes represents relatively unexplored territory. Noteworthy advancements in building-integrated wind turbines have emerged. In 2015, Park et al. proposed a wind wall turbine system integrated into facades, incorporating guide panels and small rotors for electricity generation. Computational fluid dynamics (CFD) analyses were conducted to optimize rotor shapes and the system demonstrated the capability to meet 6.3% of a residential structure's electricity demands [24]. Subsequently, in 2017, Hassanli et al. introduced a double skin facade (DSF) wind turbine system, proving its feasibility through CFD simulations [23]. Although research in this area is limited, recent studies present promising prospects for advancing building-integrated wind turbine technology.

2.1.4. Hybrid Solar-Wind Systems

This section proposes a distinctive hybrid system that synergizes thermoelectric materials, wind turbines, and solar collectors. Initially, solar heat is absorbed by the collector's absorber plate above the thermoelectric generators. The temperature difference between the hot absorber plate and a stream of fresh air is harnessed to produce energy. The thermoelectric generators heat the fresh air, causing it to ascend due to buoyancy force and the chimney effect, passing through the vertical chimney and slanted collector. Upon reaching the turbine blades, the rising air induces rotation, generating electricity generation [25]. This system encompasses a solar air collector, solar chimney, thermoelectric generators, and a Savonius wind turbine. Its integration occurs in a near-zero energy building in St. Petersburg, Russia [26]. Another employed hybrid system involves a combination of a wind turbine, PV solar panels, a tank, a compressor, a PEMEC (Proton Exchange Membrane Electrolyzer Cell) for hydrogen production with excess electricity, and a PEMFC (Proton Exchange Membrane Fuel Cell) for converting produced hydrogen into power during production deficiencies [27]. In the PEMEC, the consumption of power facilitates the conversion of water into hydrogen and oxygen. The hydrogen and oxygen generated undergo a reaction, producing water and electricity, yet another hybrid system utilizes both photovoltaic (PV) and wind technologies. This system automatically switches between photovoltaic and wind production modes based on weather conditions. It functions as a 2-in-1 wind system, featuring a device with two flexible photovoltaic panels managed by a bending mechanism. This mechanism enables the device to have two profiles [28]. In its flat and extended rectangular shape, the device maximizes sunlight absorption during the sun's dominance, producing clean electricity in PV mode. Conversely, in a half-cylindrical shape (concave and convex), it emulates the Savonius wind turbine blades' structure during wind dominance, continuing electricity production in wind mode. The device operates autonomously through an embedded electronic and artificial intelligence system. When the wind is favourable, the electro-mechanical system flexes the PV panels to transition to a semi-cylindrical mode. The PV panels extend to a flat shape in the presence of sunlight. This invention pertains to a renewable energy bi-converter system that enhances electricity generation.

2.2. *Facade Technologies*

2.2.1. Solar Paint Wall

Hydrogen presents a compelling solution to the current energy crisis and environmental challenges due to its high energy density and eco-friendly nature as a carbon-free energy source [29]. One promising method for hydrogen production is photocatalytic hydrogen evolution (PHE), a process that utilizes solar energy to split water molecules [30–32]. In this light-assisted catalysis, a newly developed solar paint exhibits the capability to split and absorb water vapour, producing hydrogen [31]. The innovative substance within the paint, synthetic molybdenum-sulfide, functions akin to silica gel but with added benefits. Unlike traditional silica gel, this novel substance acts as a semiconductor, catalyzing the separation of water molecules into hydrogen and oxygen. The subsequent step involves converting hydrogen into electricity using hydrogen fuel cells, which generate electrical energy through the combination of hydrogen and oxygen atoms [30]. The emerging class of inorganic coordination polymers, sulfur-rich molybdenum sulfides MoS_x ($x = 3^2/3$), holds significant promise for catalytic applications [30], particularly in hydrogen production. Researchers have explored the material's potential as an electrocatalyst, leveraging its quick moisture uptake and high conductivity. A catalytic ink was developed for electrolyte-free hydrogen production, avoiding the need for external power sources or complex fluid-handling machinery. To enhance water splitting efficiency, MoS_x's was combined with TiO_2 (P25) due to the former's small band gap [30]. Additionally, well-defined photocatalysts, including Al-doped $SrTiO_3$ ($SrTiO_3$:Al) loaded with a $RhCrO_x$ and CoO_y co-catalyst, were employed in a batch phase reactor using actual air samples or water vapour dosed into N_2 gas [31]. Zinc indium sulfide ($ZnIn_2S_4$) has garnered attention in PHE applications [32] owing to its outstanding semiconductor features, such as non-toxicity, a reasonable band gap, and high stability. Through electrochemical processes, fuel cells facilitate the conversion of hydrogen and oxygen's chemical energy into direct current electrical energy.

2.2.2. Photobioreactor Facade Panels

Previously, we discussed the utilization of PBRs employing microalgae for electricity production. This technology can be seamlessly integrated into building facades and even windows, as outlined in [19]. The technology resembles rooftop PBRs and can manifest in various forms, as indicated in [33]. Numerous studies have highlighted the additional benefits of incorporating PBRs into facades, serving purposes such as glazing panels [17], thermal insulation, sun-shading [11], and significantly contributing to air purification by converting CO_2 into O_2The vertical flat panels serve as a double skin facade and facilitate natural ventilation, as noted in [33]. Despite theoretical models and simulations, the practical application of this technology is challenging due to inherent problems described in [11,34]. However, there is a noteworthy real-scale application—the BIQ (bio-intelligent quotient) Building, constructed in 2013, stands as the first microalgae-powered building [18,33]. By installing vertical flat panels on two facades, the BIQ Building partially meets its energy needs [19]. Additionally, research by G. M. Elrayies et al. indicates that the Process Zero project covers 9% of the GSA office building's requirements by installing tubular PBR front panels [11]. Furthermore, integrating PBRs on both roofs and facades presents an opportunity to enhance energy production [11]. Hybrid PBRs, combining the strengths of different types, offer another avenue for maximizing benefits, as discussed in [14].

2.2.3. Microbial Biophotovoltaic Wall Technology

Microalgae have demonstrated significant potential in biotechnologies, yet they are not the sole contributors to electricity generation. Cyanobacteria, a type of bacteria, have proven to possess the ability to generate power. A specific type of microbial fuel cell, a biophotovoltaic (BPV) cell, harnesses this capability. Using water as an electron source BPVs can convert light energy into electrical output. Unlike traditional photovoltaic (PV) systems, BPV devices can produce electricity in light and darkness, making them more sustainable. Typically, the production of BPVs involves collecting cells in a liquid culture and then applying them to an electrode. However, this approach has drawbacks, primarily associated with the liquid phase. Some cyanobacterial and microalgal species, as indicated in previous studies [35], have demonstrated the ability to grow on a conducting anode without needing any organic substrate for electron transfer. The work of M. Sawa et al highlights a breakthrough in the field by showcasing the feasibility of fully printing a bioelectrode using a conventional inkjet printer [36]. The prototype featured a thin-film paper-based biophotovoltaic cell composed of a layer of cyanobacterial cells on a carbon nanotube conducting surface. A unit of nine BPVs successfully powered a digital commercial clock, cycling between 30-min "ON" periods and 30-min "OFF" intervals to recover BPV devices. Additionally, the prototype demonstrated the ability to power an LED for 60 s with one pulse every 2.5 s, providing sufficient electricity to illuminate the LED. This innovative technology is promising as a bio-solar panel during daylight hours and transforms into a bio-battery at night. The potential applications could be expanded through large-scale printing, such as creating wallpapers that generate electricity by harnessing solar energy captured during the day.

2.3. *Window Technologies*

2.3.1. Photovoltaic Glasses

The potential of fenestration systems can be significantly heightened by integrating photovoltaic (PV) technology into windows. Modern technologies utilize semi-transparent thin-film solar cells on windows, a recently developed technique that enhances daylight and thermal performance while augmenting energy generation capacity [7]. A new type of photovoltaic shutter system, known as the louvred photovoltaic window, has been introduced. This system allows for adjusting inclination angle and spacing based on solar altitude angle and weather conditions in different months [37]. Building-integrated photovoltaics (BIPVs) can also be implemented on Windows, offering the advantage of electricity generation [6]. Another potential strategy for enhancing the power output of solar cells incorporated into building windows is the Building-Integrated Concentrating Photovoltaic (BICPV) window. An innovative concept, the BICPV smart window generates energy and regulates the entry of solar heat and visible light into buildings. It features an optically switchable thermotropic layer with integrated PV cells [38]. A novel Concentrating Photovoltaic/Thermal Glazing system (CoPVTG), developed at the University of Ulster's Center for Sustainable Technologies in Belfast, UK, presents cutting-edge technology. This system consists of two glazed panels, one externally shaped to create lenses that focus solar energy onto photovoltaic cell lines. The unique characteristics of these lenses allow solar radiation to enter interior spaces during winter and be directed onto photovoltaic cells during summer, reducing solar gains while providing electricity to the building. The double-glazed panel structure of CoPVTG and CoPEG devices makes them versatile components for building glazing. The external glass panel is designed to create concentrating lenses that focus solar energy onto PV cell stripes built into the windows. Notably, the CoPVTG system facilitates heat recovery through air flowing through the air cavity, simultaneously cooling down the PV temperature and enhancing its electrical performance. Additionally, the thermal energy produced by PVs can be converted into electrical energy, with thermoelectric generators (TEGs) being one possible strategy [8].

2.3.2. Triboelectric Nanogenerator Glasses

Solar energy is commonly harnessed for electricity generation through renewable sources. Yet, an alternative approach involves tapping into mechanical energy generated by rain, mainly through utilizing triboelectric nanogenerators (TENGs). The research on TENGs, incredibly transparent ones that can be integrated into Windows, has gained significant traction. In the single-electrode mode, the friction between positively charged raindrops and the negatively charged TENG surface creates an electric current by establishing a potential difference between the system's two electrodes [39]. This technology can be coupled with a contact-mode TENG, assembled with elastic springs, to convert wind energy into electricity. This innovative approach results in a dual-mode TENG comprising a raindrop-TENG and a wind-powered-TENG, enhancing efficiency in terms of operating conditions and electrical output [40]. Two interfaces are considered: solid/solid or solid/liquid. Water (positive charge) directly contacts the SLIPS surface (negative charge) in the solid–liquid structure. On the other hand, the solid–solid system involves a triboelectric material (positive charge) obtaining the SLIPS (negative charge) when waterdrops interface with it. The liquid–solid TENG boasts a simple structure but tends to have a lower friction coefficient than the solid–solid system, which uses water as the friction material [41]. Z. Chen et al.'s work [42] demonstrates that incorporating a slippery lubricant-infused porous surface (SLIPS) into the system enhances its resilience, allowing the TENG to withstand humidity and extreme temperatures better, contributing to prolonged durability. Although the power generated by this system remains relatively low, Q. Zhou et al.'s study revealed that it can produce enough energy to light eight LEDs in series. Furthermore, after tapping on the translucent TENGs for 2.5 h, a 1000 μF capacitor was charged with a working voltage of 3 V—sufficient to power an electronic transducer for a single temperature/humidity test [43]. This transparent TENG could be a self-powered raindrop-detection sensor, automatically controlling window closure during inclement weather.

3. Discussion: SWOT Analysis Systems Coupling in the Building Envelope

To summarize the outcomes of this extensive literature review, we conducted a meticulous analysis employing SWOT analyses for each system under investigation. This strategic methodology offers an insightful view of the existing research landscape and enables a nuanced representation of both progress and obstacles. As a result, it yields valuable perspectives on the complexities essential for future studies, be they related to internal dynamics or external factors impacting the system. Table 1 shows that Multi-purpose BIPV, BAPV, or PVT systems integrated into the building offer several advantages over traditional PV systems. They generate dual energy output, exhibit higher efficiency, are flexible and efficient, and contribute to reducing fossil fuel consumption. These systems have a wide application range, are cost-effective, maintain architectural uniformity, and require less installation costs. They also require less space and help regulate indoor building temperature, utilizing excess heat. Multi-purpose photovoltaic-thermal systems offer a comprehensive solution that addresses energy production, cost-effectiveness, space efficiency, and environmental sustainability. However, these systems have the same weaknesses as conventional PV systems: the need for energy storage batteries during cloudy periods or to provide electricity in the evening. These additional components can increase the investment cost of the system as well as maintenance requirements. Indeed, the high installation costs, intermittent energy production due to weather conditions, and the necessity for energy storage to address intermittency and meet local energy demands can stop the development of these systems. Additionally, the impact of accumulated dust on reducing power output and system efficiency can be added to the difficulty of managing this system. The prospects for improving the system would involve finding solutions to electrical overproduction (via more eco-friendly storage techniques) and addressing losses due to site meteorology, orientation, and system positioning angles. Models for system deterioration or ageing should also be developed.

Table 1. SWOT analysis of the photovoltaic and thermic panels.

Strengths	Weaknesses	Opportunities	Threats
Multi-purpose: both the electricity and heat energy can be obtained from the same system [44] PVT (see line 68) system has better efficiency than the PV system [9] Flexible and efficient [44] Can help reduce fossil fuel consumption [4] Has wide application area [44] Inexpensive and convenient [44] Keeps the architectural uniformity on roofs [45] Installation cost may be reduced for the need of only one system to be installed instead of two systems [45] Lower space utilization than the two systems alone [45] Reduce the temperature of the photovoltaic panels and take advantage of the excess heat [9] Abundance of raw materials [46]	The cost of installation can be relatively high [44] The absence of the sun at night and cloudy days [47] PVT systems intermittently produce energy depending on weather [4] Need for an energy storage system to address the issues of intermittency and meet local energy needs [5] Accumulated dust can reduce power output and therefore system efficiency [5]	Improving the optical properties of the working fluid can improve efficiency [9] The better the performance of the PVT system, the higher the transmittance of visible light and solar infrared rays absorbed [9] The thermal energy generated by the system can be converted to electrical energy by the Peltier effect [33] It can be integrated into a building and forms a part of the building (BIPVT) [48] PVT systems integrated into the building envelope avoid additional land use [6] Can be integrated with other energy sources for enhanced efficiency [4] Can be coupled with another electricity production system [33] Applying PV systems to the roof can markedly decrease the heat flux through the roof [7]	Planning of site and orientation [5] Exposure to the elements and risk of premature deterioration [46] The efficiency of the modules varies significantly depending on weather conditions, climate, and the presence of shading effects [7,46] Thermal losses within the photovoltaic panel [33] Overproduction of electricity [46]

Algae (i.e., through photobioreactor facade panel systems and PBRs in Table 2) offer a promising energy source. Photobioreactors, unlike open ponds, require less space, consume less water, and are less weather-dependent. Additionally, they prevent culture evaporation, offer effective light distribution, and demonstrate climate change resistance. These factors, coupled with their ability to work at night and avoid contamination, make photobioreactors a more efficient and environmentally friendly option for energy production than traditional methods such as solar panels. The optimal conditions for algae cultivation include temperature range (16 to 27 °C), indirect middle-intensity light, necessary nutrients (salinity, CO_2, ammonia, phosphate), ideal pH (7–9), and the need for air circulation to harvest CO_2. Habibi et al. [49] shows that the initial investment required compared to an open pond is higher. Additionally, the scientific literature shows the necessity to control algae cultivation and highlights the lack of experience in building applications and the negative net present (observed) values from such photobioreactor facade panels after 15 years. Algae production presents multifaceted solutions with diverse applications outside of electricity production:

- It serves as a versatile tool for wastewater treatment, effectively cleansing water sources.
- Algae cultivation facilitates oxygen production and boasts an impressive CO_2 capture capacity, absorbing up to 85% of CO_2 content, thus aiding in carbon sequestration efforts.
- The yield of oil production from algae surpasses that of traditional sources such as soybeans by 60 times and palm by five times, promising a sustainable alternative to biofuel production.
- Algae production contributes to heat generation through innovative methods like biogas-to-electricity conversion in generators and recovering waste heat for steam supply, enhancing building energy efficiency. This algae cultivation also enables the production of food, ensuring a high-quality nutrient source compared to conventional

open pond methods. These numerous applications demonstrate that the process yields valuable by-products and offers the potential for light energy production, further diversifying its utility.

- Algae cultivation provides an unexpected benefit by offering thermal insulation, showcasing its potential as a comprehensive and sustainable solution across various domains.

However, it is imperative to adjust algae species to match specific climates and locations and to have stringent regulations in the construction domain. Indeed, it is essential to study the system's lifespan, maintenance, and cleaning requirements. We also note that the higher investment and production costs compared to open ponds render photobioreactors economically unviable. Factors like oxygen levels in water directly impact algae cultivation, whereas excessive light intensity can hinder photosynthesis. There are risks of poor or non-performance, and other renewable sources typically outperform algae in energy production. Certain algae species also pose human health risks.

Table 2. SWOT analysis of photobioreactor facade panels.

Strengths	Weaknesses	Opportunities	Threats
Generate energy [10] Algae can grow in seawater, wastewater, or harsh water [10] Algae have a high rate of growth (higher than most other productive crops) [10] More microalgae species can be developed (compared to an open pond) [10] Can produce 5 to 10 times higher yields per aerial footprint (than open pond) [10] Biogas production [10,33] Significantly decrease the building's energy demands [33] Biomass production high-efficiency (compared to open ponds) [10,33] Preventing culture evaporation [33] Effective light distribution [33] Climate change resistance [33] Tubular PBRs do not need a specific orientation for good exposure to solar light [11] Lower environmental impact than solar panel [10] Need less area (compared to an open pond) Lower water consumption (compared to an open pond) Less weather dependent (compared to an open pond) Work also during the night Avoid bacterial and dirt contamination [50] PBR design permits more effective use of light (compared to open ponds) [50]	An ideal temperature range is required for algae to bloom (being 16 to 27 °C) [10] Required indirect, middle-intensity light levels [10] Nutrients required (salinity, CO_2, ammonia, phosphate) [10] Specific pH required (7–9 is ideal) [10] Air circulation needed (harvest CO_2) [10] Initially require a higher investment (compared to an open pond) [50] Require a high control of algae cultivation [50] Lack of experience in building applications [19] Negative net present values (NPVs) after 15 years [12]	Algae production can be used for wastewater treatment [51] Oxygen production [51] CO_2-capture capacity (absorbing as much as 85% of CO_2 content) [51] The yield of oil production far exceeds that of soybeans (by 60 times) or palm (by 5 times) [51] Heat production (biogas-to-electricity conversion in the generator) [12] Recovering waste heat as steam supply [12] Able to produce food grade biomass (compared to open ponds) [10] Able to produce by-products [33] Can produce light energy [33] Provide thermal insulation [33]	A necessity to adapt algae species according to climate and location [19] Specific and tight regulations for real-life building [19] Need to study the lifetime of the system [19] Need to study the maintenance and cleaning requirements [19] Higher investment and production costs (compared to an open pond) [19] Not economically viable for the moment [19] Oxygen in the water directly affects the cultivation [10] Excessive light intensity can inhibit the photosynthesis process [33] A lack of natural light during the night causes biomass losses (25%) [33] Risks of poor or non-performance [18] Other renewables produce more energy [18] Human health risks with some algae species [18]

Table 3 shows the building-integrated wind turbine systems. The significant advantages of integrating vertical axis wind turbine (VAWT) wind walls within off-grid systems encompass several vital points, including reducing wind farm needs, particularly in off-grid settings, resulting in decreased infrastructure requirements. This reduction minimizes the necessity for cables and associated infrastructure for electricity delivery and mitigates costs and logistical challenges. Moreover, VAWT wind walls contribute to a notable decrease in energy losses, especially within off-grid systems, thereby enhancing overall efficiency. The flexibility of these wind walls, facilitated by demountable wind-harvesting panels, ensures adaptability to diverse environments. Unlike horizontal axis wind turbines (HAWTs) VAWT wind walls can capture wind from any direction without necessitating orientation. Additionally, they effectively harness turbulences, further optimizing energy capture VAWTs exhibit minimal noise production, even under low or high wind conditions, offering a quieter alternative for energy generation. The elimination of yaw mechanisms in VAWT wind walls simplifies their design and maintenance and contributes to their operational efficiency. Furthermore, compared to typical HAWTs, their lower wind startup speeds enable them to operate efficiently across varying wind conditions, solidifying their viability within off-grid systems. The inconveniences in building-integrated wind turbines are vibration and noise related to wind turbines, depending on the typology of the system. HAWTs require constant alignment with the wind direction for optimal performance. Conversely VAWTs exhibit reduced efficiency compared to traditional HAWTs and are positioned closer to the ground where wind speeds are typically lower, thus unable to harness higher wind speeds aloft. Consequently, VAWTs experience intermittent energy production influenced by varying weather conditions. Small wind turbine systems can be integrated into various structures in the building. They contribute to aesthetic design, such as in double-skin facades. Vertical axis wind turbines (VAWTs) can be positioned closer to the ground and in areas where taller structures are prohibited. Additionally, wind walls serve multiple purposes, including minimizing glare, controlling radiation, providing insulation, collecting heat, sequestering carbon emissions, and enhancing the aesthetic appeal of the building Particular avenues of research should be carried out to improve the integration of wind energy systems into the architecture of the building. Indeed, the public perceives wind turbines negatively due to visual pollution. In urban areas, turbulent and low-velocity wind conditions prevail, compounded by wind shadows caused by adjacent buildings and high urban terrain roughness. Turbines between buildings may cause discomfort for pedestrians due to high wind speeds near the ground. Additionally, buoyancy and heat effects on turbines should be considered. Early urban planning is essential in designing neighbouring buildings for turbines between them.

Hybrid solar-wind systems (see Table 4) do not rely on fossil fuels, making them more environmentally friendly and significantly reducing carbon dioxide emissions. Additionally, they require less space, have lower climate dependency, and offer better cost-effectiveness. Furthermore, they are more efficient and have a shorter payback than conventional systems. Moreover, the system includes a wind turbine that can operate during nighttime, further enhancing electricity generation and economic viability. However, hybrid technology necessitates a more significant initial investment than a singular system, such as solar panels, wind turbines, and energy storage. Combining a solar chimney with mirrors enhances the heat gain of the system. Incorporating a wind turbine and solar chimney into a PVT (photovoltaic thermal) panel system reduces the payback period and increases the potential for reducing CO_2 emissions. This configuration offers low operation and maintenance costs, generates minimal noise, and allows for integration with a storage system for both electricity and heat. Additionally, surplus power can be sold. These hybrid devices may not address all scenarios, especially in highly constrained building spaces, making installation impossible.

Table 3. SWOT analysis of the building-integrated wind turbines.

Strengths	Weaknesses	Opportunities	Threats
Reduced wind farm needs (off-grid system) [10] Limiting cables connection and infrastructure for electricity delivery [10] Decrease energy losses (off-grid system) [10] Wind walls are flexible systems (wind-harvesting panels are demountable) [49] VAWT wind walls are able to capture incoming wind from any direction (unlike HAWTs) [22] VAWT wind walls do not need to be oriented [22] VAWT wind walls can take advantage of turbulences [22] The noise is almost zero for normal winds and even for low winds with VAWTs [22] For VAWTs, no yaw mechanisms are needed [22] VAWTs have lower wind startup speeds than typical HAWTs [22]	Vibration and noise problems [22] Classic HAWTs need to be always aligned to the wind direction [22] VAWTs have decreased efficiency (more than common HAWTs) [22] VAWTs have rotors located close to the ground where wind speeds are lower [22] VAWTs cannot take advantage of higher wind speeds above [22] Intermittent energy production depending on weather [22]	Small wind turbines may be coupled with street lighting systems (smart lighting) [22] Can be paired with a photovoltaic system Can contribute to aesthetic design for the buildings (in double skin facade for instance) [22] VAWTs can be located nearer to the ground [22] VAWTs may be built at locations where taller structures are prohibited [22] Wind walls systems minimize glare and circulate air [49] Wind walls control radiation [49] Wind walls provide insulation [49] Wind walls collect heat [49] Wind walls generate energy [49] Wind walls sequester emissions [49] Wind walls provide aesthetics [49] Wind walls increase property value [49]	Wind turbines have a negative response from the public [52] Visual pollution [52] Turbulent and low-velocity wind conditions in urban areas [52] Adjacent buildings can cause wind shadow [53] Urban terrain roughness is high [53] If close to the ground, turbines between two buildings may cause discomfort for pedestrians (high wind speed) [23] Heat effects may affect the turbine (buoyancy needs to be considered) [23] Turbines between two buildings need early urban planning in the design of neighbouring buildings [23]

Table 4. SWOT analysis of the hybrid solar-wind systems.

Strengths	Weaknesses	Opportunities	Threats
Produce electricity [25–27] Do not require any fossil fuels [25] Have greater potential to reduce carbon dioxide emissions than the two systems alone [25–27] Lower climate condition dependence than the two systems alone [27] Need less area than two separated systems Better LCOE (levelized cost of electricity) [27] More environmentally friendly than the two systems alone [25,26] Better in terms of payback time than the two systems alone [27] More efficient than the two separated systems [27] The wind turbine can also rotate during the nighttime and improve the economics of the system by more electricity generation [25]	Require a larger initial investment than a unique system (solar panels, wind turbines, and energy storage) [27] Climate condition dependence [27] Intermittent production [27] Need more area than a unique solar or wind system [27]	Coupled with a solar chimney, using mirrors can increase the heat gain of the system [25] Adding a wind turbine and a solar chimney to a PVT panel system reduces payback period [25] Adding a wind turbine and a solar chimney to a PVT panel system increases the potential to reduce CO_2 emissions [25] Low operation and maintenance cost [25] Produce low noise [25] Can be equipped with a storage system for electricity and heat [25] Excess power can be sold [26]	May not be sufficient to cover all needs [26] May not fit into areas with limited space [27]

Technological features for high-efficiency, clean energy production through solar paint (see Table 5) are promising. Indeed, the adaptability of solar paint to various surfaces, the aesthetic integration into building envelopes, ease of application using a simple brush low-cost, adjustable electrochemical performance, and environmental friendliness with no emission of ozone-depleting substances are all advantages of this exceptional paint More advanced studies should be carried out at the current stage because the technology described needs to exhibit more efficiency, raising doubts about its sustainability. The solar paint technology still has room for improvement regarding its efficiency. Indeed, with a significant moisture adsorption capacity, it efficiently binds water molecules, facilitating its functioning. Moreover, its semiconductor nature ensures excellent conductivity, which is essential for its operation. Its ability to absorb light enhances its performance, whereas its high catalytic activity further contributes to its effectiveness. Additionally, its integration with standard inverter technology, akin to traditional solar cells, enables seamless connection to the electricity grid network, ensuring its compatibility and scalability within existing infrastructure. Future research perspectives would be oriented on the recent solar cell technology, requiring further investigation to determine its viability, especially in light of competition from more efficient and reliable traditional solar cells.

Table 5. SWOT analysis of solar paint.

Strengths	Weaknesses	Opportunities	Threats
High conversion efficiency [54] Produces clean energy [31] Gas phase water splitting is predicted to require less energy [31] Efficient light absorption with minimal light scattering [30] Adaptable to many surfaces [30,55] Provides aesthetic integration into the building envelope [30,55] Easy and quick application with a simple brush [56] Low-cost technology [55] Delivers an adjustable electrochemical performance [57] Environmentally friendly and emits no ozone-depleting substances after use [55]	Very low efficiency [58] Doubt regarding the sustainability of this technology [55]	A large moisture adsorption capacity for binding water molecules [30] It should be a semiconductor with good conductivity [30] Providing light adsorption capabilities [30] Features high catalytic activity [30] Utilizes the standard inverter technology employed by traditional solar cells for connecting to the electricity grid network [55]	Competition with more efficient and reliable traditional solar cells [58] Very recent technology that necessitates additional studies to ascertain its viability [58]

Table 6 underscores the remarkable capabilities of the microbial biophotovoltaic technology system under examination, showcasing its extraordinary growth potential even under limited light conditions for prolonged periods. Moreover, it significantly enhances water-use efficiency with a modest culture volume. By employing a gel as a substitute for the conventional liquid reservoir in bio-photovoltaic (BPV) devices, the system achieves notable improvements in power output compared to its counterparts. Notably, this system demonstrates remarkable endurance, sustaining electrical production for well over 100 h, starkly contrasting with the one-hour operation typically in paper-based microbial fuel cells (MFCs). Its versatility extends to delivering short power while remaining disposable and environmentally friendly, marking a significant advancement in sustainable power solutions. However, there are several critical factors affecting the performance of microbial fuel cells (MFCs):

- The notable limitation in electricity production suggests a need for further optimization.
- The printing process risks damage to cyanobacteria cells, potentially compromising their effectiveness within the MFCs.
- This study observes a decrease in power output in low-light conditions compared to well-lit environments, highlighting the dependency of MFCs on light availability.
- The Therinted Carbon Nanotube (CNT) cathode is identified as a significant bottleneck in MFC performance, indicating the necessity for alternative cathode materials or fabrication methods to enhance overall efficiency.

We underscore the complexity of MFC technology and the importance of addressing various challenges to realize its full potential in sustainable energy production. The feasibility of utilizing a low-cost commercial inkjet printer without significantly impacting the cell viability of the system has been highlighted in the scientific literature. The advantages of paper as an inexpensive and biodegradable material and the potential for miniaturizing cyanobacteria culture are also interesting avenues for improving the process. Additionally, employing high-performance carbon black (CB) could enhance power output, whereas desert CB usage might reduce material and energy expenses for scaling up. This research also proposes the development of bioenergy wallpaper and demonstrates that incorporating a hydrogel between the anode and cathode could improve power output by exposing the cathode to more air. This solar energy is an intermittent source due to its dependence on external factors such as location, weather, time of day, and seasons, resulting in inevitable drops in energy production during low light conditions. The optimization of the cell design is essential for better efficiency of the system.

Table 6. SWOT analysis of microbial biophotovoltaic technology.

Strengths	Weaknesses	Opportunities	Threats
Very great capacity for growth [36] Works in the dark (for several hours even if the range is lower) [36] Improves water-use efficiency (considering the minor volume of starting culture) [36] Uses a gel (which replaces the liquid reservoir normally used in conventional BPV devices [36] Great power output compared with conventional liquid culture-based BPV devices [36] Electrical output can be sustained for more than 100 h (paper-based MFCs can only operate for 1 h) [36] Can provide a short burst of power [36] Disposable and environmentally friendly power [36]	Low electricity production [36] Damage possibility of cyanobacteria cells during printing [36] Power output is less in the dark that in the light [36] Printed CNT cathode is a limiting factor in microbial fuel cell performance [36]	Feasibility of using an inexpensive commercial inkjet printer without (really) affecting cell viability [36] Paper is an inexpensive widespread material and biodegradable [36] The potential of miniaturization for cyanobacteria culture [36] Use of high-performance CB could increase the power output [36] Use of desert CB might reduce the material and energy costs of scale-up [36] Could be developed for bioenergy wallpaper [36] Hydrogel between anode and cathode would improve the power output (by exposing the cathode to more air) [36]	Solar energy is an intermittent energy source (inevitably drops in low light) [36] Production depends greatly on external conditions (location, weather, time of the day, and seasons of the year) [36] Optimizing cell design [36]

Integrating photovoltaic glazing and shading devices (PV devices) presents a multifaceted solution towards achieving sustainable energy practices and enhancing building efficiency (see Table 7). By harnessing clean electric energy, these innovative technologies contribute significantly to active energy conservation for windows, reducing lighting loads and overall electricity consumption. Moreover, their implementation as part of a sustainable electricity production system fosters environmental responsibility and facilitates long-term energy savings. The CoPVTG (Combined Photovoltaic and Thermal Glazing) device emerges as a standout solution, consistently delivering high energy yields while ensuring a uniform distribution of daylight. Its ability to regulate solar contribution and its economic feasibility render it a compelling option for architectural integration. Notably, CoPVTG devices not only meet the functional requirements of natural lighting but also uphold the aesthetic integrity of buildings, thereby striking a harmonious balance between sustainability and design. Furthermore, compared to alternative technologies like CoPEG (Combined Photovoltaic and Electrochromic Glazing), CoPVTG systems demonstrate superior energy performance, augmented by exploitable hot air. Adopting PV glazing and shading devices represents a pivotal step towards achieving energy efficiency and architectural excellence in contemporary construction practices. However, some negative points should be highlighted. Indeed, the climate and location of the site dramatically influence the effectiveness of photovoltaic windows. Electricity generation from BIPV systems is intermittent due to varying weather conditions. Additionally, the orientation of buildings impacts the performance of these systems. The benefits of building-integrated photovoltaic (BIPV) windows are the ability to provide adequate ventilation, reduce building cooling or heating loads, and serve as both facade windows and exterior elements. BIPV windows are noted for their insulation capabilities, with studies showing superior energy-saving performance compared to conventional insulating glass windows. Additionally, PV insulating glass units are highlighted for their more significant energy-saving potential than PV double-skin facades. The potential of Low-E coatings to minimize heat transfer through radiation is also a positive point of view for improving the system. Coloured modules can result in notable efficiency reductions, varying based on the materials and colours employed. Additionally, the duration required to recoup energy investments and the associated uncertainty regarding greenhouse gas emissions are not clarified in the scientific literature. This uncertainty underscores the challenge of competing with traditional roof PV systems.

Table 8 discusses converting ambient mechanical energy from wind impact and water droplets into electricity. This process can be utilized for a self-powered intelligent window system. The technology involved, known as TENGs (triboelectric nanogenerators), maintains transparency, ensuring that they do not obstruct or reduce the window's surface area. Their system has a high transmittance rate of over 60% and exhibits low water contact angle hysteresis when treated with SLIPSs (Slippery Liquid-Infused Porous Surfaces). Additionally, the efficiency of energy conversion is enhanced with the addition of SLIPSs, which also provide benefits such as anti-fouling, anti-icing, and drag reduction. This approach aligns with sustainability and renewable energy principles, offering advantages such as affordability, lightweight construction, and the ability to harness both wind and rain. Furthermore, introducing solid–solid/liquid–solid convertible TENGs expands the range of conditions under which energy can be generated. However, Table 8 shows the high limitations of the discussed system, emphasizing its meagre power output compared to conventional systems like photovoltaic (PV) panels and wind turbines. It also underlines the system's dependence on climate conditions, noting that variations in temperature and humidity can significantly impact its performance. In addition, the system can serve as a rain sensor or sensor for a self-powered window-closing system. For example, the component can be integrated with other electricity generation systems like photovoltaic windows. The challenges for improving the systems are increasing the durability and limiting the triboelectric nanogenerators' power. It is also important to make the system more competitive (efficient and reliable) than the existing integrated systems in the building envelope.

Table 9 shows the potential of PVTENG hybrid systems in energy production, particularly on sunny and rainy days. These systems offer advantages such as complementing individual PV and TENG components, good transparency (23.49% visible light transmittance), high colour rendering (CRI of 92), and effective window insulation. They convert ambient mechanical energy, particularly from water droplets, into electricity. Additionally, integrating SLIPSs (Slippery Liquid-Infused Porous Surfaces) leads to benefits such as low water contact angle hysteresis, increased energy conversion efficiency, and properties like anti-fouling, anti-icing, and drag reduction. Overall, these systems offer sustainable, renewable energy solutions at low cost and with lightweight construction. The difficulty observed in the hybrid system is the meagre power output and specific transmittance (i.e., the transmittance phenomenon can only be performed in a particular wavelength range). The difficulty observed in the hybrid system is the meagre power output and the specific transmittance effect of the material used in the system (i.e., the transmittance phenomenon can only be performed in a particular wavelength range). One of the main problems encountered in this hybrid system is the shading effects that impede heat transfer and lead to a decrease in air temperature, particularly in greenhouse applications where there is a high plant growth factor. Further research should be carried out to solve this shading problem. Climatic conditions, including temperature, humidity, and atmospheric pressure, can significantly impact the performance of electrical systems. These factors can influence the durability of components, potentially leading to lower overall reliability and efficiency. Additionally, in situations where short circuits occur, there may be limitations on the amount of output current that can be safely handled, which can further compromise the operational capabilities of the system. Therefore, understanding and mitigating these dependencies are crucial in ensuring the resilience and functionality of electrical infrastructure under various environmental conditions.

Table 7. SWOT analysis of photovoltaic glasses.

Strengths	Weaknesses	Opportunities	Threats
Obtain clean electric energy [37] Realize active energy saving of windows [59] The implementation of PV glazing and shading devices has the potential to decrease lighting loads and electricity consumption [7] Sustainable electricity production system [7] Integrated glazing reduces the environmental and economic impact of buildings [6] The CoPVTG device always provides the highest energy yield [60] Provide a uniform daylight distribution [7] Provide solar contribution control [7] Economically feasible [7] It can meet the needs of natural lighting while satisfying architectural aesthetics [37] CoPVTG devices provide higher energy yield than CoPEG [60] CoPVTG systems provide exploitable hot air [60]	The performance of BIPV depends highly on the climate and location site [7] Intermittent electricity production depending on weather conditions [7] Building orientation affects performances of the system [7]	Provide adequate ventilation (BIPV windows) [5] Reduce building cooling load or heat load [7,59] Can be installed as a facade window and balustrade or sloped as an exterior element [6] Capable of insulating the building [60] PV windows demonstrated superior energy-saving performance compared to conventional insulating glass windows [61] PV insulating glass units have greater energy-saving potential than PV double skin facades [61] Low-E coatings have the potential to minimize heat transfer through radiation [7]	Coloured modules can lead to significant efficiency losses depending on the materials and colours used [6] The timeframe for recovering energy investment and the associated uncertainty in greenhouse gas emissions remains unclear [6] Competition with traditional roof PV systems

Table 8. SWOT analysis of triboelectric nanogenerators.

Strengths	Weaknesses	Opportunities	Threats
Convert ambient mechanical energy (from wind impact and water droplets) into electricity [39] Can be used for a self-powered smart window system [39] TENGs are transparent (do not cover or sacrifice surface area window) [39] High transmittance of over 60% [39] Low water contact angle hysteresis with SLIPS addition [41] More efficient energy conversion with SLIPS addition [41,42] Anti-fouling, anti-icing, and drag reduction with SLIPS addition [41,42] Sustainable and renewable energy [62] Low cost [62] Lightweight [62] Take advantage of both wind and rain [39] Solid–solid/liquid–solid convertible TENG increases the conditions under which energy can be produced [40]	Very low power output compared to conventional systems such as PV panels and wind turbines [41] Climate conditions dependence [39] Temperature and humidity may affect the performances of this system [39]	Act as a rain-sensor to prevent rainwater from entering the house [41] Integrating an electrochromic device (ECD) (change colour or opacity) [39] Can be paired with other electricity-production system such as PV glasses [63] Can be used as a sensor for self-powered window-closing system [41]	Lower durability [64] Limited short circuit output current [64] Competition with more efficient and reliable systems

Table 10 provides a quantitative comparison of various systems that have been the subject of scientific articles. The comparison is conducted based on three essential criteria:

- The electrical energy production generated by the system;
- The financial cost;
- The CO_2 emission or absorption rate by the device.

It emerges that BIPV or BAPV systems are the least expensive ones (approximately 0.15 euros per kWh). This can be explained by the fact that the technology of these systems has evolved significantly in recent years, contributing to this cost reduction. Regarding PBR systems, it is noted that they are the least polluting. The explanation is that the algae used in the device need to absorb a maximum amount of CO_2 (200 m^2 of algae absorb approximately 8.5 tons of CO_2 per year) to produce this electrical energy. As for energy production, the wind turbine integrated into the building is the most profitable system (it produces approximately 0.41 kW per m^2 of habitable area).

Table 9. SWOT analysis of the photovoltaic and triboelectric nanogenerator hybrid system.

Strengths	Weaknesses	Opportunities	Threats
Energy production on sunny days and rainy days [63] PVTENG hybrid systems represent a great potential to complement vulnerable aspects of individual PV and TENG components [63] Good transparency (visible light transmittance (VLT) of 23.49%), colour rendering (CRI of 92), and window insulation [63] Convert ambient mechanical energy (from water droplets) into electricity [39] Low water contact angle hysteresis with SLIPS addition [41] More efficient energy conversion with SLIPS addition [41] Anti-fouling, anti-icing, and drag reduction with SLIPS addition [41] Sustainable and renewable energy [62] Low cost [62] Lightweight [62]	Very low power output [41] Specific transmittance (blue layer) [63]	Shading effects [63] Hampers heat transfer [63] Decreases air temperature [63] Greenhouse applications (high plant growth factor of 25.3%) [63]	Climatic conditions dependence [64] Lower durability [64] Limited short circuit output current [64]

Table 10. Quantitative comparison systems. Complementary information: positive CO_2 is when the system rejects carbon and negative CO_2 is when the system absorbs carbon in the atmosphere.

Systems	Production (in $kW \cdot m^{-2}$)	Cost (in $)	CO_2 (in $KgCO_2$/Year)	References
Photovoltaic (BIPV or BAPV) or Photovoltaic glass	0.16–0.19	0.16 (by kWh)	At + 27 until + 139	[65,66]
PBRs (bioreactor) or Microbial biophotovoltaic technology	0.06–50	1000–1500 (by m^2)	−42.5	[49,67]
Building-integrated wind turbines	0.41	767.3 (by unit)	At + $7.5.10^8$ until + 22.10^8	[68,69]
Solar paint wall	0.02–0.5	-	-	[58,70]
Triboelectric nanogenerators glasses	0.0018–0.05	-	-	[71]

4. Conclusions

The building envelope element ensures structural stability, resilience, and protection from external elements. Despite its primary functions, an opportunity exists to enhance the building's energy balance without additional surfaces. Often overlooked, the roof presents untapped potential, offering ample space and optimal exposure to harness various energy sources such as solar, rain, and wind. This makes it ideal for incorporating energy recovery devices like PVT panels, wind turbines, and PBRs for algae cultivation. In specific contexts, hybrid systems prove advantageous, generating more energy, optimizing space, and mitigating the limitations of standalone systems. Beyond energy production, specific systems offer additional functionalities; for example, algae-based systems exhibit prowess in wastewater treatment and carbon dioxide capture. Conversely, facades and windows are susceptible to climatic factors, necessitating modulating and regulating systems. Technologies like PBR facade panels and wind walls generate electricity and provide thermal and acoustic insulation, shading effects, and ventilation, contributing to reduced energy consumption. However, many of these systems require refinement and further development to validate their viability and effectiveness. Some technologies discussed in this study generate limited electrical currents, pose implementation challenges, or exist only in theoretical or simulated forms. In summary, integrating electricity production systems into the building envelope taps into the potential of existing surfaces and aligns with the imperative of meeting growing energy needs sustainably. The combination of building envelopes and energy production holds promise for creating more resilient, efficient, and environmentally conscious structures. This bibliographic study demonstrates that the evolution of electricity-producing systems integrated into the building envelope and the risks involved, if we move towards these increasingly innovative technologies, have not really been addressed in the scientific literature. Further studies should be conducted to define the economic, technical, environmental, and social implications of these electricity production systems integrated into the building envelope.

Author Contributions: Conceptualization, D.A.H.F. and E.O.; methodology, D.A.H.F.; formal analysis, D.A.H.F. and E.O.; investigation, D.A.H.F., E.O. and K.G.; resources, D.A.H.F., E.O., R.B. and K.G.; writing—original draft preparation, D.A.H.F. and E.O.; writing—review and editing, D.A.H.F. and E.O.; visualization, D.A.H.F. and E.O.; supervision, D.A.H.F.; project administration, D.A.H.F. All authors have read and agreed to the published version of the manuscript.

Funding: This research received no external funding.

Conflicts of Interest: The authors declare no conflicts of interest.

Abbreviations

The following abbreviations are used in this manuscript:

AD	Anaerobic digestion
ADEME	French Environment and Energy Management Agency
BAPV	Building-attached photovoltaic
BIPV	Building-integrated photovoltaic
BIQ	Bio-intelligent quotient
BPV	Biophotovoltaic
CFD	Computational fluid dynamics
CoPEG	Concentrating Photovoltaic Evacuated Glazing
CoPVTG	Concentrating Photovoltaic Thermal Glazing
DSF	Double skin facade
HAWT	Horizontal axis wind turbine
MFC	Microbial fuel cell
PBR	Photobioreactor
PEMEC	Polymer electrolyte membrane electrolyzer
PEMFC	Polymer electrolyte membrane fuel cell
PV	Photovoltaic
PVT	Photovoltaic thermal collector
SLIPS	Slippery lubricant-infused porous surface
ST-PSC	Semitransparent polymer solar cell
STPV	Photovoltaic Semi-Transparent
STS	Solar thermal system
TE	Thermoelectric
TEG	Thermoelectric generator
TENG	Triboelectric nanogenerator
UNEP	United Nations Environment Programme
VAWT	Vertical axis wind turbine

References

1. Agence de l'Environnement et de la Maîtrise de l'Énergie. *Climat Air et Énergie*; Agence de l'Environnement et de la Maîtrise de l'Energie—ADEME: Paris, France, 2018. Available online: https://librairie.ademe.fr/changement-climatique-et-energie/1725-climat-air-et-energie-9791029712005.html (accessed on 2 January 2024).
2. Ministères Écologie Énergie Territoires. Energie dans les Bâtiments. 2021. Available online: https://www.ecologie.gouv.fr/energie-dans-batiments (accessed on 2 January 2024).
3. Bigot, D.; Miranville, F.; Fakra, A.; Boyer, H. A nodal thermal model for photovoltaic systems: Impact on building temperature fields and elements of validation for tropical and humid climatic conditions. *Energy Build.* **2009**, *41*, 1117–1126. [CrossRef]
4. Olabi, A.G.; Shehata, N.; Maghrabie, H.M.; Heikal, L.A.; Abdelkareem, M.A.; Rahman, S.M.A.; Shah, S.K.; Sayed, E.T. Progress in Solar Thermal Systems and Their Role in Achieving the Sustainable Development Goals. *Energies* **2022**, *15*, 9501. [CrossRef]
5. Vodapally, S.N.; Ali, M.H. A Comprehensive Review of Solar Photovoltaic (PV) Technologies, Architecture, and Its Applications to Improved Efficiency. *Energies* **2023**, *16*, 319. [CrossRef]
6. Kuhn, T.E.; Erban, C.; Heinrich, M.; Eisenlohr, J.; Ensslen, F.; Neuhaus, D.H. Review of technological design options for building integrated photovoltaics (BIPV). *Energy Build.* **2021**, *231*, 110381. [CrossRef]
7. Taşer, A.; Koyunbaba, B.K.; Kazanasmaz, T. Thermal, daylight, and energy potential of building-integrated photovoltaic (BIPV) systems: A comprehensive review of effects and developments. *Sol. Energy* **2023**, *251*, 171–196. [CrossRef]
8. Farhani, R.; Ennawaoui, C.; Hajjaji, A.; Boughaleb, Y.; Rivenq, A.; El Hillali, Y. Photovoltaic-thermoelectric (PV-TE) hybrid system for thermal energy harvesting in low-power sensors. *Mater. Today Proc.* **2022**, *66*, 223–228. [CrossRef]
9. Maythem, A.; Al-Shamani, A.N. Using Hybrid System Photovoltaic Thermal/Phase Change Materials/Thermoelectric (PVT/PCM/TE): A Review. *For. Chem. Rev.* **2022**, 1365–1400. Available online: http://www.forestchemicalsreview.com/index.php/JFCR/article/view/1241 (accessed on 2 January 2024).
10. Biloria, N.; Thakkar, Y. Integrating algae building technology in the built environment: A cost and benefit perspective. *Front. Archit. Res.* **2020**, *9*, 370–384. [CrossRef]
11. Elrayies, G.M. Microalgae: Prospects for greener future buildings. *Renew. Sustain. Energy Rev.* **2018**, *81*, 1175–1191. [CrossRef]
12. Bender, A. Cost and Benefit Analysis of Implementing Photobioreactor System for Self-Sustainable Electricity Production from Algae. *Master Work* **2017** . [CrossRef]
13. Bohutskyi, P.; Bouwer, E. Biogas production from algae and cyanobacteria through anaerobic digestion: A review, analysis, and research needs. In *Advanced Biofuels and Bioproducts*; Springer: Berlin/Heidelberg, Germany, 2013; pp. 873–975. [CrossRef]
14. Sirohi, R.; Pandey, A.K.; Ranganathan, P.; Singh, S.; Udayan, A.; Awasthi, M.K.; Hoang, A.T.; Chilakamarry, C.R.; Kim, S.H.; Sim, S.J. Design and applications of photobioreactors—A review. *Bioresour. Technol.* **2022**, *349*, 126858. [CrossRef]
15. Schenk, P.M.; Thomas-Hall, S.R.; Stephens, E.; Marx, U.C.; Mussgnug, J.H.; Posten, C.; Kruse, O.; Hankamer, B. Second generation biofuels: High-efficiency microalgae for biodiesel production. *Bioenergy Res.* **2008**, *1*, 20–43. [CrossRef]
16. Berzin, I. Photobioreactor and Process for Biomass Production and Mitigation of Pollutants in Flue Gases. 2002. Available online: https://patents.google.com/patent/US20050260553A1/en?oq=+patent+US+20050260553 (accessed on 2 January 2024).
17. Araji, M.T.; Shahid, I. Symbiosis optimization of building envelopes and micro-algae photobioreactors. *J. Build. Eng.* **2018**, *18*, 58–65. [CrossRef]
18. Wilkinson, S.; Biloria, N.; Ralph, P. The technical issues associated with algae building technology. *Int. J. Build. Pathol. Adapt.* **2020**, *38*, 673–688. [CrossRef]
19. Öncel, S.; Köse, A.; Öncel, D. Façade integrated photobioreactors for building energy efficiency. In *Start-Up Creation*; Elsevier: Amsterdam, The Netherlands, 2016; pp. 237–299. [CrossRef]
20. Drew, D.; Barlow, J.; Cockerill, T. Estimating the potential yield of small wind turbines in urban areas: A case study for Greater London, UK. *J. Wind. Eng. Ind. Aerodyn.* **2013**, *115*, 104–111. [CrossRef]
21. Yang, A.-S.; Su, Y.-M.; Wen, C.-Y.; Juan, Y.-H.; Wang, W.-S.; Cheng, C.-H. Estimation of wind power generation in dense urban area. *Appl. Energy* **2016**, *171*, 213–230. [CrossRef]
22. Casini, M. Small vertical axis wind turbines for energy efficiency of buildings. *J. Clean Energy Technol.* **2016**, *4*, 56–65. [CrossRef]
23. Hassanli, S.; Hu, G.; Kwok, K.C.; Fletcher, D.F. Utilizing cavity flow within double skin façade for wind energy harvesting in buildings. *J. Wind. Eng. Ind. Aerodyn.* **2017**, *167*, 114–127. [CrossRef]
24. Park, J.; Jung, H.J.; Lee, S.W.; Park, J. A new building-integrated wind turbine system utilizing the building. *Energies* **2015**, *8*, 11846–11870. [CrossRef]
25. Kazemi, K.; Ebrahimi, M.; Lahonian, M.; Babamiri, A. Micro-scale heat and electricity generation by a hybrid solar collector-chimney, thermoelectric, and wind turbine. *Sustain. Energy Technol. Assessments* **2022**, *53*, 102394. [CrossRef]
26. Deymi-Dashtebayaz, M.; Baranov, I.V.; Nikitin, A.; Davoodi, V.; Sulin, A.; Norani, M.; Nikitina, V. An investigation of a hybrid wind-solar integrated energy system with heat and power energy storage system in a near-zero energy building—A dynamic study. *Energy Convers. Manag.* **2022**, *269*, 116085. [CrossRef]
27. Shahverdian, M.H.; Sohani, A.; Pedram, M.Z.; Sayyaadi, H. An optimal strategy for application of photovoltaic-wind turbine with PEMEC-PEMFC hydrogen storage system based on techno-economic, environmental, and availability indicators. *J. Clean. Prod.* **2023**, *384*, 135499. [CrossRef]

28. Lahlou, Y.; Abdelghani, H.; Mohammed, A.; Mohammed, I. Système Compact de Production d'énergie Renouvelable Hybride PV-Eolien. 2021. Available online: https://patentscope.wipo.int/search/fr/detail.jsf?docId=WO2021137680 (accessed on 2 January 2024).
29. Elçiçek, H. Determination of critical catalyst preparation factors (cCPF) influencing hydrogen evolution. *Int. J. Hydrogen Energy* **2023**, *48*, 3824–3837. [CrossRef]
30. Daeneke, T.; Dahr, N.; Atkin, P.; Clark, R.M.; Harrison, C.J.; Brkljaca, R.; Pillai, N.; Zhang, B.Y.; Zavabeti, A.; Ippolito, S.J.; et al. Surface water dependent properties of sulfur-rich molybdenum sulfides: Electrolyteless gas phase water splitting. *ACS Nano* **2017**, *11*, 6782–6794. [CrossRef] [PubMed]
31. Shearer, C.J.; Hisatomi, T.; Domen, K.; Metha, G.F. Gas phase photocatalytic water splitting of moisture in ambient air: Toward reagent-free hydrogen production. *J. Photochem. Photobiol. A Chem.* **2020**, *401*, 112757. [CrossRef]
32. Zheng, X.; Song, Y.; Liu, Y.; Yang, Y.; Wu, D.; Yang, Y.; Feng, S.; Li, J.; Liu, W.; Shen, Y.; et al. $ZnIn_2S_4$-based photocatalysts for photocatalytic hydrogen evolution via water splitting. *Coord. Chem. Rev.* **2023**, *475*, 214898. [CrossRef]
33. Talaei, M.; Mahdavinejad, M.; Azari, R. Thermal and energy performance of algae bioreactive façades: A review. *J. Build. Eng.* **2020**, *28*, 101011. [CrossRef]
34. Talaei, M.; Mahdavinejad, M. Probable cause of damage to the panel of microalgae bioreactor building façade: Hypothetical evaluation. *Eng. Fail. Anal.* **2019**, *101*, 9–21. [CrossRef]
35. McCormick, A.J.; Bombelli, P.; Scott, A.M.; Philips, A.J.; Smith, A.G.; Fisher, A.C.; Howe, C.J. Photosynthetic biofilms in pure culture harness solar energy in a mediatorless bio-photovoltaic cell (BPV) system. *Energy Environ. Sci.* **2011**, *4*, 4699–4709. [CrossRef]
36. Sawa, M.; Fantuzzi, A.; Bombelli, P.; Howe, C.J.; Hellgardt, K.; Nixon, P.J. Electricity generation from digitally printed cyanobacteria. *Nat. Commun.* **2017**, *8*, 1327. [CrossRef]
37. Chen, H.; Cai, B.; Yang, H.; Wang, Y.; Yang, J. Study on natural lighting and electrical performance of louvered photovoltaic windows in hot summer and cold winter areas. *Energy Build.* **2022**, *271*, 112313. [CrossRef]
38. Liu, X.; Wu, Y. Design, development and characterisation of a Building Integrated Concentrating Photovoltaic (BICPV) smart window system. *Sol. Energy* **2021**, *220*, 722–734. [CrossRef]
39. Yeh, M.-H.; Lin, L.; Yang, P.-K.; Wang, Z.L. Motion-driven electrochromic reactions for self-powered smart window system. *ACS Nano* **2015**, *9*, 4757–4765. [CrossRef]
40. Chung, J.; Cho, H.; Yong, H.; Heo, D.; Rim, Y.S.; Lee, S. Versatile surface for solid–solid/liquid–solid triboelectric nanogenerator based on fluorocarbon liquid infused surfaces. *Sci. Technol. Adv. Mater.* **2020**, *21*, 139–146. [CrossRef] [PubMed]
41. Zeng, Y.; Luo, Y.; Lu, Y.; Cao, X. Self-powered rain droplet sensor based on a liquid–solid triboelectric nanogenerator. *Nano Energy* **2022**, *98*, 107316. [CrossRef]
42. Chen, Z.; Lu, Y.; Li, R.; Orlando, R.J.; Manica, R.; Liu, Q. Liquid-solid triboelectric nanogenerators for a wide operation window based on slippery lubricant-infused surfaces (SLIPS). *Chem. Eng. J.* **2022**, *439*, 135688. [CrossRef]
43. Zhou, Q.; Park, J.G.; Kim, K.N.; Thokchom, A.K.; Bae, J.; Baik, J.M.; Kim, T. Transparent-flexible-multimodal triboelectric nanogenerators for mechanical energy harvesting and self-powered sensor applications. *Nano Energy* **2018**, *48*, 471–480. [CrossRef]
44. İlhan Volkan, Ö.; Yeşilyurt, M.K.; Yilmaz, E.Ç.; Ömeroğlu, G. Photovoltaic thermal (PVT) solar panels. *Int. J. New Technol. Res.* **2016**, *2*, 13–16.
45. Jia, Y.; Alva, G.; Fang, G. Development and applications of photovoltaic–thermal systems: A review. *Renew. Sustain. Energy Rev.* **2019**, *102*, 249–265. [CrossRef]
46. Segura, E.; Belmonte, L.M.; Morales, R.; Somolinos, J.A. A Strategic Analysis of Photovoltaic Energy Projects: The Case Study of Spain. *Sustainability* **2023**, *15*, 12316. [CrossRef]
47. Allouhi, A.; Rehman, S.; Buker, M.S.; Said, Z. Recent technical approaches for improving energy efficiency and sustainability of PV and PV-T systems: A comprehensive review. *Sustain. Energy Technol. Assessments* **2023**, *56*, 103026. [CrossRef]
48. Yu, G.; Yang, H.; Yan, Z.; Ansah, M.K. A review of designs and performance of façade-based building integrated photovoltaic-thermal (BIPVT) systems. *Appl. Therm. Eng.* **2021**, *182*, 116081. [CrossRef]
49. Habibi, S.; Valladares, O.P.; Peña, D.M. Sustainability performance by ten representative intelligent Façade technologies: A systematic review. *Sustain. Energy Technol. Assessments* **2022**, *52*, 102001. [CrossRef]
50. SCHOTT. Open Raceway Pond, Plastic Bags or Glass Tube Photobioreactor. 2018. Available online: https://knowledge.schott.com/art_resource.php?sid=qhwy.2rri80a (accessed on 2 January 2024).
51. Proksch, G. Growing sustainability-integrating algae cultivation into the built environment. *Edinb. Archit. Res. J.* **2013**, *33*, 147–162.
52. Yi, Y.K.; Kang, B. Integrating a wind turbine into a parking pavilion for generating electricity. *J. Build. Eng.* **2020**, *32*, 101471. [CrossRef]
53. Walker, S.L. Building mounted wind turbines and their suitability for the urban scale—A review of methods of estimating urban wind resource. *Energy Build.* **2011**, *43*, 1852–1862. [CrossRef]
54. Ravi, P.; Noh, J. Photocatalytic Water Splitting: How Far Away Are We from Being Able to Industrially Produce Solar Hydrogen? *Molecules* **2022**, *27*, 7176. [CrossRef]
55. Sharma, S.; Baral, R. Solar Photovoltaic Paint for Future: A Technical Review. *Adv. J. Eng.* **2022**, *1*, 18–23. [CrossRef]

56. Abbas, M.A.; Basit, M.A.; Yoon, S.J.; Lee, G.J.; Lee, M.D.; Park, T.J.; Kamat, P.V.; Bang, J.H. Revival of solar paint concept: Air-processable solar paints for the fabrication of quantum dot-sensitized solar cells. *J. Phys. Chem. C* **2017**, *121*, 17658–17670. [CrossRef]
57. Saleem, M.; Albaqami, M.D.; Bahajjaj, A.A.A.; Ahmed, F.; Din, E.; Arifeen, W.U.; Ali, S. Wet-Chemical Synthesis of TiO_2/PVDF Membrane for Energy Applications. *Molecules* **2022**, *28*, 285. [CrossRef] [PubMed]
58. Genovese, M.P.; Lightcap, I.V.; Kamat, P.V. Sun-believable solar paint. A transformative one-step approach for designing nanocrystalline solar cells. *ACS Nano* **2012**, *6*, 865–872. [CrossRef] [PubMed]
59. Wang, H.; Lin, C.; Hu, Y.; Zhang, X.; Han, J.; Cheng, Y. Study on indoor adaptive thermal comfort evaluation method for buildings integrated with semi-transparent photovoltaic window. *Build. Environ.* **2023**, *228*, 109834. [CrossRef]
60. Barone, G.; Buonomano, A.; Chang, R.; Forzano, C.; Giuzio, G.F.; Mondol, J.; Palombo, A.; Pugsley, A.; Smyth, M.; Zacharopoulos, A. Modelling and simulation of building integrated Concentrating Photovoltaic/Thermal Glazing (CoPVTG) systems: Comprehensive energy and economic analysis. *Renew. Energy* **2022**, *193*, 1121–1131. [CrossRef]
61. Wang, M.; Peng, J.; Li, N.; Yang, H.; Wang, C.; Li, X.; Lu, T. Comparison of energy performance between PV double skin facades and PV insulating glass units. *Appl. Energy* **2017**, *194*, 148–160. [CrossRef]
62. Walden, R.; Kumar, C.; Mulvihill, D.M.; Pillai, S.C. Opportunities and challenges in triboelectric nanogenerator (TENG) based sustainable energy generation technologies: A mini-review. *Chem. Eng. J. Adv.* **2021**, *9*, 100237. [CrossRef]
63. Liu, T.; Zheng, Y.; Xu, Y.; Liu, X.; Wang, C.; Yu, L.; Fahlman, M.; Li, X.; Murto, P.; Chen, J.; et al. Semitransparent polymer solar cell/triboelectric nanogenerator hybrid systems: Synergistic solar and raindrop energy conversion for window-integrated applications. *Nano Energy* **2022**, *103*, 107776. [CrossRef]
64. Singh, S.; Tripathi, R.K.; Gupta, M.K.; Dzhardimalieva, G.I.; Uflyand, I.E.; Yadav, B. 2-D self-healable polyaniline-polypyrrole nanoflakes based triboelectric nanogenerator for self-powered solar light photo detector with DFT study. *J. Colloid Interface Sci.* **2021**, *600*, 572–585. [CrossRef] [PubMed]
65. Amoruso, F.M.; Schuetze, T. Carbon Life Cycle Assessment and Costing of Building Integrated Photovoltaic Systems for Deep Low-Carbon Renovation. *Sustainability* **2023**, *15*, 9460. [CrossRef]
66. Bigot, D.; Miranville, F.A.; Ingar, I.; Guichard, S.; Boyer, H. Thermal performance of photovoltaic systems integrated in buildings. In *Solar Collectors and Panels, Theory and Applications*; InTech: Lodon, UK, 2010; pp. 405–428.
67. Vajdi, S.; Aslani, A. Design and techno-economic analysis of direct CO_2 capturing with integrated photobioreactors as a building façade. *Sustain. Energy Technol. Assessments* **2023**, *56*, 103068. [CrossRef]
68. Loganathan, B.; Chowdhury, H.; Mustary, I.; Rana, M.M.; Alam, F. Design of a micro wind turbine and its economic feasibility study for residential power generation in built-up areas. *Energy Procedia* **2019**, *160*, 812–819. [CrossRef]
69. Dayan, E. Wind energy in buildings: Power generation from wind in the urban environment-where it is needed most. *Refocus* **2006**, *7*, 33–38. [CrossRef]
70. Khan, S.A.; Rahman, A. Efficiency of thin film photovoltaic paint: A brief review. *Int. J. Recent Technol. Eng.* **2019**, *7*, 163–169.
71. Hasan, M.A.M.; Zhang, T.; Wu, H.; Yang, Y. Water Droplet-Based Nanogenerators. *Adv. Energy Mater.* **2022**, *12*, 2201383. [CrossRef]

sustainability MDPI

Review

Social Life-Cycle Assessment in the Construction Industry: A Review of Characteristics, Limitations, and Challenges of S-LCA through Case Studies

Prisca Ayassamy * and Robert Pellerin

Department of Mathematical and Industrial Engineering, Polytechnique Montréal, Montréal, QC H3T 1J4, Canada; robert.pellerin@polymtl.ca
* Correspondence: prisca.ayassamy@hotmail.com

Abstract: The paper aims to examine how researchers have operationalized social impact assessment in construction projects over the last ten years. A systematic review was used to investigate case studies in the Social Life-Cycle Assessment (S-LCA) to analyze the application of the methodology. In total, 19 articles published between 2012 and 2023 were classified according to their scope, functional unit measure, S-LCA indicators used, and the main challenges. Our findings revealed limitations in both qualitative and quantitative aspects of measuring social indicators, primarily stemming from difficulties associated with scoring and assessment methodologies. Additionally, we observed deficiencies in social data within the S-LCA framework. This suggests that potential social impacts may be inadequately addressed and evaluated due to various challenges that have been highlighted in the existing literature.

Keywords: S-LCA; social impacts; literature review; challenges; operationalization

Citation: Ayassamy, P.; Pellerin, R. Social Life-Cycle Assessment in the Construction Industry: A Review of Characteristics, Limitations, and Challenges of S-LCA through Case Studies. *Sustainability* **2023**, *15*, 14569. https://doi.org/10.3390/su151914569

Academic Editors: Zora Vrcelj and Malindu Sandanayake

Received: 10 July 2023
Revised: 25 September 2023
Accepted: 5 October 2023
Published: 8 October 2023

1. Introduction

The construction industry is one of the major industries in any national economy, regardless of its level of development (Ilhan and Yobas, 2019 [1]). It is responsible for substantial material and resource consumption and its impact on climate change (Balasbaneh et al., 2018 [2]). In 2018, the building and construction sectors were responsible for 40% of global greenhouse gas emissions (Larsen et al., 2022 [3]), and 36% of final energy use contributed to climate change effects and negatively impacted health (UNEP, 2019 [4]). With a lack of consideration for waste management and waste reduction in the early phases of projects, there tends to be waste generated by construction and demolition through the life cycle of buildings (Esa et al., 2017 [5]), with a remarkable impact of 50% at the end of the life of a project (Kibert, 2016 [6]). It also has a reputation for its high consumption rate of natural resources, which generates between two and three billion tonnes of building waste per year (Jain, 2021 [7]).

Fortunately, the construction sector has started to adopt life-cycle assessments (LCA) to conduct environmental assessments. On the other hand, Social Life-Cycle Assessment (S-LCA) has not gained as much popularity despite being recognized as key in designing processes and sustainable products (Vitorio and Kripka, 2021 [8]). S-LCA is a methodology to assess the social impacts of products and services throughout their life cycle, from raw resource extraction to their final disposal. It is based on the UNEP/SETAC guidelines (UNEP, 2020 [9]). While LCA involves material, energy, and economic flows in production and consumption that impact stakeholders, S-LCA provides a systematic assessment framework combining quantitative and qualitative data to support social and socio-economic decision-making (UNEP, 2020 [9]). S-LCA comprises four phases: goal and scope definition, inventory analysis, impact assessment, and interpretation (Ramirez et al., 2014 [10]). It includes these steps: characterization, normalization, and weighting. According to Dong et al. (2016) [11], characterization is converting social information into interpretable indicators that reflect a list of impacts; normalization is rescaling the characterization results into a comparable range; and weighting is modifying the normalization results according to the importance of subcategories. The two strengths of the S-LCA are (1) its focus on the product and (2) the definition of social impacts, which encompasses a company's behavior and socioeconomic perspective (Zamagni et al., 2011 [12]).

However, little research focuses on S-LCA in the construction industry. Larsen et al. (2022 [3]) said that S-LCA is neither considered nor applied in the building industry to evaluate the impact of construction and refurbishing buildings on the social aspect. However, social value should be considered in the construction industry, as social value tends to increase or improve the social image of stakeholders (Daniel and Pasquire, 2019 [13]). Considering current challenges, there is a need to enhance social indicators within the building sector. While specific solutions have been identified with commendable social characteristics, it is crucial to acknowledge that studies might overlook the social advantages inherent in these solutions (Ostermeyer et al., 2013 [14]). Most research focuses on technology and neglects social and human needs (Fan et al., 2018 [15]). Tokede et Traverso (2020 [16]) pointed out that the challenge with S-LCA is defining wellbeing, which should provide a holistic understanding of the human condition and aspirations. It is worth noting that most studies in this field rely on qualitative and semi-qualitative data, which can present challenges when attempting to draw definitive conclusions from the obtained results. (Huertas-Valdivia et al., 2020 [17]). On the other hand, concerns revolve around the methodological operationalization and measurability of social indicators, which pose limitations on data gathering and stakeholder identification (Tokede and Traverso, 2020 [16]). Though the social aspect is important, no standardized methodologies exist for S-LCA (Larsen et al., 2022 [3]).

In this sense, this paper analyzes how the literature addresses the operationalization of S-LCA in the construction industry by assessing to what extent Social Life-Cycle Assessment has been reported in case studies in the last decade.

This paper tries to answer the following questions:

RQ1—What is the scope of the S-LCA case studies?
RQ2—What is the functional unit measure studied in the case studies?
RQ3—What is the nature of the S-LCA indicators used in the selected case studies?
RQ4—What are the main challenges of Social Life-Cycle Assessment in the literature in case studies presented in the construction industry?

The paper's structure is organized as follows: Section 2 serves as the methodological section, encompassing descriptions of data selection, the research protocol overview, the classification framework, and four key research questions. Section 3 delves into the presentation of our literature review findings, while Section 4 is dedicated to the discussion of these results. Finally, the paper reaches its conclusion in Section 5.

2. Methodology

This paper seeks to understand to what extent the operationalization of S-LCA has been reported by researchers in the literature. To do so, this research assessed articles published during the last ten years (2012–2023), focusing on case studies in the construction industry.

The research methodology in this literature review includes (1) the data collection protocol and (2) the classification of selected papers. Each part is described in detail in the following subsections.

2.1. Data Collection and Research Screening Overview

The search was based on bibliographic databases and electronic libraries such as Web of Science and Google Scholar. Web of Science was used as it has selective, balanced, and complete coverage of the world's leading research, covering around 34,000 journals (Birkle et al., 2020 [18]). On the other hand, Google Scholar was used as well, and it provides an instant method to build on a digital snowball to retrieve literature (Zientek et al., 2018 [19]). A systematic literature protocol was used to evaluate our findings. From these, four research questions were defined to pursue our research. To analyze the S-LCA literature, the inclusion criteria were as follows in Table 1: (a) the period was set between 2012 and 2023, inclusive, and focused on literature written in English; (b) documents were limited to journal articles; and (c) early access articles that had a focus on case studies of S-LCA in the construction industry were also included. Additional articles were added to our article.

Table 1. Inclusion criteria.

Inclusion Criteria
Studies published between 2012 and 2023
Articles addressing social impacts
Articles presenting case studies
Articles focusing on the construction industry

The following search strings and keyword combinations of terminologies and abbreviations were used: ("SLCIA" OR "Social life cycle impact assessment" OR "SLCA" OR "S-LCA" OR "Social life-cycle assessment" OR "Social LCA" (Topic) AND ("construction" OR "building" OR "AEC") (Topic) AND ("case studies" OR "case study" OR "use case") (Topic) AND between 2012 and 2023 (Publication Years) AND Article (Document Types) and Article (Document Types) AND Article or Early Access (Document Types) AND English (Languages).

The first screening of databases resulted in 32 publications and contained scientific publications (peer-reviewed journal articles). To ensure that the articles mentioned both case studies and S-LCA and the construction industry for the literature portfolio, all abstracts screened through Endnotes 20 and 19 were proven irrelevant based on the inclusion and exclusion criteria. After we screened out the articles, 13 were removed as 9 were not in the construction industry, and 4 were not on social assessment, leaving us with 19 relevant articles.

Figure 1 summarizes the research screening process and the criteria for classifying the articles. The following section describes the classification framework used to analyze the selected papers.

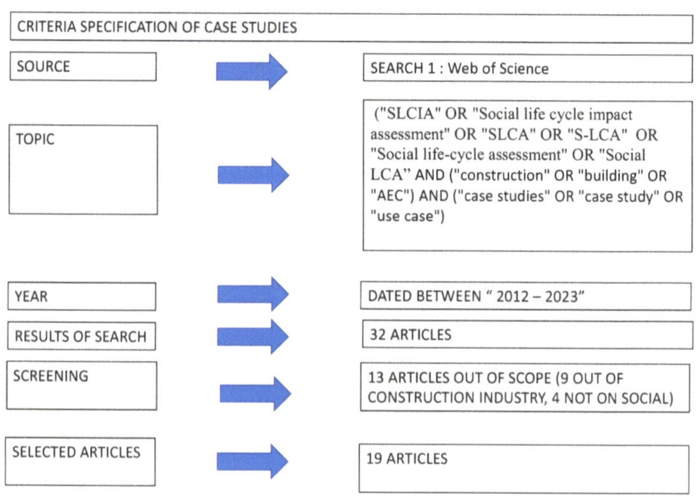

Figure 1. Research screening process.

2.2. Classification Framework

To classify our paper, we first categorized our articles according to their location, type of infrastructure, model, and the stakeholders involved. The location is the country where the research has been performed and studied. It is prescriptive and helps to understand if the results are biased and influenced because of their location in a specific part of the world. The type of infrastructure determines if it is a building, a part of the building, or a particular model. The actors involved help identify the types of stakeholders/individuals involved.

A specific classification scheme was then developed to answer each research question, as described in the following subsections.

2.2.1. Scope

A scope's objective is to identify and "define the object of the study and to delimit the assessment" (Jørgensen et al., 2008 [20]). According to the UNEP (2020 [9]) Guidelines, some elements are included or excluded depending on the study's goal. In those, we shall find (1) the definition of the object of the study (product, function, or service), the number of materials needed to produce the product or output, and the steps, activities, and organizations to comply with the functional unit; (2) the identification of parts of the production system in the assessment (the system boundaries); (3) the variable(s) to determine the importance of different activities in the product system; the stakeholders included and affected (workers, value chain actors, society, consumers, government, construction enterprises, real estate developments, community); the type of impact assessment method, and impact categories and/or subcategories included; the data collection strategies (inventory indicators, data type, and data collection) and data quality requirements.

To answer this question, we looked at the scope of the study presented in each article. Moreover, four elements' criteria were used: (a) the type of construction (building, route, etc.), (b) the scope of infrastructure studied (entire building or parts of buildings), (c) the type of case studies (single or multiple case studies), (d) the organizational type (public or private), and (e) the type of stakeholders involved (workers, local community, society, consumers, value chain actors).

2.2.2. Functional Unit and System Boundary

Functional units are the "quantified performance of a product system for use as a reference unit in a Life-Cycle Assessment study "(UNEP, 2009 [21]). Its purpose is to provide a reference to the relationship between inputs and outputs (Tokede and Traverso, 2020 [16]). It is a critical issue in S-LCA as it is difficult to identify (Fan et al., 2018 [15]). However, it needs to be consistent with the goal and scope of the study (UNEP, 2020 [9]) in which it is involved.

The system boundary "determines parts of the product system that will be included in the system assessed" (UNEP, 2020 [9]). They are defined according to the life cycle stages from upstream processes (i.e., resource use, purchase of goods, and services) to downstream processes (i.e., distribution use and end-of-life products). According to the guidelines (UNEP, 2020 [9]), it is defined as (a) the full life cycle of products and services (cradle-to-grave; from resource extraction to end-of-life); (b) the supply chain of the product (cradle-to-gate; exclude use phase and end-of-life); and (c) parts of the life cycle (gate-to-gate or gate-to-grave).

To answer this question, we separated the functional units for S-LCA, the boundary (cradle to grave or cradle to gate), and the functional units used for LCA.

2.2.3. S-LCA Indicators

Social indicators can be described as "evidence, subjective or objective, qualitative, quantitative, or semi-quantitative, being collected to facilitate concise, comprehensive, and balanced judgments about the condition of specific social aspects concerning a set of values and goals" (UNEP, 2020 [9]). It includes (a) approaches such as impact pathways (mentioned in Question 3), (b) social topics as stakeholders and impact categories, (c) characterization models and types of impact pathways used for assessment, and (d) the weighting approach (UNEP, 2020 [9]).

The character of assessment for social indicators is divided into three types: (a) qualitative, which is nominative and will use words (description); (b) quantitative, which will use a numerical description of the issue (physical units); and (c) semi-qualitative, which will have results expressed in a yes/no form or a scale (scoring system) (UNEP, 2020 [9]).

UNEP (2009) [21] defines the impact pathways as "social LCI results and/or social impact categories" and the impact categories as "logical groupings of S-LCA results related to social issues of interest to stakeholders and decision-makers". The impact pathway is an important part, as it provides an assessment framework. With the identification of indicators, it will give a better assessment of impacts throughout the life cycle (Tokede et Traverso, 2020 [16]). It can be qualitative, which will cover social topics and categories, or quantitative, which will focus on measurable numbers and targets (UNEP, 2020 [9]). It is referred to as a cause-and-effect relationship between the midpoint and endpoint (Jørgensen et al., 2008 [20]). The midpoint is the parameters in the social mechanism network (UNEP, 2020 [9]). The endpoint is the determined damage levels (UNEP, 2020 [9]). The impact is mostly linked to the midpoint and endpoint impact pathways. The impact categories cover specific social issues of interest to stakeholders and decision-makers. They can be grouped as subcategories results (UNEP, 2020 [9]). It is separated between additive and descriptive according to the criteria of their functional units in the case studies.

To answer this question, we looked at the stakeholder categories and their impact categories linked to their activities, the social indicators, the character of the assessment process through impact pathways, impact effects, and impact and stakeholder categories. The impact categories are linked to an indicator, and the indicator is a way to relate to the identified impact (Mathe, 2014 [22]). Indicators are direct measurements of social issues, and they act as a bridge to link the data with subcategories and impact categories to guide the data collection process (Wu et Chen, 2014 [23]).

2.2.4. Challenges

Jørgensen et al. (2008 [20]) described the impact assessment as the phase where inventory information is translated into impacts. Characterization considers the inventory results within the same impact category (Jørgensen et al., 2008 [20]). It is required to translate results into value for an impact indicator at the midpoint or endpoint (UNEP, 2020 [9]). This is the final stage of the S-LCA, where the results are checked and discussed. They are broken down into life cycle phases, impact categories, impact subcategories, and stakeholder categories (UNEP, 2020 [9]).

This section comprises data for stakeholders and impact categories. Three approaches are used to prioritize data collection: (a) conduct a literature review to highlight key potential social impacts to identify specific unit processes for which data should be collected; (b) explore the intensity of different unit processes in a product's life cycle and determine a variable; and (c) identify hotspots in the product's life cycle (UNEP, 2020 [9]). The approach to prioritizing data is to identify the hotspot. It highlighted that hotspots are linked to social issues, and impact subcategories cover these. Hotspots can generally be evaluated at the country's level, but for case-specific S-LCA, more precise geographic information is needed (Hosseinijou et al., 2014 [24]). Benoit et al. (2010 [25]) claimed that the impact assessment is underdeveloped as the guidelines provide a general structure with a set of categories and subcategories.

To answer this question, our approach involved a comprehensive examination of each article's methodology regarding data collection and its accessibility, impact assessment procedures, and the subsequent interpretation of results. Specifically, we scrutinized aspects such as data collection methods and availability (drawing from literature, assessing unit intensity, and identifying hotspots), the nature of impact assessment (whether midpoint or endpoint), and the nuances within result interpretations (considering life cycle phases, impact categories, and subcategories). This meticulous analysis enabled us to discern and identify the challenges elucidated across the 19 articles.

In our next section, we discussed our results and findings through the respective tables associated with each question.

3. Results

Our protocol brought 19 articles (Table 2). A numerical value is attributed to each article and will be used in the remaining tables and figures.

Table 2. Selected papers.

Paper	Authors	Paper Objective	Type of Construction	Location
1	Dong et Ng, 2015 [26]	Develop a S-LCA model for building construction projects	Precast Façade, Semi Precast Slab, and Precast Staircase	Hong Kong
2	Fan et al., 2018 [15]	Evaluate the social impact of a green building district within its designed service life	Green Building	China
3	Balasbaneh et al., 2018 [2]	Analyze the stakeholder toward contribution to economic development	Timber and Concrete structure	Malaysia
4	Dong et Ng, 2016 [11]	Develop a LCSA framework and a case study of LCSA	Building	Hong Kong
5	Hosseinijou et al., 2014 [24]	Assess and compare socio-economic impacts of materials in their life cycle.	Material (steel and iron)	N/A
6	Fauzi et al., 2022 [27]	Propose a method and case study based on the S-LCA framework and guidelines	Wood, Concrete, Steel, Gravel, Aluminum, Gypsum, and Brick	Ontario, Canada; Quebec, Canada; China
7	Santos et al., 2019 [28]	Collect and analyze information to assist the decision-making process	Construction and Demolition waste	Not specific
8	Balasbaneh et al., 2021 (Life cycle...) [29]	Assess three different construction techniques	Materials (concrete) and Single-story residential building	Malaysia
9	Zheng et al., 2020 [30]	Evaluate the social impacts on pavement	Hot-Mix Asphalt Pavement	China
10	Hossain et al., 2018 [31]	Assess the social implications and sustainability of construction materials with a comparative rating model	Recycled Materials	Hong Kong
11	Bezama et al., 2021 [32]	Compare two different demonstrator systems for lightweight wood-based concrete elements	Wood-Based Concrete Elements	Germany
12	Liu et Qian, 2019 (Towards) [33]	Propose an integrated building-specific sustainability assessment model	PPVC Project Semi-Prefab	Singapore
13	Liu et Qian., 2019 (Evaluation...) [34]	Develop a methodological framework for social life cycle assessment	Social Sustainability Projects	Singapore
14	Zheng et al., 2019 [35]	Compare LCSA	Pavement	China
15	Vitorio et Kripka, 2021 [8]	Verify the evolution of social indicators in the sectors	Single-Family Residences	Brazil
16	Balasbaneh et al., 2020 [36]	Evaluate carbon dioxide (CO_2) emissions and the cost and social impacts	Windows in School Building	Malaysia

Table 2. *Cont.*

Paper	Authors	Paper Objective	Type of Construction	Location
17	Ostermeyer et al., 2013 [14]	Address the potential of LCSA in the built environment	Refurbishment of Building	Europe
18	Balasbaneh et al., 2021 (Applying...) [37]	Evaluate the sustainability performance of different flooring systems	Flooring Systems	Malaysia
19	Osorio-Tejada et al., 2022 [38]	Analyze the social performance of companies involved in the supply chain of road transport	Road Freight service	Latin America, Europe, and Asia

3.1. Scope of the S-LCA Case Studies

For scope, we classified the articles as per their contribution to research. Table 2 showed that out of the 19 articles, three aimed to develop and assess, four were to evaluate, one was to collect, compare, and verify S-LCA, and two were to propose and analyze S-LCA. Studies were mostly performed on Asian countries, with Hong Kong (three), China (four), Malaysia (four), Singapore (two), and Asia in general (1). Four studied buildings, and fifteen were in built-in environments.

In Table 3, we review the types of case studies. They were mostly single-case studies rather than multiple. The organizational type was mostly public (16). For the different kinds of stakeholders, the focus was on workers, value chain actors, consumers, and the local community.

We can also note that authors focused on (1) developing a model and methodological or LCSA framework; (2) assessing social impacts and sustainability of building materials; (3) evaluating the social cost of carbon dioxide emissions in green buildings and their materials; (4) comparing LCSA and systems of construction materials; (5) verifying the evolution of social indicators; and (6) analyzing stakeholder's contribution and the social performance of companies.

Mathe (2014 [22]) highlighted the need to consider the range of actors and indicators chosen and the individuals concerned. As such, one category in Table 3 is the type of stakeholders involved. Stakeholders should be present during the development of the S-LCA analysis and are considered from the impact analysis's point of view (Arcese et al., 2013 [39]). However, although S-LCA aims to validate an assessment and its consequences, some factors are personal. In the sense that when a company carries out its activity, it focuses on the product and not on the "behavior" (Zamagni et al., 2011 [12]). Indeed, certain changes cannot be directly linked to the specific individuals affected, as noted by Jorgensen et al. (2013 [40]). For instance, consider the evaluation of forced labor, where its existence might be recognized but not quantified in relation to the number of T-shirts within the product system, as illustrated by the example provided in UNEP (2020 [9]).

3.2. Functional Units and System Boundaries in Case Studies

Among the 19 articles reviewed, the functional units employed in S-LCA exhibited significant variation, as shown in Table 4. These included assessments ranging from worst to best in two articles, percentages in three articles, square meters (m^2) in three articles, scoring ranging from unconcerned to very strong priority in one article, a linkage with national and international laws with scoring from strongly positive to strongly negative in one article, numeric values, categorical distinctions, yes/no indicators, and hours each featured in one article, while another article used weight as a specification. Furthermore, three articles utilized cost evaluation as their functional unit, one considered the level of risk, and finally, two articles did not specify the functional unit employed in their assessments.

Table 3. Review of the case studies.

Paper	Type of Case Studies		Organization Type		Type of Stakeholders Involved:				Methods and Models Used
	Single	Multiple	Public	Private	Workers	Value Chain Actors	Consumers	Local Community	
1	X		X		X			X	Questionnaires and surveys on social performance of environmental friendly practices
2	X			X			X	X	Questionnaires for green residential districts
3		X	X		X				Interviews and research assessment on contribution to economic development for numbers of job creation
4	X		X	X	X			X	Interviews on material stages
5	X		X		X	X	X	X	MFA (tool) to identify hotspots within communities, companies and employment
6	X		X		X	X	X	X	Multi-level Analysis on unit, company, sector and country
7	x		X		X	X	X	X	Model identifies stakeholders' perspectives, experiences in waste management.
8		X	X		X	X	X	X	Multi-criteria decision-making on sustainable Flooring (type of flooring)
9	X		X		X		X	X	Framework on raw materials, production, construction, use and maintenance
10	X		X		X	X	X	X	Interviews on challenges and recycled materials
11	X		X		X				RESPONSA model on indicators and organizational learning
12	X		X		X		X	X	Model on structural design and well-being of stakeholders
13	X		X		X		X	X	Interviews and indicators on cause-effect relationships (soundproof issue)
14	X		X		X		X	X	Interviews and questionnaires on social impacts for pavement
15	X			X	X	X	X	X	WFWP method-raw materials and workers' wages
16	X		X		X		X	X	Multi-criteria decision-making on user satisfaction and indoor noise and parameters
17	X			X			X		A multidimensional Pareto optimization methodology on Building materials, Workers and job conditions
18		X	X		X		X	X	Interviews and Multi-criteria decision-making on judgement of stakeholders on types of construction
19		X	X		X		X	X	Interviews on labor rights

Table 4. Functional units S-LCA, boundary, and functional unit LCA.

#	FU-S-LCA	Boundary	FU-LCA
1	Scale of −1 to 1, and −1 is the worst and 1 is the best social	Cradle to grave	Scale of 1 to −1
2	None	Cradle to gate	None
3	%	Cradle to grave	Cost per MYR
4	−1 to 1, with 1 being the best social performance; the score ranges from −5 to 5, with 5 being the best social performance	Cradle to grave	Kg
5	T, m^2	Cradle to grave	None
6	M^2	Cradle to grave	None
7	None	Cradle to grave	None
8	9 is unconcerned, 7 is moderate priority, 5 is strong priority, 3 is solid priority, and 1 is very strong priority	Cradle to grave	M^2
9	%	Cradle to gate	None
10	1.00 Strongly positive, fully agreed, very highly related, highly compatible 0.75 Mostly positive, moderately agreed, highly related, moderately compatible 0.50 Neutrally affected, agreed, neutrally related, compatible 0.25 Mostly negative, partially disagreed, moderately negative, negatively compatible 0.00 Strongly negative, fully disagreed, highly unrelated, incompatible	Cradle to grave	None
11	Number, category, percent, yes and no, hour	Cradle to gate	Kg, mm
12	Weights	Cradle to gate	none
13	Weights	Cradle to gate	Weight
14	HHCP, milli-DALYS eq, S, nox eq)	Cradle to gate	T, m^3
15	R$/month	Cradle to grave	Kg, m^2
16	Weight	Cradle to grave	US $
17	Cost	Cradle to gate	Investment cost
18	Cost	Cradle to grave	Cm
19	Low, medium, high, very high risk	Cradle to gate	None

In terms of the functional unit concerning LCA, the analysis of the 19 articles revealed diverse approaches. Specifically, two articles were cost-related, three utilized weight specifications, one employed centimeters (cm), one used kilograms (kg), one utilized cubic meters (m^3), another employed tons (T), one used square meters (m^2), and one adopted the worst and best criteria. Remarkably, seven articles did not specify the functional unit used.

Regarding the system boundary, it was observed that eight articles focused on the cradle-to-gate perspective, while eleven adopted the cradle-to-grave approach. This distinction can be attributed to the fact that four of the 19 articles were exclusively concerned with the building itself, whereas the remaining articles encompassed broader project or material assessments.

We can see in Table 4 that the unit processes to fulfill the functional unit are set up for both S-LCA and LCA. However, even if so, this approach is not feasible in S-LCA as the measures are mostly on the socioeconomic impacts, which are related to the company's behavior instead of the product's function unit (Jørgensen, 2013 [40]). As such, if S-LCA is applied to assess a product by focusing on the product system itself, the behavior will not be caught because the supplier will be held responsible for only the part of production included in the product system (Zamagni et al., 2011 [12]). Thus, the impacts cannot be expressed through the functional unit. In addition, S-LCA works with data on attributes and characteristics of processes that cannot be expressed per unit (Hosseinijou et al., 2014 [24]).

Lagarde and Macombe (2013 [41]) also mentioned that the concepts used to describe the systems and the boundaries are unclear in the literature since authors do not clearly explain their models and criteria for making their choice. It has been pointed out that to support management decisions, it is sufficient to include only parts of the life cycle that are directly influenced by companies (Hosseinijou et al., 2014 [24]). In our study, the articles are from cradle to gate or cradle to grave, meaning that all life cycle stages were involved.

3.3. Nature of the S-LCA Indicators Used in the Selected Case Studies

We have organized five distinct stakeholder categories and aligned them with their respective impact categories, as outlined in the UNEP guidelines. This arrangement provides a structured overview of the relationships between stakeholders and the specific impact categories relevant to their interests and concerns.

As Chan and Oppong (2017 [42]) discussed, a stakeholder has different attributes (power, legitimacy, and urgency), and it is important to understand their effect on construction projects. For workers, the categories associated were fair salary, working hours, forced labor, equal opportunities, health and safety, and discrimination. For value chain actors, the subcategories of capacity for job creation and local employment were of paramount importance. Consumers, on the other hand, exhibited a distinct concern related to consumer privacy. The community stakeholder category was associated with a broader range of impact subcategories, encompassing local job creation, respect for indigenous rights, land use, cultural heritage, safe living conditions, community engagement, human health, public commitment, and technology development.

However, it is noteworthy that in our analysis of the 19 articles, we encountered challenges in obtaining certain data, primarily due to difficulties in identifying the relevant stakeholders. This observation is somewhat surprising, given that the UNEP (2020 [9]) guidelines emphasize that the initial step in the analysis should be the identification of stakeholders. Nonetheless, we identified a recurring pattern whereby specific impact categories were consistently linked to particular stakeholder categories, as illustrated in Figure 2.

The subcategories were capacity for job creation and local employment. For consumers, consumer privacy was one of their concerns. The community was linked to local job creation and respect for indigenous rights, land use, cultural heritage, safe living conditions, community engagement, human health, public commitment, and technology development.

In Table 5, our analysis of the 19 articles revealed diverse characteristics of the assessment approaches. Specifically, we found that nine articles employed a quantitative assessment methodology, five utilized a qualitative approach, and five adopted a semi-quantitative method. Regarding the consideration of midpoint and endpoint criteria, only three articles exclusively focused on midpoint (two) or endpoint (one), while the remaining sixteen articles considered both criteria. Additionally, seven articles assessed direct impacts, while eleven focused on indirect impacts. The majority of articles, precisely fifteen out of nineteen, presented descriptive assessments, while the remaining four followed an additive approach.

Furthermore, our examination of the impact pathways, which illustrate the cause-and-effect relationship between midpoint and endpoint indicators, revealed that many articles established links, particularly between fair wages (midpoint) and human health (endpoint). However, as noted by Hosseinijou et al. (2014 [24]), the precise measurement of this process remains a challenge. In line with the observations of Neugebauer et al. (2014) [43], it became evident that the boundaries between impact indicators and inventory indicators are not distinctly defined.

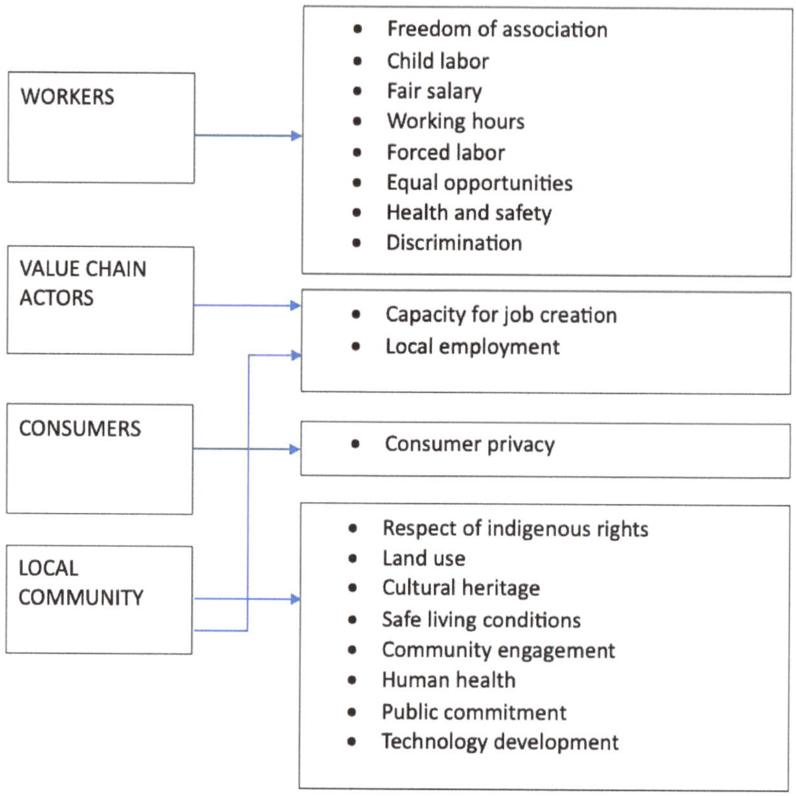

Figure 2. Stakeholders' categories and subcategories from UNEP guidelines.

3.4. Main Challenges of Social Life-Cycle Assessment

Addressing research question 4, this section categorizes methodological challenges into four distinct types: data quality, data uncertainties, data measurement, and missing data. We found that six studies employed surveys, five relied on interviews, two used literature reviews, two employed models, three utilized multi-criteria decision-making approaches, and five employed statistical methods and matrix analysis.

Within the data quality category, Dong and Ng (2015 [26]) emphasized that data quality is a significant concern, particularly because certain indicators cannot be effectively measured, rendering it impossible to estimate scores accurately. Furthermore, the scoring method itself is problematic, given the absence of a widely accepted and standardized scoring system, as noted by Fan et al. (2018 [15]).

Concerning data uncertainties, despite the collection of information via questionnaires, it was recognized that these data may still exhibit uncertainties. Additionally, Balasbaneh et al. (2020 [36]) highlighted the difficulty in identifying certain stakeholders, particularly in governance, making it challenging to collect data from them.

In the context of data measurement, the quantification of social impact was deemed challenging overall. Sustainability, as highlighted in Article 5, was described as an underdeveloped aspect. Moreover, Vitorio and Kripka (2021 [8]) underscored that the construction industry still lacks comprehensive social data inventories.

Addressing missing data, it was noted that some stakeholders were excluded due to data gaps, especially in the case of value chain actors and governance, where obtaining information proved difficult (Balasbaneh et al., 2018 [2]; Balasbaneh et al., 2020 [36]).

Table 5. Character of assessment, impact assessment, and result interpretations.

Paper	Assessment Process			Impact Assessment						Result Interpretations			
	Qualitative	Qualitative	Semi-Qualitative	Midpoint	Endpoint	Classification	Characterisation	Normalization	Weighting	Impact Effects		Impact Categories	
										Direct	Indirect	Addictive	Descriptive
1			X	X	X	X		X	X	X		X	
2	X			X	X	X	X		X	X		X	
3			X	X	X	X	X				X		X
4				X	X	X	X	X	X		X	X	
5		X	X	X	X	X	X		X		X	X	
6	X	X		X	X	X				X			X
7		X		X	X	X	X			X			X
8	X			X	X	X	X	X	X	X			X
9	X	X		X	X		X			X			X
10		X		X	X	X	X	X	X		X		X
11	X	X		X	X	X	X	X	X		X		X
12	X	X		X	X		X	X	X		X		X
13		X	X	X	X	X	X	X	X		X		X
14	X		X	X	X	X	X	X	X		X		X
15	X			X	X	X	X		X	X			X
16	X	X		X	X		X	X	X		X		X
17	X	X		X	X		X	X	X		X		X
18	X			X	X	X	X	X	X		X		X
19		X		X	X	X	X	X	X		X		X

Lastly, we observed that among the studies, eight focused on the cradle-to-gate perspective, while eleven adopted the cradle-to-grave approach. This choice of system boundary in life cycle analysis was influenced by the challenges associated with missing data, as highlighted in Table 4.

As pointed out by Fauzi et al. (2022 [27]), it is crucial to encompass the entire product life cycle because impacts occurring throughout various phases are frequently overlooked due to missing data. This underscores the need for comprehensive data collection and analysis across the entire life cycle of a product.

Moreover, Ostermeyer et al. (2013 [14]) emphasize that these missing data represent essential information gaps that should be addressed through future research efforts, highlighting the importance of ongoing data generation to enhance the effectiveness of S-LCA. Zanchi et al. (2018 [44]) have also identified several key elements that can significantly influence the application of S-LCA. These elements encompass the perspective from which affected stakeholders are considered, the methodology's selection and prioritization of indicators, the critical role of the functional unit linked with social inventory information, the definition of system boundaries that may not always encompass all relevant unit processes, and the potential disparities in data collected when assessing social impacts on-site within a company compared to data collected at a national or geographic level. These considerations underscore the complexity and multifaceted nature of social life-cycle assessment.

In summary, Social Life-Cycle Assessment continues to pose significant challenges within the construction sector. These challenges (In Table 6) are primarily driven by issues related to missing data, difficulties in measurement, uncertainties, and the inconsistent quality of available data. Addressing these challenges is essential to advancing the effectiveness and reliability of Social Life-Cycle Assessment in construction projects.

Table 6. Challenges in data collection.

Paper	Challenges of Data Collection			
	Data Quality	**Data Uncertainties**	**Data Measurement**	**Missing Data**
1	X	X	X	
2	X	X	X	X
3			X	X
4	X		X	X
5		X		
6		X	X	X
7				X
8	X			
9				X
10	X		X	
11		X		X
12	X	X	X	X
13	X		X	X
14		X	X	
15		X	X	
16		X	X	
17			X	X
18	X	X		X
19	X			

4. Discussion

This paper had the overarching goal of examining how researchers have put social impact assessment into practice within construction projects over the past decade. The study encompassed various facets, including the development, assessment, evaluation, comparison, and verification of methodologies and frameworks. These encompassed activities such as (1) the creation of models and methodological frameworks for Life Cycle Sustainability Assessment (LCSA); (2) the assessment of social impacts and sustainability concerning building materials; (3) the evaluation of the social cost associated with carbon dioxide emissions in green buildings and their materials; (4) the comparison between LCSA and construction material systems; (5) the verification of the evolution of social indicators; and (6) the analysis of stakeholder contributions and the social performance of companies.

The study predominantly featured single-case studies rather than multiple-case studies, with a focus on evaluating and assessing a total of three private organizations and sixteen public organizations. In terms of stakeholders, the research centered on workers, value chain actors, consumers, and the local community, shedding light on the intricate dynamics and implications of social impact assessment in the construction sector.

Furthermore, in terms of functional units, the measurements and scoring employed varied widely, encompassing scales such as worst to best in two cases, percentages in three cases, square meters (m^2) in three cases, scoring from unconcerned to very strong priority in one case, linkage with national and international laws with scoring from strongly positive to strongly negative in one case, numeric values, categorical distinctions, yes/no indicators, and hours each featured in one case, with weight used as a specification in one case. Additionally, three articles utilized cost evaluation as their functional unit, one considered the level of risk, and finally, two articles did not specify the functional unit employed in their assessments.

These diverse scoring methods, as pointed out by Jørgensen (2013 [40]), raise questions about the appropriateness of measuring socio-economic impacts based on a company's behavior rather than focusing on the product's functional unit. Zamagni et al. (2011 [12]) similarly argued that behavior may not be adequately captured, as suppliers are held responsible only for the part of production included in the product system. Moreover, S-LCA often deals with data on attributes and characteristics of processes that cannot be expressed per unit, as highlighted by Hosseinijou et al. (2014 [24]). This complexity arises from the fact that social impacts are intricately linked to human well-being, making it challenging to establish a direct connection to a specific physical unit, as noted by Zheng et al. (2020 [30]). These considerations underscore the complexity and nuances associated with the choice of functional units in Social Life-Cycle Assessments.

Thirdly, the nature of indicators within the case studies spanned five stakeholder categories. These case studies were characterized as quantitative in nine instances, qualitative in five cases, and semi-quantitative in five instances. As outlined by Zanchi et al. (2018 [44]), various elements can significantly influence S-LCA applications, with a particular emphasis on the perspective of affected stakeholders. Stakeholders possess distinct attributes, including power, legitimacy, and urgency, and comprehending their impact on construction projects is vital, as noted by Chan and Oppong (2017 [42]). Consequently, for management decision support, it may be sufficient to consider only those parts of the life cycle directly influenced by companies, as highlighted by Hosseinijou et al. (2014 [24]).

Within the context of the case studies, the impact on affected stakeholders became evident. Workers were associated with indicators related to fair salaries, working hours, forced labor, equal opportunities, health and safety, and discrimination. Value chain actors were concerned with indicators related to capacity for job creation and local employment, while consumers prioritized indicators such as consumer privacy. The local community was linked to a broad range of indicators, including local job creation, respect for indigenous rights, land use, cultural heritage, safe living conditions, community engagement, human health, public commitment, and technology development.

It is noteworthy that the articles in our study encompassed a cradle-to-gate or cradle-to-grave perspective, indicating the involvement of all life cycle stages in the assessments. This comprehensive approach allows for a holistic understanding of social impacts throughout the entire life cycle of construction projects.

Additionally, the authors noted that the predominant challenges revolved around data quality and uncertainties. There exists an intricate interplay between scope, assessments, and measurements, rendering methodological evaluation challenging. Notably, S-LCA exhibits deficiencies in terms of comparability and transparency, as highlighted by Pollok et al. (2021) [45]. It is important to acknowledge that assessing the same item produced in different locations can yield disparate impacts, thereby influencing the indicators, as discussed by Zamagni et al. (2011 [12]). Furthermore, collecting data on-site within a company or at a country level within geographical zones introduces additional complexities, as underscored by Zanchi et al. (2018 [44]).

Connecting the inventory results of the social dimension to functional units is another challenge (Zheng et al., 2019 [35]; Dong and Ng, 2015 [26]). For the impact assessment, the weighting and scoring of social issues remain a challenge (Hosseinijou et al., 2014 [24]). For example, in the study of Dong et al. (2016 [11]), interviewees suggested leaving the choice of weighting to the users. In the study of Hossain et al. (2018 [31]), an indirect scoring system is used based on indirect weighing based on respondents' opinions as ratings. In Liu and Qian (2019 [22]), equal weights were assigned to four stakeholders because of experts' opinions regarding ethical issues. Knowing that, when the weighting of any criteria is higher, there is a chance for the related criteria to over-influence decision-making (Balasbaneh et al., 2021 [37]). As such, an uncertainty analysis of the scoring and weighting models is needed. Since social impacts are mostly associated with human well-being there is a risk of inevitable subjectivity (Zheng et al., 2020 [30]). Characterization models should be able to translate inventory results into impacts in a comparative way (Hosseinijou et al., 2014 [17]). Still, there is a knowledge gap with the social indicators to characterize social issues (Liu et Qian, 2019 [34]). Dong et al. (2015 [26]) said that calculation is no longer part of characterization, while normalization and weighting are the quantification steps. Characterization models are not able to link the impact results to the functional unit (Hosseinijou et al., 2014 [24]). Therefore, the process of the scoring system usually includes normalization, which is a quantifying process with a lack of scientific method (Dong et al., 2015 [26]). Therefore, a specific set of indicators needs to be developed depending on the goal and scope definition as well as data accessibility (Liu et Qian, 2019 [22]). An international and multidiscipline expert panel could help solve these issues (Zheng et al., 2019 [35]).

We have identified and analyzed the specific methods that authors have employed to operationalize Social Life-Cycle Assessment (S-LCA) in construction projects. Our findings have unveiled numerous issues within the S-LCA assessment process. These challenges have been categorized into four distinct types: data quality, data uncertainties, data measurement, and missing data. Notably, the primary concern that emerged from our study pertains to the quality of the data, with certain indicators proving difficult or impossible to measure accurately. Therefore, it becomes nearly impossible to estimate the score (Dong and Ng, 2015 [26]).

The scoring method is also problematic, as there is a lack of a well-accepted scoring system (Fan et al., 2018 [15]). Concerning data uncertainties, even though information is gathered through questionnaires, it may still exhibit certain degrees of uncertainty. This inherent uncertainty complicates the identification of some stakeholders, particularly those involved in governance, making it challenging to gather data from them, as highlighted by Balasbaneh et al. (2020 [36]). As highlighted by Backes et Traverso (2021 [46]), the principal challenges identified revolved around the selection and quantification of social criteria and indicators.

Regarding data measurement, the quantification of social impact remains a challenging endeavor and is still underdeveloped in terms of sustainability, as noted by Hosseinijou et al. (2014 [24]). The complex and multifaceted nature of social impact assessment presents difficulties in precisely quantifying these impacts, particularly within the context of sustainability.

Concerning missing data, challenges persist, and some stakeholders are omitted due to data gaps, especially among value chain actors and governance, where obtaining information proves to be particularly challenging, as highlighted by Balasbaneh et al. (2018 [2] and 2020 [36]). Inventory data collection also presents a hurdle due to the limited availability of databases, as noted by Zheng et al. (2020 [30]). These data limitations underscore the ongoing issues associated with social impact assessment in the construction sector, where comprehensive and reliable social data inventories remain lacking, as highlighted by Vitorio and Kripka (2021 [44]).

Furthermore, the absence of international consensus on the social life cycle impact assessment method is a noteworthy challenge, as indicated by Fan et al. (2018 [15]). The lack of standardized criteria, particularly in the domain of social culture, further complicates the development and implementation of a cohesive approach, as observed by Balasbaneh et al. (2018 [2]). These challenges highlight the need for continued research and standardization efforts in the field of Social Life-Cycle Assessment. Pollock et al. (2021 [45]) even argued that the complication is in the complex testing and verification of social impact pathways and social issues' facets that connect to different disciplines and theories.

To sum up, the limitation of the implementation of the S-LCA is the selection of different stakeholder categories, impact subcategories, indicators, and weighting methods (Hossain et al., 2018 [31]). Upon juxtaposing our discoveries with those of Pollock et al. (2021 [45]), it becomes evident that analogous concerns have surfaced in other industries, notably in the automotive sector. Furthermore, we touched upon the behavioral dimension as a noteworthy factor, stemming from the inherent challenge of quantifying social aspects. The authors of the study similarly acknowledged this challenge, emphasizing its role as a causal factor for uncertainty, particularly in relation to Environmental Life Cycle Inventory (E-LCI) data. Although they addressed issues related to indicators, our selected case studies did not center on a specific site or location. In contrast, Backes and Traverso (2023 [46]) meticulously curated a list of indicators and linked them to a subsequent hotspot analysis focused on production countries. This approach introduces the possibility of variations in data collection, whether conducted on-site at individual companies or aggregated per country, especially within distinct geographic zones, as noted by Zanchi et al. (2018 [44]).

5. Conclusions

This paper conducted a comprehensive scoping review of case studies published in the past decade to gain insights into the current trends in Social Life-Cycle Assessment within the construction industry. Our primary objective was to explore the operationalization of S-LCA (Social Life-Cycle Assessment). The review encompassed various aspects, including the scope, functional unit, system boundaries, nature of indicators used, and challenges encountered in S-LCA applications. However, it is important to acknowledge that we cannot encompass every aspect of operationalization comprehensively. Notably, we did not delve into specific details such as the individuals responsible for conducting the analyses, the requisite skill sets, or the software tools employed, among other factors. These particulars are infrequently documented in the articles we reviewed. As such, it is important to acknowledge that there are numerous other facets of operationalization that warrant further investigation. Nevertheless, the findings from this review shed light on several noteworthy observations:

Flaws in S-LCA: It became evident that S-LCA faces certain shortcomings related to the quality of measurement, scoring methods, and the availability of social data. These deficiencies are not exclusive to the construction sector but are prevalent across various industries.

Measurement Challenges in Construction: Particularly within the construction industry, there is a pressing need to develop more suitable measurement approaches for construction output. The assessment of impacts in this context can be complex and challenging.

Focus on Products: The majority of S-LCA assessments focus on the product level, which may not adequately capture behavioral aspects (social) that are integral to understanding social impacts. Social impacts are often attributed primarily to companies rather than individual processes and materials.

Materials-Centric Assessments: A significant portion of the reviewed articles primarily focused on materials rather than the construction of buildings themselves. This emphasis on materials could contribute to the prevalence of flaws and challenges encountered in S-LCA within the construction sector.

Standardization and Quantification Challenges: The field of social culture in S-LCA lacks standardization, and quantifying social impacts remains a complex task. This subjectivity is particularly pronounced in scoring and weighting during assessments.

In summary, this review underscores the need for continued research and development efforts to enhance the effectiveness and reliability of S-LCA, both within the construction industry and across other sectors. Addressing measurement challenges, standardizing social culture aspects, and considering the broader behavioral context are key steps in advancing the field of social impact assessment.

This research has its limitations, primarily stemming from the constraints imposed during the article selection process. The main limitations include:

Limited Article Pool: The study's scope was narrowed down by focusing exclusively on scientific articles within the construction sector that feature case studies. This approach excluded potentially valuable insights from non-academic sources and gray literature, which may contain relevant case studies conducted by industry professionals. Consequently, the findings may not provide a comprehensive view of all relevant S-LCA case studies in the construction industry. The decision to exclude gray literature, while upholding academic rigor, may have inadvertently omitted valuable real-world case studies and practical applications of S-LCA conducted by non-academic stakeholders in the construction sector.

Confidentiality and Ethical Considerations: Some case studies or industry-specific data may be subject to confidentiality agreements or ethical constraints, limiting their inclusion in the analysis. These constraints can potentially result in an incomplete representation of available S-LCA case studies within the construction industry.

Despite these limitations, the research provides valuable insights into the current state of S-LCA within the construction sector based on the available academic literature. Future research endeavors may seek to address these limitations by exploring a broader range of sources and considering alternative methodologies to capture a more comprehensive view of S-LCA case studies and applications in the field.

Author Contributions: Methodology, P.A. and R.P.; EndNote software, P.A.; validation, R.P.; formal analysis, P.A. and R.P.; investigation, P.A.; resources, P.A.; data curation, P.A.; writing—original draft preparation, P.A.; writing, P.A.; Review and editing—P.A. and R.P.; supervision, R.P.; funding acquisition, R.P. All authors have read and agreed to the published version of the manuscript.

Funding: This research was funded by Jarislowsky/Atkins Réalis Research Chair.

Institutional Review Board Statement: Not applicable.

Informed Consent Statement: Not applicable.

Data Availability Statement: Not applicable.

Conflicts of Interest: The authors declare no conflict of interest.

References

1. Ilhan, B.; Yobas, B. Measuring construction for social, economic and environmental assessment. *Eng. Constr. Archit. Manag.* **2019**, *26*, 746–765. [CrossRef]
2. Balasbaneh, A.T.; Bin Marsono, A.K.; Khaleghi, S.J. Sustainability choice of different hybrid timber structure for low-medium cost single-story residential building: Environmental, economic, and social assessment. *J. Build. Eng.* **2018**, *20*, 235–247. [CrossRef]
3. Larsen, V.G.; Tollin, N.; Sattrup, P.A.; Birkved, M.; Holmboe, T. What are the challenges in assessing circular economy for the built environment? A literature review on integrating LCA, LCC and S-LCA in life cycle sustainability assessment, LCSA. *J. Build. Eng.* **2022**, *50*, 104203. [CrossRef]
4. United Nation Environment Programme [UNEP]. Global Status Report for Buildings and Construction Sector. In 2019 Global Status Report for Buildings and Construction: Towards a Zero Emission, Efficient and Resilient Buildings, and Construction Sector. [PDF]. 2019. Available online: https://www.unep.org/resources/publication/2019-global-status-report-buildings-and-construction-sector (accessed on 1 December 2022).
5. Esa, M.R.; Halog, A.; Rigamonti, L. Developing strategies for managing construction and demolition wastes in Malaysia based on the concept of circular economy. *J. Mater. Cycles Waste Manag.* **2017**, *19*, 1144–1154. [CrossRef]
6. Kibert, C.J. *Sustainable Construction: Green Building Design and Delivery*; John Wiley & Sons: Hoboken, NJ, USA, 2016.
7. Jain, M.S. A mini review on generation, handling, and initiatives to tackle construction and demolition waste in India. *Environ. Technol. Innov.* **2021**, *22*, 101490. [CrossRef]
8. Vitorio, P.C.; Kripka, M. Fair wage potential as a tool for social assessment in building projects. *Eng. Constr. Archit. Manag.* **2021**, *28*, 1295–1318. [CrossRef]
9. United Nation Environment Programme [UNEP]. Guidelines for Social Life Cycle Assessment of Products 2020. [PDF]. 2020. Available online: https://www.lifecycleinitiative.org/library/guidelines-for-social-life-cycle-assessment-of-products-and-organisations-2020/ (accessed on 9 July 2023).
10. Ramirez, P.K.S.; Petti, L.; Haberland, N.T.; Ugaya, C.M.L. Subcategory assessment method for social life cycle assessment. Part 1: Methodological framework. *Int. J. Life Cycle Assess.* **2014**, *19*, 1515–1523. [CrossRef]

11. Dong, Y.H.; Ng, S.T. A modeling framework to evaluate sustainability of building construction based on LCSA. *Int. J. Life Cycle Assess.* **2016**, *21*, 555–568. [CrossRef]
12. Zamagni, A.; Amerighi, O.; Buttol, P. Strengths or bias in social LCA? *Int. J. Life Cycle Assess.* **2011**, *16*, 596–598. [CrossRef]
13. Daniel, E.I.; Pasquire, C. Creating social value within the delivery of construction projects: The role of lean approach. *Eng. Constr. Archit. Manag.* **2019**, *26*, 1105–1128. [CrossRef]
14. Ostermeyer, Y.; Wallbaum, H.; Reuter, F. Multidimensional Pareto optimization as an approach for site-specific building refurbishment solutions applicable for life cycle sustainability assessment. *Int. J. Life Cycle Assess.* **2013**, *18*, 1762–1779. [CrossRef]
15. Fan, L.; Pang, B.; Zhang, Y.R.; Zhang, X.J.; Sun, Y.W.; Wang, Y.F. Evaluation for social and humanity demand on green residential districts in China based on SLCA. *Int. J. Life Cycle Assess.* **2018**, *23*, 640–650. [CrossRef]
16. Tokede, O.; Traverso, M. Implementing the guidelines for social life cycle assessment: Past, present, and future. *Int. J. Life Cycle Assess.* **2020**, *25*, 1910–1929. [CrossRef]
17. Huertas-Valdivia, I.; Ferrari, A.M.; Settembre-Blundo, D.; García-Muiña, F.E. Social Life-Cycle Assessment: A Review by Bibliometric Analysis. *Sustainability* **2020**, *12*, 6211. [CrossRef]
18. Birkle, C.; Pendlebury, D.A.; Schnell, J.; Adams, J. Web of Science as a data source for research on scientific and scholarly activity. *Quant. Sci. Stud.* **2020**, *1*, 363–376. [CrossRef]
19. Zientek, L.R.; Werner, J.M.; Campuzano, M.V.; Nimon, K. The Use of Google Scholar for Research and Research Dissemination. *New Horiz. Adult Educ. Hum. Resour. Dev.* **2018**, *30*, 39–46. [CrossRef]
20. Jørgensen, A.; Le Bocq, A.; Nazarkina, L.; Hauschild, M. Methodologies for social life cycle assessment. *Int. J. Life Cycle Assess.* **2008**, *13*, 96–103. [CrossRef]
21. United Nation Environment Programme [UNEP]. Guidelines for Social Life Cycle Assessment of Products. [PDF Report]. 2009. Available online: https://wedocs.unep.org/bitstream/handle/20.500.11822/7912/-Guidelines%20for%20Social%20Life%20Cycle%20Assessment%20of%20Products-20094102.pdf?sequence=3&amp%3BisAllowed= (accessed on 1 December 2022).
22. Mathe, S. Integrating participatory approaches into social life cycle assessment: The SLCA participatory approach. *Int. J. Life Cycle Assess.* **2014**, *19*, 1506–1514. [CrossRef]
23. Wu, R.; Yang, D.; Chen, J. Social Life Cycle Assessment Revisited. *Sustainability* **2014**, *6*, 4200–4226. [CrossRef]
24. Hosseinijou, S.A.; Mansour, S.; Shirazi, M.A. Social life cycle assessment for material selection: A case study of building materials. *Int. J. Life Cycle Assess.* **2014**, *19*, 620–645. [CrossRef]
25. Benoît, C.; Norris, G.A.; Valdivia, S.; Ciroth, A.; Moberg, A.; Bos, U.; Prakash, S.; Ugaya, C.; Beck, T. The guidelines for social life cycle assessment of products: Just in time! *Int. J. Life Cycle Assess.* **2010**, *15*, 156–163. [CrossRef]
26. Dong, Y.H.; Ng, S.T. A social life cycle assessment model for building construction in Hong Kong. *Int. J. Life Cycle Assess.* **2015**, *20*, 1166–1180. [CrossRef]
27. Fauzi, R.T.; Lavoie, P.; Tanguy, A.; Amor, B. On the possibilities of multilevel analysis to cover data gaps in consequential S-LCA: Case of multistory residential building. *J. Clean. Prod.* **2022**, *355*, 131666. [CrossRef]
28. Santos, A.C.; Mendes, P.; Teixeira, M.R. Social life cycle analysis as a tool for sustainable management of illegal waste dumping in municipal services. *J. Clean. Prod.* **2019**, *210*, 1141–1149. [CrossRef]
29. Balasbaneh, A.T.; Sher, W. Life cycle sustainability assessment analysis of different concrete construction techniques for residential buildings in Malaysia. *Int. J. Life Cycle Assess.* **2021**, *26*, 1301–1318. [CrossRef]
30. Zheng, X.Y.; Easa, S.M.; Ji, T.; Jiang, Z.L. Modeling life-cycle social assessment in sustainable pavement management at project level. *Int. J. Life Cycle Assess.* **2020**, *25*, 1106–1118. [CrossRef]
31. Hossain, M.U.; Poon, C.S.; Dong, Y.H.; Lo, I.M.C.; Cheng, J.C.P. Development of social sustainability assessment method and a comparative case study on assessing recycled construction materials. *Int. J. Life Cycle Assess.* **2018**, *23*, 1654–1674. [CrossRef]
32. Bezama, A.; Hildebrandt, J.; Thran, D. Integrating Regionalized Socioeconomic Considerations onto Life Cycle Assessment for Evaluating Bioeconomy Value Chains: A Case Study on Hybrid Wood-Concrete Ceiling Elements. *Sustainability* **2021**, *13*, 4221. [CrossRef]
33. Liu, S.Y.; Qian, S.Z. Towards sustainability-oriented decision-making: Model development and its validation via a comparative case study on building construction methods. *Sustain. Dev.* **2019**, *27*, 860–872. [CrossRef]
34. Liu, S.Y.; Qian, S.Z. Evaluation of social life-cycle performance of buildings: Theoretical framework and impact assessment approach. *J. Clean. Prod.* **2019**, *213*, 792–807. [CrossRef]
35. Zheng, X.Y.; Easa, S.M.; Yang, Z.X.; Ji, T.; Jiang, Z.L. Life-cycle sustainability assessment of pavement maintenance alternatives: Methodology and case study. *J. Clean. Prod.* **2019**, *213*, 659–672. [CrossRef]
36. Balasbaneh, A.T.; Yeoh, D.; Abidin, A.R.Z. Life cycle sustainability assessment of school window renovations against noise pollution in tropical climates. *J. Build. Eng.* **2020**, *32*, 101784. [CrossRef]
37. Balasbaneh, A.T.; Yeoh, D.; Juki, M.I.; Gohari, A.; Abidin, A.R.Z.; Bin Marsono, A.K. Applying three pillar indicator assessments on alternative floor systems: Life cycle study. *Int. J. Life Cycle Assess.* **2021**, *26*, 1439–1455. [CrossRef]
38. Osorio-Tejada, J.L.; Llera-Sastresa, E.; Scarpellini, S.; Morales-Pinzon, T. Social Organizational Life Cycle Assessment of Transport Services: Case Studies in Colombia, Spain, and Malaysia. *Sustainability* **2022**, *14*, 10060. [CrossRef]
39. Arcese, G.; Lucchetti, M.; Merli, R. Social Life Cycle Assessment as a Management Tool: Methodology for Application in Tourism. *Sustainability* **2013**, *5*, 3275–3287. [CrossRef]
40. Jørgensen, A. Social LCA—A way ahead? *Int. J. Life Cycle Assess.* **2013**, *18*, 296–299. [CrossRef]

41. Lagarde, V.; Macombe, C. Designing the social life cycle of products from the systematic competitive model. *Int. J. Life Cycle Assess.* **2013**, *18*, 172–184. [CrossRef]
42. Chan, A.P.C.; Oppong, G.D. Managing the expectations of external stakeholders in construction projects. *Eng. Constr. Archit. Manag.* **2017**, *24*, 736–756. [CrossRef]
43. Neugebauer, S.; Traverso, M.; Scheumann, R.; Chang, Y.J.; Wolf, K.; Finkbeiner, M. Impact pathways to address social well-being and social justice in SLCA—Fair wage and level of education. *Sustainability* **2014**, *6*, 4839–4857. [CrossRef]
44. Zanchi, L.; Delogu, M.; Zamagni, A.; Pierini, M. Analysis of the main elements affecting social LCA applications: Challenges for the automotive sector. *Int. J. Life Cycle Assess.* **2018**, *23*, 519–535. [CrossRef]
45. Pollok, L.; Spierling, S.; Endres, H.-J.; Grote, U. Social Life Cycle Assessments: A Review on Past Development, Advances and Methodological Challenges. *Sustainability* **2021**, *13*, 10286. [CrossRef]
46. Backes, J.G.; Traverso, M. Application of Life Cycle Sustainability Assessment in the Construction Sector: A Systematic Literature Review. *Processes* **2021**, *9*, 1248. [CrossRef]

sustainability MDPI

Review

A Review on the Way Forward in Construction through Industrial Revolution 5.0

Muhammad Ali Musarat [1,2], Muhammad Irfan [3], Wesam Salah Alaloul [1,*], Ahsen Maqsoom [4] and Maria Ghufran [5]

1 Department of Civil and Environmental Engineering, Universiti Teknologi PETRONAS, Bandar Seri Iskandar 32610, Perak, Malaysia; ali.musarat@utp.edu.my
2 Offshore Engineering Centre, Institute of Autonomous System, Universiti Teknologi PETRONAS, Bandar Seri Iskandar 32610, Perak, Malaysia
3 Civil Engineering Department, HITEC University, Taxila 47080, Pakistan; irfan7235@yahoo.com
4 Department of Civil Engineering, COMSATS University Islamabad Wah Campus, Wah Cantt 47040, Pakistan; ahsen.maqsoom@ciitwah.edu.pk
5 Department of Construction Engineering and Management, National University of Sciences and Technology (N.U.S.T.), Islamabad 44000, Pakistan; mghufran.cem17nit@student.nust.edu.pk
* Correspondence: wesam.alaloul@utp.edu.my

Abstract: The growing concept of Industry 5.0 (IR 5.0) has enhanced the study horizon of the technology-centered Industry 4.0 (IR 4.0) to an intelligent and balanced socioeconomic change powered mutually by people and technologies. The role of humans in the technological revolution is largely focused on IR 5.0, which is already a future trend. IR 4.0's cyber–physical systems revolution has evolved into IR 5.0, or in other words, from machine-to-machine integration to human-to-machine integration, which is radically altering how people live, work, and interact with one another. Therefore, the current study aims to comprehensively review transformation through industrial revolutions and provide a way forward in the construction industry with the incorporation of IR 5.0. This study has used a narrative-based research methodology in which multiple databases such as Scopus, Web of Sciences, Google Scholar, and Science Direct have been utilized for extracting articles related to the subject area of the current study. Moreover, through narrative-based methodology, which is a generic-based review technique, the information gathered from multiple sources has been summarized and synthesized. The findings of the review indicate that resilience, human-centricity, economic efficiency, and sustainable development are the key characteristics of IR 5.0. Moreover, the adoption of IR 5.0 in the construction industry also faces some major challenges such as a shortage of IR 5.0-related technical skills, investment-hesitancy among investors, security, and cultural concerns for human-to-machine integration, and an unavailability of data for effective decision-making for governments and stakeholders. The study results also highlight that with selective technology adoption, project teams embracing IR 5.0 for improved collaboration and coordination, more environmentally friendly technology adoption through human-to-machine collaboration, and stakeholders leveraging the power of human knowledge and innovative proficiency through machines, reforms can be brought into the construction industry through the incorporation of IR 5.0. It is also important to keep in mind that adopting IR 4.0 is still difficult in some areas and it may seem like achieving IR 5.0 will require years of effort and significant cultural change; however, it needs to be considered right away. The effects of disruptive technologies on Industry 4.0 are covered in several studies; however, IR 5.0 is a novel idea that is still in its early stages, thus its consequences have not been well examined in the construction industry. Therefore, the current study has expanded the body of knowledge on this important subject in detail and has comprehensively explained the transformation by providing a way forward for the adoption of IR 5.0 in the construction industry.

Keywords: revolution; IR 5.0; construction; sustainable development

Citation: Musarat, M.A.; Irfan, M.; Alaloul, W.S.; Maqsoom, A.; Ghufran, M. A Review on the Way Forward in Construction through Industrial Revolution 5.0. *Sustainability* **2023**, *15*, 13862. https://doi.org/10.3390/su151813862

Academic Editors: Zora Vrcelj and Malindu Sandanayake

Received: 5 August 2023
Revised: 7 September 2023
Accepted: 15 September 2023
Published: 18 September 2023

1. Introduction

The construction sector plays a pivotal role in a country's economy [1] as it accounts for about 6% of global gross domestic product (GDP) [2]. GDP is important in the market since it helps to balance various industries. Over time, the construction sector has encountered a massive transformation in the use of technology in projects [3]. The construction industry promotes the economic activities that encompass the entire construction process, from the production of raw construction materials and components, through the provision of professional solutions such as design and project management, to actual construction site activities [4,5].

Considering the significance of the construction sector, governmental stakeholders globally agree that boosting the infrastructure and real estate development industries, as well as providing affordable housing, is a common economic goal [6,7]. However, setting the wheel of the construction industry into motion faces numerous problems and obstacles These include legal frameworks, organizational culture barriers, economic incentives, tax regimes, funding programs, procurement, supply chain tools and methods, and a system of checks and balances that must all be developed and made to function to cover both providers and purchasers. When a system is constructed from the ground up, the job becomes substantially more complex. That requires the utilization of more advanced tools, techniques, processes, and optimized coordination among teams, which is difficult to attain with obsolete methods [8,9].

Most governments are investing in research and development to promote environmentally friendly building practices. As a result, modernizing the construction sector will secure long-term developments. In terms of modernization prospects, the market size for Industry 4.0 was estimated to be USD 114.55 billion in 2021. Artificial intelligence, 3D printing, augmented and virtual reality, the Internet of Things (IoT), and blockchain are just a few examples of digital technologies that are disrupting every step of the value chain, including product design, supply chains, manufacturing techniques, and customer experiences. Strategic attempts by governments to digitize production processes across multiple industries were also supporting worldwide manufacturing industry giants like Germany, the U.S., France, and Japan. Since the inception of industrial revolutions, humans have consistently been evolving, i.e., the transformation from IR 1.0 to IR 5.0 [10,11].

Furthermore, it has long been argued that the construction sector is labor-intensive and sluggish in embracing new technologies. However, IR 5.0, which emphasizes the integration of humans with machines through automation, artificial intelligence (AI), and data analytics, is starting to take shape and has the potential to drastically change the way construction is realized. The current review study explores several topics within the context of transformational evolution through different industrial revolutions with the potential to revolutionize construction practices by raising standards for productivity, efficiency, and safety. The drive of this review is to examine the significance of this transformation in the context of different industrial revolutions, from IR 1.0 to IR 5.0, and provide a way forward for IR 5.0 adoption from the perspective of the construction industry. Perhaps this review study makes a persuasive case for the adoption of IR 5.0 in the construction sector as a way of advancing it into a more productive and technologically sophisticated era. Moreover, the current review will assist government stakeholders, key top management personnel, policymakers, and major players in the construction industry worldwide in addressing the industry's key challenges in the form of skilled workers in the face of a knowledge economy [12]. Furthermore, the concept of IR 5.0 is novel in the context of the construction industry and to expand and enrich the body of knowledge, a framework for decision-makers to successfully implement IR 5.0 in the construction industry has been proposed.

2. History of Industrial Revolutions

The Industrial Revolution (IR) was a significant turning point in history that had some sort of impact on practically every element of daily living [13]. The most significant outcome of the IR, according to some economists, was that this was the first time in history that the average person's standard of living in the West started to rise steadily. However, other economists believe that this advancement did not occur until the late nineteenth and early twentieth centuries [14]. Before the IR and the creation of the modern capitalist economy, GDP per capita remained mostly steady. However, with the IR, capitalist economies entered a period of per capita economic growth. Since the domestication of animals and plants, economic historians concur that the start of the IR was a significant event in human annals [15]. The exact beginning and conclusion of the IR, as well as the rate of societal and economic change, is still up for debate among historians [16]. T. S. Ashton claimed that the IR took place roughly between 1760 and 1830, while Eric Hobsbawm claimed that it started in Britain in the 1780s and was not felt until the 1830s or 1840s [17]. In the nineteenth century, mechanized textile production spread from the United Kingdom to continental Europe and the United States, with important textile, iron, and coal centers emerging in Belgium and the United States, and later textiles in France [18]. Figure 1 illustrates the transformation in different industrial revolutions.

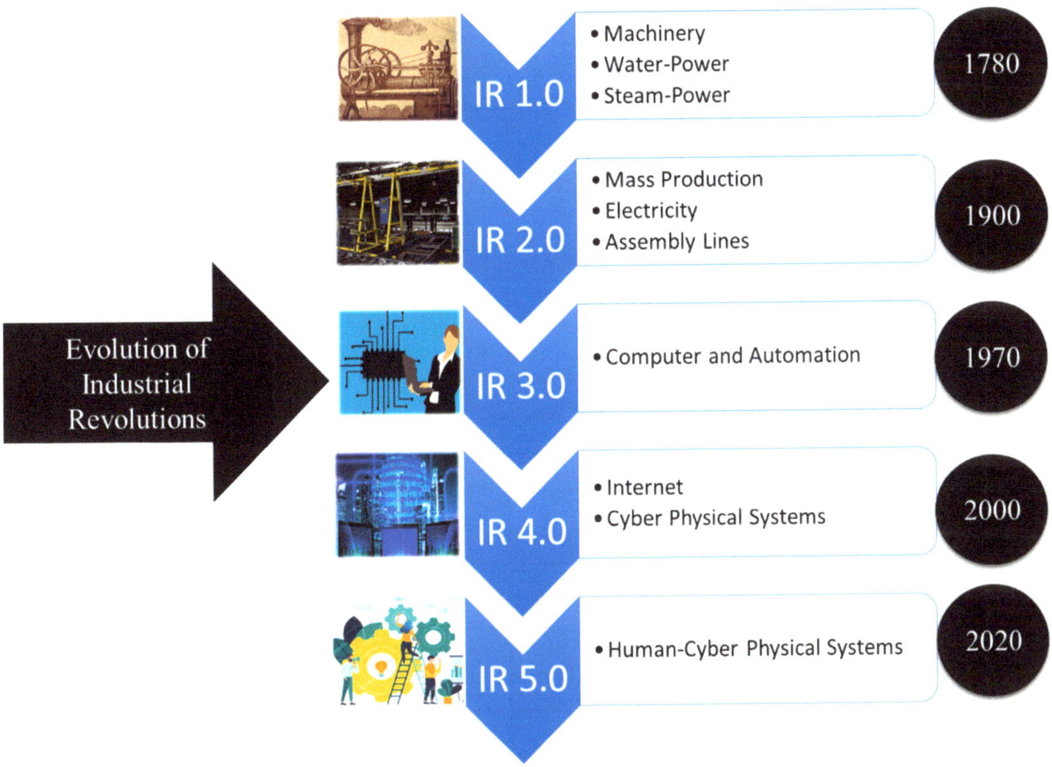

Figure 1. Industrial Revolution History.

2.1. *Industrial Revolution 1.0*

IR 1.0 had a significant impact on society, transforming the way goods were produced and leading to an increase in productivity and economic growth. However, it also led to social and environmental problems such as poor working conditions and pollution. The lessons learned from this period continue to shape the way we think about industrialization and its impact on society. The system improved [19]. Also, banking, and other financial systems improved to run the industries and business firms smoothly. Child and infant mortality rates decreased and fertility rates increased. As a result, population growth had changed intensely. This research concluded that IR 1.0 completely changed the history of human beings [20].

2.2. Industrial Revolution 2.0

From the late nineteenth century to the early twentieth century, the second IR, also known as the Technological Revolution, was a period of rapid scientific discovery, standardization, mass production, and industrialization. The first IR, which ended in the middle of the nineteenth century, was marked by a slowdown in significant inventions before the Second Industrial Revolution, which began in 1870 [21]. The development of the railway network allowed for the faster and cheaper transportation of goods, leading to the growth of international trade. The telegraph and telephone revolutionized communication, allowing for faster and more efficient communication between individuals and businesses. The assembly line, invented by Henry Ford, revolutionized the production of goods, allowing for the mass production of cars and other consumer goods at a much faster rate than before [22].

2.3. Industrial Revolution 3.0

Industry 3.0 utilized a newly recorded micro database for real advancement yield, the main impetuses, and mechanical interdependencies [23]. The combined forces of information technology derived the third IR, which has altered not just the way we work but also how we perceive the world and how we define it. A global village replaced society after the Third Industrial Revolution. People were empowered by technology and information providers to locate, retrieve, exchange, and use data in ways that improve their lives [24].

The word "technology" encompasses both dimensions of innovation. Due to China's rapid e-commerce growth, third-party payment technology has advanced significantly [20]. The payment business of the traditional financial industry, represented by commercial banks, is expanding in both breadth and depth thanks to this significant technological advancement, which was started by emerging internet enterprises. In the meantime, there is also a significant degree of substitution, competition, and crowding out among these banks in terms of the potential consumers, traditional intermediary firms, deposit and loan services, and basic payment and settlement activities of the traditional financial industry. However, an organization's technology transformation and acceptance are driven by entrepreneurial activities to develop new technologies. The value proposition that the company offers to the consumer has been impacted by technological progress in the financial industry. Understanding the sources of technical innovation is essential for maximizing investment and market potential. The market benefits from technological innovation, as well as inventors and early adopters [25].

2.4. Transformation of IR 3.0 to 4.0

The Third Industrial Revolution (TIR) was led by high-tech innovations in manufacturing, distribution, and energy factors. The TIR was global, but it was also local, giving rise to the term 'glocal'. The TIR was set to change the way we work, produce, and entertain. It fundamentally changed the way we plan and manage cities and regions [26]. IR 3.0 had a profound impact on the ICT, knowledge, defense, health, education, advanced manufacturing, financial, and administrative sectors [22].

2.5. Industrial Revolution 4.0

Industry 4.0 or IR 4.0 was established in Germany to modernize industrial manufacturing by utilizing emerging technologies and digitalization to their fullest extent [27]. Due to its capacity to increase operational effectiveness, efficiency, and the availability of new prospects, digitalization has become a widely accepted concept around the globe today [28]. A realistic and sustainable production system is the main focus of IR 4.0 [29]. The goal of Industry 4.0 is to digitalize industrial processes to establish a broad, adaptable network for production and services. The frameworks of IR 4.0, which are controlled by artificial intelligence, increase the usefulness of the human–machine interface [30]. The four main drivers underpinning Industry 4.0, which help transform manufacturing into a fully digital and intelligent process, are the Internet of Things (IoT), the Industrial Internet of Things (IIoT), cloud-based manufacturing, and smart manufacturing [31].

BIM (within the planning domain) can help overcome the digital gap that currently exists in the construction sector and continue to have an impact on upcoming building practices when combined with Industry 4.0 (the production domain) [32]. IR 4.0 technologies, as shown in Figure 2, provide better control and forecast of beneficial outcomes not only for business but also for society and the environment [33]. Industry 4.0 is a different field for stakeholders to adopt in the construction industry because they are not gaining a clear view of implementation [34], and its acceptance in the construction industry is challenging due to the advanced technology; therefore, stakeholders cannot invest. IR 4.0, also known as Industry 4.0, is the ongoing revolution that builds on the Digital Revolution. It is described by the incorporation of sophisticated technologies such as robotics, 3D printing, and blockchain into the production process [24].

Figure 2. IR 4.0 Technologies.

As a result, IR 4.0 in the system for the construction sector concentrates on the shift from analog to digital [35] and, finally, the digital-to-physical transition to provide better coordination, planning, and execution of built environment infrastructure [36]. As previous scholars have stated, the idea of construction 4.0 is still evolving, and it is informed by its predecessors' conception of industry 4.0 [37]. As per [38], building a digital construction site with a variety of tools to track progress throughout a project's life cycle is the main goal of building 4.0. By using IR 4.0, the construction process as well as corporate and project frameworks would change, integrating the fragmented construction industry through cyber–physical systems [39] as shown in Figure 3.

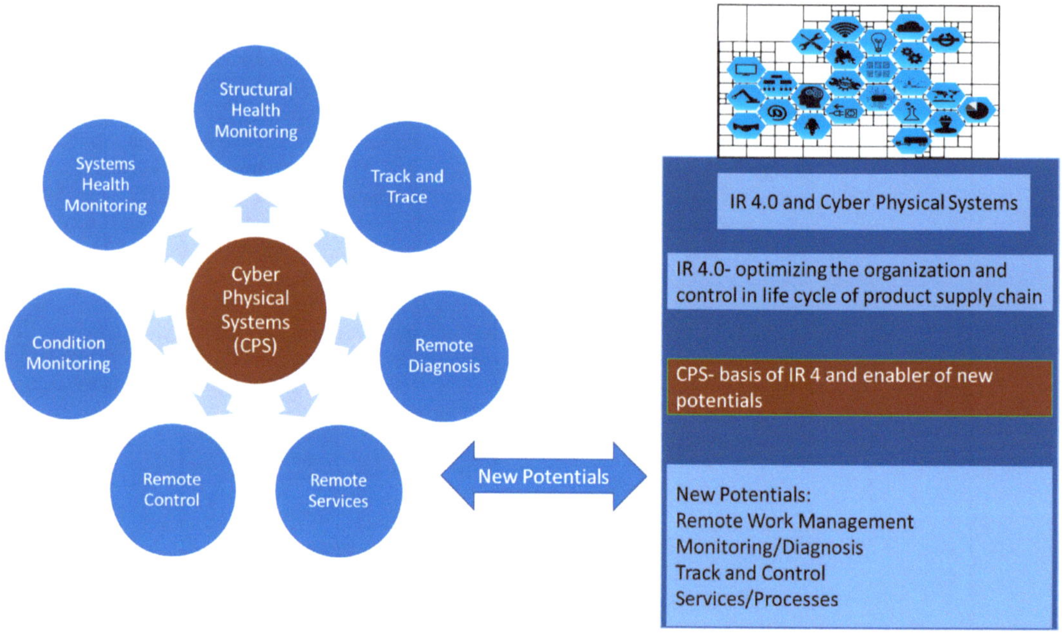

Figure 3. Cyber–Physical System in IR 4.0.

Construction is one of the most lucrative businesses, but it also has one of the lowest levels of R&D intensity [40]. Similarly, employment growth in the AEC has slowed over time despite nearly doubling in other industries [41]. In an era of Industry 4.0, the role of human resources is changing from machine operator to strategic decision-maker [42]. Robots assist humans with difficult, risky, and time-consuming activities, but for successful human–machine cooperation, humans must be effectively trained for these duties [36]. Construction is a labor-intensive industry; thus, there is a considerable possibility to boost productivity through technological innovation (such as the usage of robots), especially for potentially dangerous human labor. With the digital building platform, robots are only employed sparingly for tasks like 3D printing, wall construction, installing rebar, welding, using drones, etc. [43]. Industry 4.0 technologies are anticipated to have the biggest effects on businesses around the world [44]. Among these technologies, IoT and AI have the most drastic impacts of 72% and 68%, respectively [45].

2.6. *Transformation of IR 4.0 to 5.0*

To improve efficiency and productivity, IoT devices were created in Industry 4.0 and, as a result, increased efficiency and mass production and reduced costs [46]. Cooperation between humans and robots was challenging in Civilization 4.0 because knowledge and information exchange were insufficient and desired. The transformation from Industrial Revolution 4.0 to 5.0 is currently an ongoing process and is characterized by the integration of advanced technologies such as artificial intelligence, virtual and augmented reality, and advanced robotics [47]. The primary focus of Industrial Revolution 5.0 is on developing intelligent systems that can work collaboratively with humans and improve the efficiency of industrial processes. This includes the development of autonomous systems that can make decisions and take actions based on data analytics and machine learning algorithms.

One of the key areas of focus in Industrial Revolution 5.0 is the development of a more sustainable industrial system. This includes the use of renewable energy sources, the reduction of waste, and the development of closed-loop systems that minimize the use of resources. The use of advanced analytics and sensors is also critical in enabling real-time monitoring and decision-making to optimize resource utilization and reduce environmental impact. Another key aspect of Industrial Revolution 5.0 is the development of a more inclusive industrial system that is accessible to all. This includes the use of advanced technologies such as virtual and augmented reality to provide training and support to workers, as well as the development of adaptive manufacturing processes that can accommodate workers with different abilities [48].

Overall, the transformation from Industrial Revolution 4.0 to 5.0 represents a significant shift in the way industrial processes are designed and managed, with a greater focus on sustainability, inclusivity, and intelligent systems that work collaboratively with humans. This transformation is likely to bring about new opportunities and challenges, and organizations and individuals need to adapt to these changes to remain competitive and productive in the years to come. Industry 5.0, which incorporates the integration of human intelligence and cognitive computing, was developed to enhance the methods and efficiency of production [49]. With collaborative operations, Industry 5.0 aims to combine these cognitive computing capabilities with human intelligence and resourcefulness [50], as shown in Figure 4.

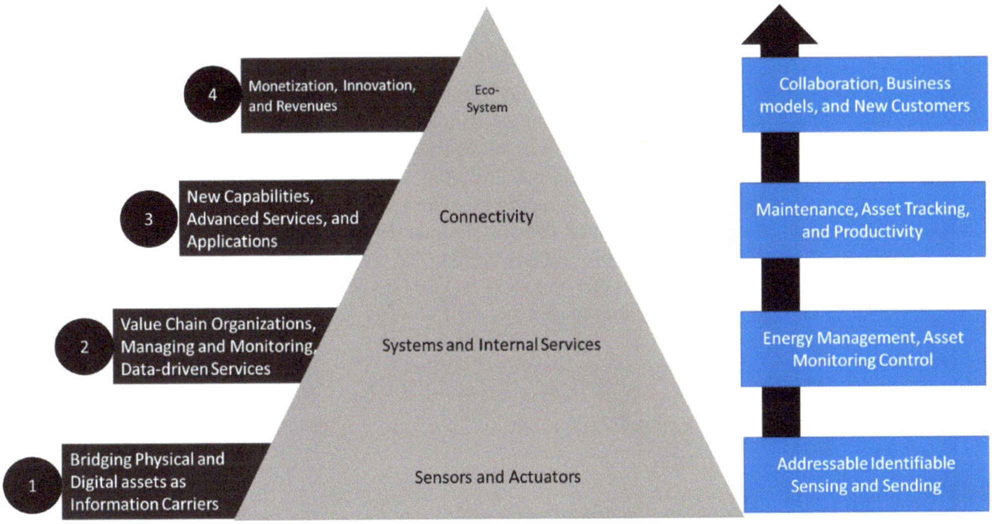

Figure 4. Industrial Transformations.

The fundamental tenet of "an orthogonal exit" is that events in hyper-connected networks have no bearing on the orthogonal departure pathways, hence the phrase "safe exit strategy". Such integration has a positive impact on Industry 5.0 as well, for instance: optimization of human efficiency and liberty of design [51]. Increasing the safety of the employees increases customer satisfaction and loyalty. In the construction industry, such integration of Industry 5.0 may alleviate barriers such as social polarization of unemployment, increased cyber security threats, huge amounts of investment, accountability, customer subjectivity, and monopoly [52]. The summary of these transformations is depicted in Figure 5.

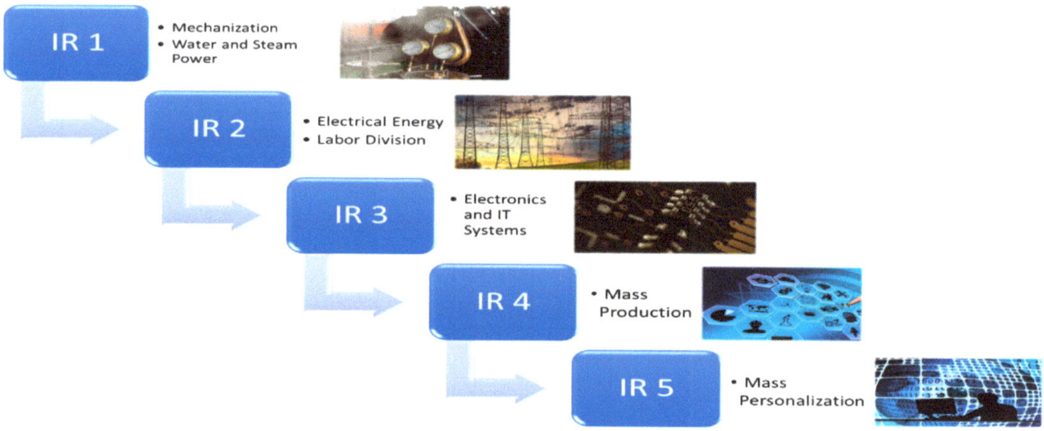

Figure 5. Summary of Industrial Revolution Transformation.

3. Industrial Revolution 5.0

The concept of the Fourth Industrial Revolution was unveiled at the Hannover Messe trade show in Germany in 2011, which is regarded as the year when Industry 4.0 first gained popularity. The initial vision of Industry 4.0 was also created largely for the German industrial automation and smart manufacturing markets, as well as for other nations that are members of the EU and are subject to EU legislation [53]. Industry 4.0 was also once thought to be primarily important to the European corporate climate and policies, even though it quickly became a global movement [54]. Whereas IR 5.0 compliments IR 4.0 philosophy by integrating humans with machines instead of replacing them. The transformational differences from IR 4.0 to IR 5.0 are depicted in Figure 6, while Figure 7 shows the technology enablers for IR 5.0.

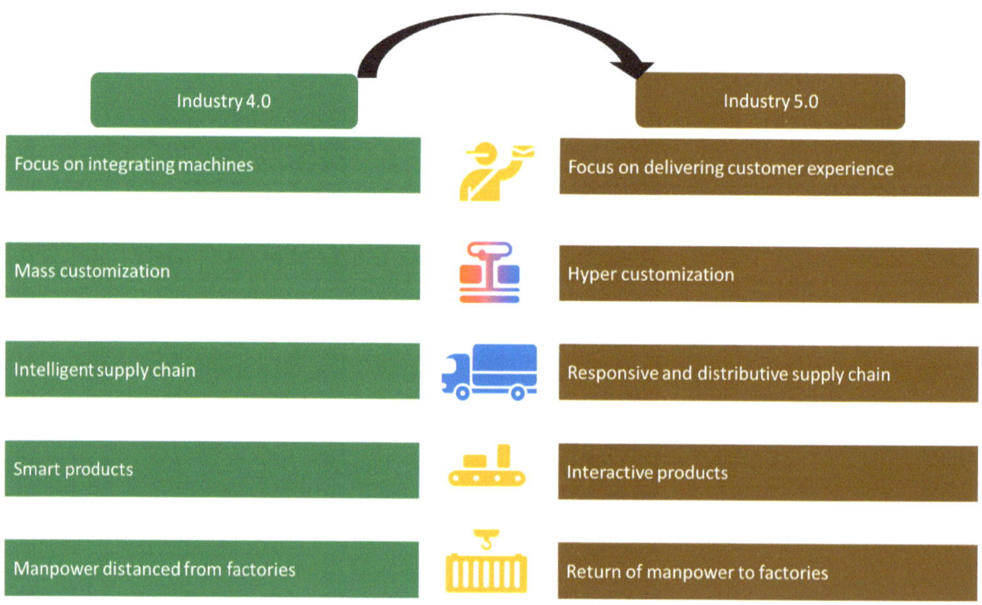

Figure 6. Transformation from IR 4.0 to IR 5.0.

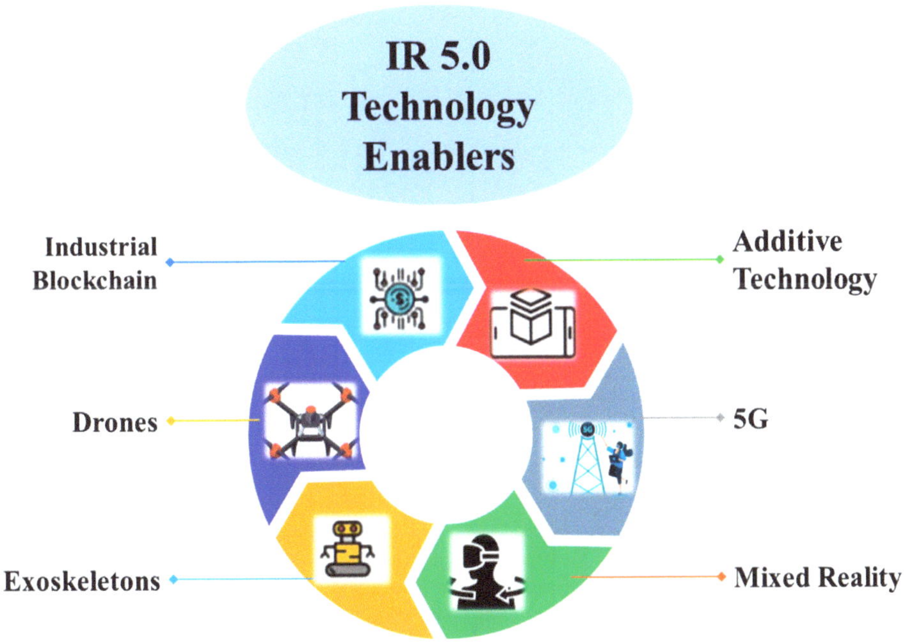

Figure 7. Technology Enablers for IR 5.0.

Industry 5.0 is a notion for the next stage of industrialization that is characterized by the return of labor to factories, distributed manufacturing, intelligent supply chains, and hyper-customization to constantly give a personalized consumer experience [55]. The key characteristics of IR 5.0 include:

3.1. Sustainability and Resiliency

In addition to empowering people, Industry 5.0 pays close attention to several resilience and sustainability-related issues. The impression is that Industry 5.0 proposes a framework that ought to strike a balance between the initial vision's competitiveness and business efficiency and a focus on sustainability, sensitivity to the environment, and deliberation on the processes of industrial automation on the environment [56]. Relying on adaptable, flexible, and agile technology is another aspect of Industry 5.0's push for resilience [57].

3.2. Collaboration between Machines and People

The emphasis is now on cooperative interactions between machines and humans, even though Industry 5.0 does not minimize the critical role that robots and automated equipment play in the new industrial revolution [58]. In addition to recognizing the well-known benefits and characteristics of robotic automation, such as the fact that these systems are more accurate, reliable, and productive than human workers, Industry 5.0 also recognizes all the drawbacks of excessive automation. For instance, the limited adaptability of highly automated solutions to shifting needs and specifications [58].

3.3. Enhancing Client Experience and Going All out on Personalization

The new idea is focused on offering better customer experiences—instead of just achieving high performance by linking machines and software when it comes to the consumer-facing side of industrial automation—products and services produced by Industry 5.0 solutions [59]. The capacity of businesses to provide clients with even better choices and product customization options while still reducing production costs due to robots automation, and other cutting-edge technologies is what brings hyper customization and hyper-personalization to consumers [60].

3.4. Reasons to Adopt IR 5.0 Enabling Technologies

Most of the main technologies used in Industry 4.0 systems are also important mechanisms of the Industry 5.0 paradigm [61]. Industry 5.0's enabling technologies are typically broken down into six broad groups, as shown in Figure 7 [62]. The following are the key reasons to adopt IR 5.0 enabling technologies:

1. The use of augmented reality (AR) and virtual reality (VR) for inclusivity, training and industrial testing;
2. Sophisticated safety gear and working tools that improve human skills through robotic tools and data communication;
3. Automatic speech and gesture detection:

Using AI and other cutting-edge technology to enhance human workers' cognitive capacities. Tracking gadgets to keep tabs on their physical and mental well-being.

4. Digital twins and simulations;
5. Robots that work together:

Since Industry 5.0 mainly relies on collaborative robots, or robots (also known as universal robots) that operate alongside humans rather than completely autonomous robotic solutions and industrial equipment, it is important to explore this area as well. In comparison to Industry 4.0, the significance of robots and all other types of human–machine interaction technologies and solutions is significantly greater in Industry 5.0 environments.

6. Human–machine integration:

Robots are made to interact directly with human workers, unlike industrial robots, which are often made to operate autonomously or as a part of a wider network of machines.

7. High adaptability:

The Industry 4.0 concepts' industrial robots are often created as specialized instruments that can perform one or more tasks. Without significant coding and engineering improvements, reading such machines is either impossible or highly challenging. On the other hand, robots are intended to be adaptable, simple-to-use machines that may be used for a variety of purposes.

8. Reduced prices and smaller sizes:

Robots are considerably more compact than standard Industry 4.0 industrial robotic systems. Additionally, because they are far more affordable to produce, this technology is more readily available to institutions and companies who do not have the funds to invest in expensive industrial machinery.

9. Easy to use and intuitive:

Additionally, collaborative robots are moving away from the excessive complexity of traditional industrial automation solutions. They are often made to be user-friendly and simple to program to make it easier to employ robots by human workers to increase productivity as well as to redeploy robots throughout a production area, for example. According to the Industry 5.0 paradigm, installing, programming, and calibrating collaborative robotic systems for each user's unique demands should not be prohibitively expensive.

10. Application to carry out a wider range of duties:

Robots are meant to be able to handle a variety of light activities as well, such as pick and place, material handling, and other jobs where they can help human employees but cannot fully replace humans, unlike industrial robots which are primarily made for heavy production.

11. Increased safety:

Additionally, robots are more frequently used for operations that have a high risk of accidents and injury for human workers. Utilizing them would enhance security, lessen accidents, and improve working conditions for human workers.

3.5. Transforming Construction Process with Construction 5.0 through IR 5.0

Construction 5.0 is an innovative approach that makes use of technology with the integration of human creativity to revolutionize industrial processes through the incorporation of the IR 5.0 concept [63]. To improve efficiency, productivity, effectiveness, and safety in construction projects, Construction 5.0, an enhanced iteration of conventional construction practices where technology compliments human creativity instead of replacing it, incorporates cutting-edge technology including artificial intelligence (AI), the Internet of Things (IoT), BIM, and robotics [64]. Stakeholders can significantly enhance project planning, implementation, and management by utilizing digital twins, unmanned aerial vehicles, smart sensors, and autonomous machines inside an interconnected system made possible by IR 5.0. Analysis of data in real time for better decision-making while eliminating mistakes and cost overruns is made possible by this transformational approach. Additionally, Construction 5.0 supports sustainability by encouraging eco-friendly products and procedures, minimizing waste production, and lowering carbon footprints during a project [65]. Construction professionals can unleash enormous value and foster innovation in a market environment that is becoming more competitive by embracing this paradigm change towards Construction 5.0 through IR 5.0.

By utilizing technology like drones, Internet of Things (IoT) sensors, and building information modeling (BIM), the construction industry may achieve human-to-machine integration. BIM makes team planning and design possible, and IoT sensors offer real-time information on tools, equipment, and worker safety [44]. Aerial photography from drones is available for site assessments and progress tracking. Through the continuous exchange of data between humans and machines, made possible by these technologies, the efficiency, safety, and decision-making of building projects are all improved. The construction industry has been gradually adopting technological advancements to increase efficiency, reduce costs, and improve sustainability. Some potential directions that the industry could take as part of a hypothetical Industrial Revolution 5.0 in construction might include:

(a) Using Advanced Automation and Robotics in Construction

The construction sector is being revolutionized by advanced automation and robotics, which increase productivity, security, and accuracy. These technologies include a wide range of applications, including robotic bricklaying, autonomous heavy machinery, and drones for site inspection [66]. Construction projects can be finished more rapidly, with lower labor costs, and with lessened human risk by automating repetitive operations and utilizing AI for project management. Additionally, these developments make it possible to build intricate structures that were previously thought to be unfeasible, ushering in a new era of eco-friendly and futuristic architecture [67]. The use of cutting-edge automation and robotics speeds up the construction process and provides better outcomes, making it a crucial part of the industry's continued development.

(b) 3D-Printed Building Modules and Additive Manufacturing

The construction industry has seen the emergence of ground-breaking technologies like 3D-printed building modules and additive manufacturing, which are fundamentally changing how we design and build structures [68]. These techniques enable the production of highly customized and complicated building components, from walls and façades to complete modules, by precisely stacking materials like concrete, plastic, or metal. This method not only vastly shortens building deadlines but also lowers labor and material costs, allowing architects and engineers to test out cutting-edge, sustainable solutions that were previously difficult to realize. Furthermore, 3D printing can be used to construct structures on-site quickly and efficiently in remote or disaster-stricken places [69]. The ability of these technologies to transform the construction sector and promote environmentally friendly building techniques is growing as they develop.

(c) Digital Twin and Building Information Modelling (BIM)

In the design and construction sectors, key advances include digital twins and building information modeling (BIM). Using digital twin technology, physical buildings can be digitally replicated in real time, enabling continuous monitoring and analysis throughout a building's lifetime. Conversely, BIM entails the development of a comprehensive 3D model of a construction project that includes information about its systems, materials, and components [70]. Together, these technologies give planners, designers, and contractors the ability to precisely optimize planning, design, and construction processes. Through continuous insights into a building's performance, digital twins enable predictive maintenance and increases in energy efficiency. BIM improves stakeholder coordination and collaboration, minimizing mistakes and delays. In the construction industry, they usher in a new era of data-driven decision-making, sustainability, and efficiency [71].

(d) Smart Materials and Sustainable Construction

Sustainable construction techniques are being advanced by smart materials. These cutting-edge materials have special qualities that react to outside stimuli, like temperature, light, or stress, allowing structures to adapt and perform at their best in the moment [72]. They contribute to sustainability by improving energy efficiency by using self-regulating insulation or photovoltaic materials to capture solar energy. In addition, replacement can increase a structure's longevity and durability, reducing the need for routine upkeep and replacement, and lowering resource consumption and waste production. The modern search for sustainable and resilient construction solutions is perfectly aligned with the incorporation of smart materials into construction, which not only make buildings more environmentally friendly but also improve occupant comfort and overall operating efficiency [73].

(e) Renewable Energy Integration

One of the most important steps towards a sustainable and environmentally responsible future is the incorporation of renewable energy resources into our energy infrastructure. The utilization of clean, nearly endless renewable energy sources, such as solar, wind, hydro, and geothermal energy, can drastically lower greenhouse gas emissions and our reliance on fossil fuels. We can generate electricity effectively and reliably while reducing the negative consequences of climate change by integrating these sources into our energy systems and infrastructure [74]. Additionally, improvements in grid management systems and energy storage technology are making it easier to integrate renewable energy, guaranteeing a reliable and resilient energy supply for communities and companies around the world. In addition to addressing the urgent need for carbon reduction, this shift towards renewable energy also promotes economic growth [67].

(f) IoT and Connectivity

Connectivity and the Internet of Things (IoT) have ushered in a revolutionary era across many industries, providing previously unattainable levels of data-driven insights and automation. The Internet of Things (IoT) is a large network of networked gadgets and sensors that can gather, distribute, and analyze data instantly. IoT-enabled devices are improving efficiency, productivity, and decision-making across a variety of industries, from manufacturing and healthcare to transportation and smart cities [75]. IoT provides proactive maintenance, remote monitoring, and better resource management by facilitating a constant flow of information. Additionally, convenience and personalization are improved by the incorporation of IoT into daily life through wearable technology and smart houses. The potential for IoT and connectivity to create a more effective, sustainable, and integrated future continues to expand as our world becomes more networked, with ripple effects.

(g) AI and Predictive Analytics

By utilizing the power of data to make proactive, data-driven decisions, AI and predictive analytics are revolutionizing numerous industries. Artificial intelligence (AI) systems can analyze large datasets, spot patterns, and predict future trends and outcomes thanks to sophisticated machine learning algorithms [76]. Because it makes it possible to assess risk, detect fraud, provide individualized advice, and estimate demand, this technology is invaluable in fields including banking, healthcare, marketing, and supply chain management. AI increases productivity and accuracy by automating difficult data analysis and prediction activities, which ultimately spurs innovation and competition. As AI and predictive analytics advance, they could change business models, streamline processes, and influence how decisions are made in the future for a variety of applications [77].

(h) Modular Off-Site Construction

Modular off-site construction, often referred to as off-site or prefabricated construction, is a revolutionary approach to building that involves manufacturing building components or modules in a controlled factory environment and then transporting and assembling them on-site. This method offers numerous advantages, including reduced construction time, increased cost-effectiveness, and improved quality control. It minimizes weather-related delays and site disruptions, making it especially appealing for projects with tight schedules or challenging environmental conditions [78]. Additionally, modular construction promotes sustainability by reducing material waste and energy consumption during construction. Its versatility allows for innovative architectural designs and can be applied across various building types, from residential and commercial structures to healthcare facilities and educational institutions. In an era of increasing demand for efficient, eco-friendly, and adaptable construction solutions, modular off-site construction is poised to play a pivotal role in shaping the future of the industry.

3.6. Role of IR 5.0 in the Manufacturing Industry

Industry 5.0, a wave of cognitive computing infrastructure and apps that are redefining how commodities are created, is transforming the manufacturing sector [79]. Process manufacturing will experience yet another upheaval as Industry 5.0, a new wave of cognitive computing applications and infrastructure, transforms the production of chemicals, pharmaceuticals, and biotechnology products. It is critical to comprehend the initiative's definition and all its ramifications. The first aspect of Industry 5.0 is how people will collaborate with robots and intelligent machinery [56]. Of course, encouraging individuals in manufacturing plants to share information is a necessary part of shift-to-shift communication. However, in Industry 5.0, robots will be assisting humans in their work by using the IoT and big data [80]. The industry should emphasize the requirement for a personal human touch more than Industry 4.0 pillars of productivity and automation [81].

Robots have regularly performed physically demanding, dull, or dangerous work like welding in auto plants or loading and unloading heavy objects from trucks in warehouses. Industry 5.0 makes strides beyond Industry 4.0 that will enable collaborative operations that combine cognitive computer skills with human intelligence and resourcefulness. Industry 4.0 introduced smarter and more connected robots to the workplace [82]. Industry 5.0 will soon provide the acceptability and acknowledgment required to integrate people's creative and cognitive abilities with the speed and accuracy of the technology. As a result, the system will be stronger and more competitive. People will have more freedom to use their innate cognitive abilities to contribute even more value to the plant floor because of coexistence, which will open a wide range of exploratory prospects, including novel and interesting employment opportunities [83]. Even the creation of new social contracts and improved communication on the production floor are possible outcomes of Industry 5.0. There will be more opportunities for human-to-human collaboration in more significant ways, invoking information that will promote resilience and responsibility, and ensure compliance, in addition to communication between humans and their robots. Human-led activities, for instance, can react to unfavorable situations or assist in reducing the risk from large-scale catastrophes like the pandemic. Let's face it, when all is said and done, people, not machines, are in charge.

The question is not whether a company can gain from having humans collaborate with robots, but rather how they might use AI-enabled tools to achieve the best results from human–machine interactions. Businesses that leverage technology to enable employees to harness their innate skills and abilities to boost productivity will succeed [84]. Industry 5.0 is expected to provide a situation where people and robots may interact effectively. This will make it possible to address the problems of complicated production in the future, including increased customizations through automated manufacturing processes that are optimized, which will call for a lot of cooperation [85]. All plant functions will benefit from increased transparency, dependability, and visibility because of which teams will be better able to interact and provide the best results. Manufacturing companies may increase productivity, cost-effectiveness, quality, and safety thanks to people-centric technologies [86]. All in all, Industry 5.0 is a concept that is starting to gain traction and that places the Industrial Revolution's power to positively affect society at its core.

3.7. Role of IR 5.0 in the Construction Industry

3.7.1. Executing a Project with a Human Focus

Industrial projects are very intricate. When automation systems are fully implemented, human ingenuity can be repressed according to the mass automation paradigm. Incremental improvements are frequently impossible once the system achieves a steady state. Changing an operating system frequently calls for extensive design, new hardware or software, and financial expenditure [87]. As a result, innovation may be discouraged or downright forbidden for a while (until capital expenditures can be recovered). Contrarily, using humans in automated operations can be discouraged, mostly because it requires investing in costly software. Think about the implementation of an automated pipe drafting system that chooses the best piping routing for a designer, as an example. Even if the designer does not think the routing is ideal, if the designer is forced to accept it because of an overreliance on technology, we can perceive the gap between people and technology [88].

People are empowered by Industry 5.0. To spur innovation and industrial progress, it blends its creative potential with cutting-edge technology. Furthermore, with capital projects continuing to grow in complexity, this innovation is more important than ever. Digital literacy is crucial because human capital is highly valued in Industry 5.0 [20]. Programs for developing technical skills and digital literacy will both receive increased funding from project teams. To decrease costs and shorten deployment schedules, many teams are currently choosing to slash training expenses. This will no longer be a workable budgeting choice in Industry 5.0 because human innovation is essential to success [89]. Greater inventiveness in the initial stages of project planning will result from Industry 5.0. To enhance overall project delivery models and construction execution tactics, innovative ideas will be sought after and fostered. To guarantee that important project goals are created and met, engineering design programs will incorporate strategic planning and tactical planning sessions for constructability and sustainability [90].

Transparency in material sourcing and production processes, near real-time visibility of fabrication status, and a material tracking protocol, are more in line with what we have come to expect from online purchasing thanks to shipping and logistics data (continuous alerts and real-time delivery tracking). Engagement and empowerment using cutting-edge technology will significantly increase construction efficiency instead of a persistent concentration on highly automated operations. Construction sites are not manufacturing facilities; the success of a project depends just as much on good manufacturing and design as it does on inventive field execution and dynamic planning protocols. Furthermore, if the success of any new project is solely determined by the reduction in time spent using the tools (a poor efficiency model), the operational model will be swiftly superseded by the 5.0 revolution. Instead of a constant race to automate, construction innovation and originality will become fundamental drivers of Industry 5.0 models [91]. In a 5.0 model, measuring productive time is outcome-oriented rather than reliant on how much time a craft professional spends holding a wrench.

3.7.2. Adaptive to Collaboration

Unlike Industry 4.0, which encouraged teams to work in partially connected virtual project delivery models, Industry 5.0 encourages teams to work together again in co-located or immersive virtual environments rather than in siloed organizations. Teamwork is essential, and both in-person and virtual teams need to be able to communicate constantly [92]. To facilitate stakeholder involvement in this new paradigm, even between contractors who might typically be viewed as rivals, virtual collabs that bring together stakeholders from engineering, sourcing, construction, commissioning, and project management teams are essential. Collabs move rapidly, collaborate, make comprehensive decisions, and have a beneficial impact on project trajectories thanks to advanced data and technology use. Being adaptive to collaboration also benefits organizations. In today's fast-paced and competitive business environment, organizations need to be able to respond quickly and effectively to changing circumstances [83]. This requires a high degree of collaboration and adaptability, as teams need to be able to work together to achieve a common goal, even in the face of unexpected challenges or obstacles. Organizations that promote collaboration and adaptability are more likely to succeed, as they are better able to leverage the collective skills and knowledge of their employees to achieve their objectives [93].

3.7.3. The Tech That Gives Power

Industry 5.0 wants systems to be resilient rather than stable. As more black swan events (climate- or health-related, and political) take place and disrupt regular project operations, technology systems and solutions that favor stability over resilience will depreciate [94]. In this scenario, technology that promotes system resilience will be increasingly common, and project teams will place less emphasis on mass-automation solutions (automated and heavily scripted process enablers). Rapidly deployable software with adaptable workflow choices will therefore be favored in the industrial project arena over extremely inflexible, process-driven technologies [95]. Digital twins are essential to initiatives that use 5.0. Being a hub for knowledge, they link people and technology in ways that foster creativity. The democratization of data will enable project teams, including engineers, project managers, construction team members, and operations staff, to collaborate easily within digital twins. In Industry 5.0, edge computing will become increasingly prevalent to meet the demands of project teams for large-scale data collecting and processing [96]. The adoption of machine learning and AI will grow, and teams will manage more data instead of just big data. Systems thinking will be crucial for developing and deploying cyber–physical systems as well as for designing and operationalizing systems and solutions [97]. Cobots will become more commonplace. Cobots operate and interact directly with humans, unlike robots, which are physically separated from their human counterparts [59]. These increase work productivity while enabling both mass manufacturing and mass personalization of work outputs. In fabrication facilities, on-site laydowns, and support spaces like tool cribs and field offices, cobots will be commonplace [98].

3.7.4. Focus on Sustainability

Projects under Industry 5.0 are heavily focused on the environment. Any project's success depends on sustainability, which should not just be a top concern. Teams in Industry 5.0 assess the environmental effects of their actions and collaborate to reduce environmental risks [89]. They also place a strong emphasis on resource efficiency and design, create project plans that reduce jarring changes in resource requirements, incorporate sustainable production techniques, and factor in embodied carbon when setting design parameters. Teams working on Industry 5.0 projects see sustainability as a responsibility, and they genuinely care about producing results that have a beneficial influence on the built and social surroundings [99].

3.8. Barriers and Challenges in the Adoption of IR 5.0

Industry 5.0 advances now face challenges that must be overcome for a company to succeed. The following are the main obstacles to the implementation of Industry 5.0:

1. Using state-of-the-art technology requires more time and effort from human workers In addition to specialized software-connected factories, Industry 5.0 demands the utilization of collaborative robotics, artificial intelligence, real-time data, and the Internet of Things [58].
2. Investments in cutting-edge technologies are necessary. Adopting Industry 5.0 is expensive since smart machines and highly skilled personnel are required to increase production and efficiency [59].
3. To communicate with a variety of devices and defend against potential quantum computing applications when deploying IoT nodes, authentication has been utilized in the industry. Since ICT systems lie at the heart of Industry 5.0 applications, strict security standards are required to prevent security risks [50].
4. Currently, the economic impacts of Industry 5.0 in the construction industry have not been researched comprehensively and coverage of the literature on this aspect is missing. The main reason lies in the fact that it is a novel concept and has not been accepted widely by major stakeholders. The economic impacts will be studied once it is implemented globally, which is currently a piece missing from this puzzle.

3.9. Reforms with IR 5.0 in the Construction Industry

The Fifth Industrial Revolution has already begun and, because of the overlap with the Fourth Industrial Revolution, the transition will take longer and require more conscious effort to accept than the changes companies had to make during previous industrial revolutions as shown in Table 1. Due to the exponential growth of technological capability, the time between industrial revolutions has historically gotten shorter; yet, this is the first time that an industrial revolution has overlapped in recorded history. For many, embarking on this 5.0 journey entails improving upon or extending the present 4.0 experience [98]. It will result in a full reframing of operating strategies for some. It is difficult to move from 4.0 to 5.0. However, this transition is merely another step along the path for engineering and construction teams that embrace the potential for positive change. Forward-thinking teams are in a great position to start moving toward version 5.0 right away for an unavoidable revolution [100].

Table 1. Case Studies (Practical Implementation).

Sr.No	Case Study	IR 4.0 Incorporation	Improvement through IR 5.0	Reference
1	**Apis or 3D-Printed House in Russia** A modern method of building is the integration of the IR 4.0 construction protocol into the Russian Apis or 3D-Printed House project. This protocol uses Industry 4.0 technologies to automate and optimize the building of homes, including 3D printing, IoT sensors, advanced robotics, and data analytics. The result is a building process that is more effective and economical, wastes less labor and materials, and offers more design flexibility. This protocol offers a substantial advance in modernizing the building sector and addressing housing difficulties with creative, sustainable solutions by merging digital design, real-time monitoring, and automated construction technologies.	3D printing, IoT sensors, Advanced Robotics, and Data Analytics	Utilization of interactive products and hyper-customization could further enhance the construction process and customer experiences.	[101]

Table 1. *Cont.*

Sr.No	Case Study	IR 4.0 Incorporation	Improvement through IR 5.0	Reference
2	**TECLA house construction in Italy** The implementation of cutting-edge Industry 4.0 technology in the construction industry is demonstrated by the IR 4.0 construction protocol used on the TECLA house construction project in Italy. This project embodies efficiency and sustainability because it uses a distinctive 3D printing technique and materials that are acquired locally. Real-time monitoring using IoT sensors and data analytics ensures accuracy and quality control during the construction process. The TECLA project shows how cutting-edge robotics, automation, and digital design have the potential to revolutionize the building sector and make it possible to build eco-friendly, individualized, and affordable housing solutions that meet contemporary needs.	3D printing and IoT sensors	With the incorporation of a responsive and distributive supply chain, resource efficacy can be improved. Moreover, through human–machine interactions, the creativity of humans and the efficiency of machines can further improve construction operations.	[102]
3	**Shanghai Chenshan Botanical Garden in China** Technology like IoT sensors for plant monitoring, data analytics for resource optimization, and automation for maintenance and guest services might all be used in a botanical garden to implement Industry 4.0 principles.	IoT sensors, Data Analytics, and Automation	Instead of completely employing machines, humans working with intelligent devices and robots should be adopted. It is about using cutting-edge technology like the Internet of Things (IoT) and big data to enable robots to assist people in working more effectively and quickly.	[103]

People are given more authority by Industry 5.0, which uses the resources at hand to produce more significant value. Industry 5.0 offers a holistic project delivery paradigm that incorporates key issues of concern to present to future project leaders, which is a significant advance above IR 4.0. Although the evolution to IR 5.0 is unavoidable, there are certain significant advantages that CEOs should consider when determining whether to hasten the transition inside their businesses and project teams:

1. Teams that are empowered by Industry 5.0 leverage human knowledge and innovation more frequently;
2. Using technology more selectively, teams switch from mass automation to flexible solutions that can automate, or mass customize, enabling performance at scale under a variety of scenarios;
3. Improved cooperation—teams use virtual co-location technology and seamless communication channels to work more effectively together;
4. More environmentally friendly construction techniques—project teams place a greater emphasis on sustainability, incorporating environmental and social effects into planning and decision-making.

4. Conceptual Framework

Human–economy-centric, sustainable, and resilient development are the three leading stakes of Industry 5.0 as proposed in the framework shown in Figure 8. Industry 4.0

was primarily focused on using technology to remove humans from the manufacturing process and only entrust them with supervision and control tasks, which caused production workers to perceive increasing automation and digitalization as a severe danger. The core of Industry 5.0 is the interaction of three sectors: technological, social, and ecological. Industry 5.0 emphasizes the importance of technology in driving economic growth (business). However, achieving corporate goals also involves achieving social goals both inside and outside of the workplace, such as those related to human–machine relations (social and ecological responsibility). System designers must create projects in Industry 5.0 with a "human-centered" rather than a "technology-centered" mindset. The potential for complex judgments to be made outside of the control of humans raises ethical concerns regarding the repercussions of the development of artificial intelligence. For human values and ethical considerations to be addressed as design needs rather than expenses in freshly created cyber–physical systems, it is necessary to analyze them in advance.

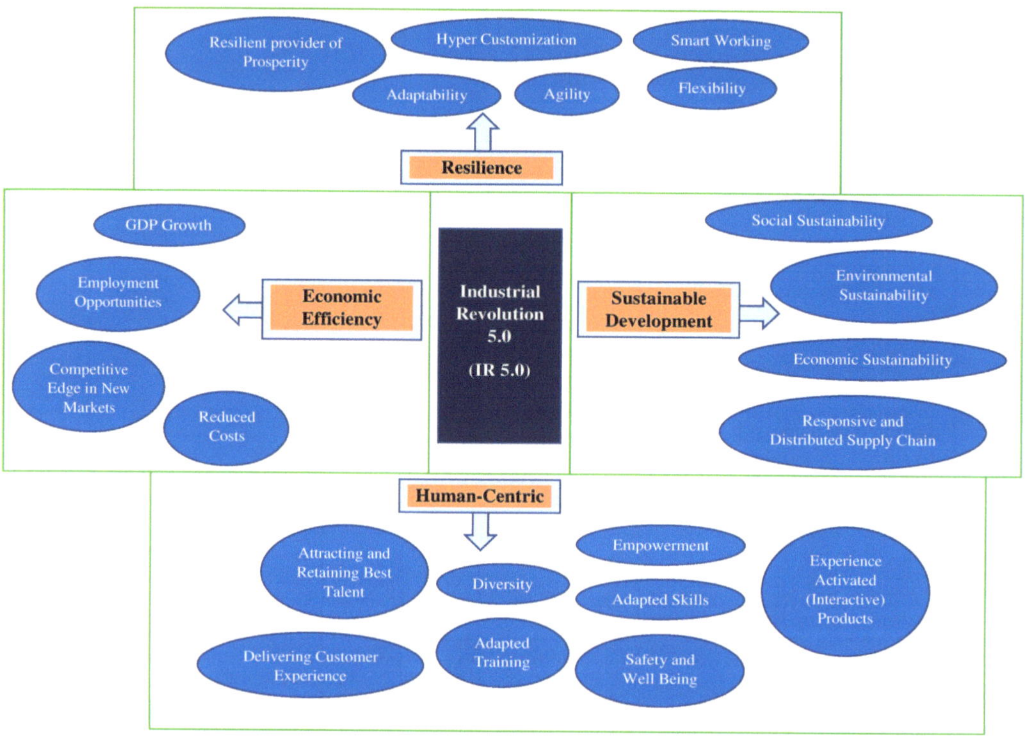

Figure 8. Framework for IR 5.0 Adoption.

Industrial Revolution 5.0 technologies are essential for increasing productivity and competitiveness in the manufacturing and construction sector. An illustration of this is the use of the Seru production system, a lean manufacturing strategy created in Japan. To optimize its production processes, Seru mainly relies on automation, IoT sensors, and data analytics. These technologies enable manufacturers to create highly adaptable and effective manufacturing cells where employees and machines work together seamlessly. Real-time data collected by sensors enables predictive maintenance and quick problem solutions. This improves product quality while simultaneously decreasing downtime. Manufacturing is now able to switch from mass production to highly customized and adaptable production methods, satisfying changing market demands while remaining cost-effective.

4.1. *IR 5.0 and Sustainability in Construction*

The Fifth Industrial Revolution, which will emphasize collaborative interaction between machines and people, is now upon us. Professionals are free to provide clients with value-added tasks because of personalization and the incorporation of collaborative robots. This most recent iteration includes greater resilience, a human-centric strategy, and an emphasis on sustainability in addition to manufacturing and construction methods. By actively pursuing initiatives to bring about change, Industry 5.0 expands sustainability beyond merely reducing, minimizing, or adapting against climate impact. This objective sometimes referred to as "Net Positive," strives to improve the planet by having businesses contribute to the solution rather than contributing to the problem or merely paying lip-service to objectives of sustainability through greenwashing. IR 5.0 has the potential to profoundly impact the economy. The first way that IR 5.0 can increase production is by streamlining procedures and lowering human error rates using robotic and automated equipment. This may lead to reduced project costs and quicker project completion schedules. Furthermore, AI-powered analytics facilitate better decision-making by offering immediate information on the efficiency of projects, the distribution of resources, and risk control. Additionally, the adoption of these technologies opens up opportunities for workers to upgrade their skills and acquire the new technical abilities needed to operate alongside sophisticated systems. It is important to understand that IR 5.0 encourages job growth in industries like software development and maintenance of these complex systems, despite concerns about job displacement brought on by automation.

Consequently, construction operations can be improved to cut down on waste, energy use, and environmental effects with the use of IR 5.0 technology. AI-enabled systems may analyze data and identify possible bottlenecks or defects in designs, maximizing resource efficiency and reducing material waste. Real-time monitoring of energy use and atmospheric conditions on building sites is also made possible by IoT devices, improving resource control and efficiency. Robotics is essential in automating monotonous or dangerous building operations for workers while assuring precision and minimizing human error. Industry experts may help create a greener infrastructure that supports sustainable development and a sustainable future by incorporating IR 5.0 concepts in construction practices with a sustainability lens.

4.1.1. Sustainable Development

By encouraging eco-friendly and effective practices, Industrial Revolution 5.0 has the potential to greatly contribute to sustainable development. Industries may cut back on resource use, cut emissions, and improve supply chains by integrating technology like AI, IoT, and renewable energy. Smart grids, for instance, can improve energy efficiency, and AI-driven logistics can reduce carbon footprints associated with transportation. Additionally, IR 5.0's emphasis on circular economies and sustainable materials promotes responsible resource management and lower waste production, bringing industrial advancement and long-term environmental and social sustainability goals into alignment.

4.1.2. Human-Centric

Strong emphasis is placed on human-centric design and technologies in Industrial Revolution 5.0. It combines automation, robotics, and AI to improve workers' productivity and well-being. For instance, wearable technology and exoskeletons can improve safety and comfort in physically demanding industries, while AI-driven systems can assist people with repetitive activities, decreasing workload and errors. A more inclusive and adaptable work environment is fostered through the creation of adaptive workspaces, where technology adjusts to human requirements and preferences, fostering a better work-life balance and overall job satisfaction. In IR 5.0, technology is used to enhance and empower the workforce, ensuring that innovation and advancement are focused on bettering people's lives.

4.1.3. Economic Efficiency

Using cutting-edge technology like artificial intelligence (AI), automation, and data analytics, Industrial Revolution 5.0 promotes economic efficiency. These developments improve resource allocation, lower operational costs, and optimize production processes For instance, AI-driven supply chain management assures just-in-time inventory and cost savings, while predictive maintenance enabled by IoT sensors reduces downtime and maintenance costs. Additionally, IR 5.0's customizable features help companies effectively satisfy the demands of each customer. This emphasis on accuracy, adaptability, and data-driven decision-making increases productivity while also fostering competition in a global economy that is continually changing.

4.1.4. Resilience

Industrial Revolution 5.0 encourages flexibility and risk management, which increases the resilience of industries and economies. Businesses may monitor and respond to interruptions in real time using cutting-edge technologies like AI, IoT, and digital twins, assuring business continuity. Predictive analytics, for instance, can spot possible supply chain problems, enabling early actions to safeguard essential resources. Additionally, by lowering sensitivity to environmental and energy-related problems, smart infrastructure, and the use of renewable energy boost resilience. Businesses and societies are better prepared to endure and recover from diverse crises thanks to IR 5.0, which encourages a more flexible and responsive approach to both planned and unforeseen disturbances.

4.2. IR 5.0 and Managerial Insights

The development of cutting-edge technology, including automation, artificial intelligence, and big data analytics, has had an enormous effect on managerial insights. Today's managers can view enormous volumes of real-time data that may be examined to learn a great deal about consumer behavior, market dynamics, and operational effectiveness. This enables them to execute successful strategies that spur growth and enhance organizational efficiency while making data-driven decisions [59]. Furthermore, IR 5.0 empowers managers to make proactive decisions by utilizing predictive analytics and machine learning systems to estimate future events more accurately. Overall, IR 5.0 equips managers with useful information from tech-based systems and resources, enabling quick decisions that may be adjusted to business contexts that are evolving frequently.

5. Conclusions

The impact of the fifth revolution on many industries, such as the construction industry, cannot yet be determined due to the unavailability of adequate resources and skills. Although the world is shifting from IR 4.0 to IR 5.0, the concept of IR 5.0 is still in its early stages as industry professionals are working to streamline and ease the integration of humans with machines instead of replacing them. Considering the significance of the adoption of IR 5.0 in the construction industry, the current review study is conducted to assess the transformational changes within different industrial revolutions and a way forward for the adoption of IR 5.0 in the construction industry. Furthermore, through a policy framework for decision-makers, the present study has tried to fill an important role by providing a way forward for the construction industry professionals to transform themselves in the era of IR 5.0

The findings of the current study highlight that Industry 5.0 can promote innovation in how we interpret large data by incorporating changes into the design of future innovation ecosystems. Without compromising the viability and security of an innovative ecosystem and its constituent parts, Industry 5.0 envisions a world of linked networks. This revolution aims to provide accountability while utilizing the most automation and big data analysis possible. Robots have historically played significant roles in manufacturing and production facilities, but the most recent generation of collaborative robots is fitted with sensors that allow them to perform tasks other than mechanical and repetitive ones. The results of the current review also indicate that in some applications, robots collaborate with people to promote a higher level of product customization. The integration of human intelligence and cognitive computing with machines will be at the heart of the Fifth Industrial Revolution. Innovation in construction has been critical to this transformation, with new materials, techniques, and tools being developed to improve productivity and reduce environmental impact. For example, the use of 3D printing in construction is enabling faster and more efficient construction of buildings and infrastructure, while also reducing waste and energy consumption.

Similarly, the study has also identified that another area of innovation in construction is the development of smart buildings that use sensors and data analytics to optimize energy use and improve the comfort and safety of occupants. The use of advanced materials such as biodegradable composites and self-healing concrete is also enabling more sustainable construction practices. The interaction between people and machines alters many aspects of production and has an impact on the economy and ecosystems. Manufacturers are under pressure to reduce costs due to fierce competition, which can be achieved by producing zero waste. Zero waste production, which emphasizes the human component of manufacturing, helps maintain a healthy environment. It is time to go from Industry 4.0 to Industry 5.0, where robots and human intelligence will enhance cyber–physical systems, considering the societal implications of Industry 4.0. The key differences between IR 4.0 and IR 5.0 include the shift of focus from integrating machines to delivering customer experiences, mass customization to hyper customization, intelligent supply chains to responsive and distributive supply chains, smart products to interactive products, and manpower distanced from factories to the return of manpower to their workstations. Thus, future work will be conducted in virtually every industry and will be based on the massive amounts of data supplied by these IoT devices. The industrial process is reinforced with value addition that promotes mass personalization by reintroducing human interaction with collaborative robots. Although the fifth revolution is still in its early stages, businesses are attempting to go forward with it sooner because they want to outperform their rivals. As a result, the knowledge gained from the analysis of Industry 5.0 is further distilled to define a new research program.

5.1. Practical Implications

A review of the potential effects Industrial Revolution 5.0 can have on the construction industry suggests some real-world implications, including the adoption of cutting-edge technologies like AI, IoT, and automation for increased project efficiency, the incorporation of digital twins and BIM for better project management, the use of smart materials for sustainable construction, and the use of renewable energy sources for eco-friendly infrastructure. Adopting these advances can result in more effective, environmentally friendly, and technologically sophisticated construction practices, ensuring the sector is competitive and prepared for the future.

5.2. Managerial Implications

A study on the future of building through Industrial Revolution 5.0 has important managerial ramifications. Construction managers and leaders must invest in the training and development of their employees to be ready to react to rapid technological advances. Additionally, they want to reevaluate conventional project management strategies and switch to more collaborative, data-driven techniques to take advantage of innovations like BIM, digital twins, and predictive analytics. Additionally, as smart materials, renewable energy integration, and modular building methods become crucial elements of future projects, innovation and sustainability should be at the center of their strategic planning. Construction managers can position their businesses for success in the changing construction landscape, fueled by Industrial Revolution 5.0 technology, by embracing these changes.

5.3. Recommendations and Limitations for Future Research

It has been highlighted already that the concept of IR 5.0 is novel and is in its early phase of development. Through this review study, IR 5.0 adoption has been discussed comprehensively, however, the following are various recommendations and future agenda items in the context of the current subject area:

1. It is recommended to explore the education and training workshop requirements for preparing the workforce of the future to be equipped with the relevant skills and knowledge in the era of IR 5.0;
2. The adoption of IR 5.0 in the context of the construction industry should be explored with a greater focus on the economy, society, and the environment;
3. Government agencies, policymakers, decision-makers, stakeholders, and investors should collaborate to increase investments and minimize the regulatory hurdles in the adoption or implementation of IR 5.0-enabling technologies;
4. The concept of IR 5.0 should be explored with the inclusion of real-world case studies Although, at this moment, due to the unavailability of resources and skills, it would be a difficult job, but still, making some progress is better than no progress at all;
5. The concept of IR 5.0 has not been assessed adequately in the context of sustainability. The challenges and limitations are abundant, but they can be removed through the engagement of major industry stakeholders from public and private organizations.

Author Contributions: Conceptualization, M.A.M. and M.I.; methodology, M.A.M., M.I., W.S.A., A.M. and M.G.; formal analysis, M.A.M., M.I., W.S.A. and M.G.; investigation, M.A.M., M.I. and M.G.; resources, M.A.M., M.I. and W.S.A.; data curation, M.A.M., M.I., W.S.A., A.M. and M.G.; manuscript preparation, M.A.M., M.I. and M.G.; review and editing of the draft, M.A.M., M.I., W.S.A., A.M. and M.G.; supervision and project administration, M.A.M., W.S.A. and A.M. All authors have read and agreed to the published version of the manuscript.

Funding: This research received no external funding.

Institutional Review Board Statement: Not applicable.

Informed Consent Statement: Not applicable.

Acknowledgments: The authors would like to thank Universiti Teknologi PETRONAS (UTP) for the support provided in this research.

Conflicts of Interest: The authors declare no conflict of interest.

References

1. Zamani, S.H.; Rahman, R.; Fauzi, M.; Yusof, L. Effect of COVID-19 on building construction projects: Impact and response mechanisms. In *IOP Conference Series: Earth and Environmental Science*; IOP Publishing: Bristol, UK, 2021; Volume 682, p. 012049.
2. Craveiroa, F.; Duartec, J.P.; Bartoloa, H.; Bartolod, P.J.J.S.D. Additive manufacturing as an enabling technology for digital construction: A perspective on Construction 4.0. *Sustain. Dev.* **2019**, *4*, 251–267. [CrossRef]
3. Ilic, D.; Markovic, B.; Milosevic, D. Strategic business transformation: An industry 4.0 perspective. *Econ. Law* **2017**, *7*, 49.
4. Musarat, M.A.; Alaloul, W.S.; Liew, M. Impact of inflation rate on construction projects budget: A review. *Ain Shams Eng. J.* **2021**, *12*, 407–414. [CrossRef]
5. Musarat, M.A.; Alaloul, W.S.; Liew, M.; Maqsoom, A.; Qureshi, A.H. Investigating the impact of inflation on building materials prices in construction industry. *J. Build. Eng.* **2020**, *32*, 101485. [CrossRef]
6. Alawag, A.M.; Alaloul, W.S.; Liew, M.; Musarat, M.A.; Baarimah, A.O.; Saad, S.; Ammad, S. Critical success factors influencing total quality management in industrialised building system: A case of malaysian construction industry. *Ain Shams Eng. J.* **2023**, *14*, 101877. [CrossRef]
7. Musarat, M.A.; Alaloul, W.S.; Cher, L.S.; Qureshi, A.H.; Alawag, A.M.; Baarimah, A.O. Applications of Building Information Modelling in the Operation and Maintenance Phase of Construction Projects: A Framework for the Malaysian Construction Industry. *Sustainability* **2023**, *15*, 5044. [CrossRef]
8. Locatelli, G.; Mancini, M.; Romano, E. Systems engineering to improve the governance in complex project environments. *Int. J. Proj. Manag.* **2014**, *32*, 1395–1410. [CrossRef]
9. Musarat, M.A.; Alaloul, W.S.; Irfan, M.; Sreenivasan, P.; Rabbani, M.B. Health and safety improvement through Industrial Revolution 4.0: Malaysian construction industry case. *Sustainability* **2022**, *15*, 201. [CrossRef]

10. Tavares, M.C.; Azevedo, G.; Marques, R.P. The Challenges and Opportunities of Era 5.0 for a More Humanistic and Sustainable Society—A Literature Review. *Societies* **2022**, *12*, 149. [CrossRef]
11. Musarat, M.A.; Alaloul, W.S.; Hameed, N.; Qureshi, A.H.; Wahab, M.M.A. Efficient Construction Waste Management: A Solution through Industrial Revolution (IR) 4.0 Evaluated by AHP. *Sustainability* **2022**, *15*, 274. [CrossRef]
12. Alaloul, W.S.; Liew, M.S.; Zawawi, N.A.W.A.; Mohammed, B.S. Industry revolution IR 4.0: Future opportunities and challenges in construction industry. In *MATEC Web of Conferences*; EDP Sciences: Castanet-Tolosan, France, 2018; Volume 203, p. 02010.
13. Harley, C.K. Reassessing the industrial revolution: A macro view. In *The British Industrial Revolution*; Routledge: Abingdon, UK, 2018; pp. 160–205.
14. Chen, Q.; de Soto, B.G.; Adey, B.T. Construction automation: Research areas, industry concerns and suggestions for advancement. *Autom. Constr.* **2018**, *94*, 22–38. [CrossRef]
15. Hossain, M.A.; Nadeem, A. Towards digitizing the construction industry: State of the art of construction 4.0. In Proceedings of the ISEC, Riverside, CA, USA, 8 September 2019; Volume 10.
16. Sherwani, F.; Asad, M.M.; Ibrahim, B.S.K.K. Collaborative robots and industrial revolution 4.0 (ir 4.0). In Proceedings of the 2020 International Conference on Emerging Trends in Smart Technologies (ICETST), Karachi, Pakistan, 26–27 March 2020; IEEE: New York, NY, USA, 2020; pp. 1–5.
17. More, C. *Understanding the Industrial Revolution*; Routledge: Abingdon, UK, 2002.
18. Sawhney, A.; Riley, M.; Irizarry, J. (Eds.) *Construction 4.0: An Innovation Platform for the Built Environment*; Routledge: Abingdon, UK, 2020; ISBN 9780429398100.
19. Kamaruzaman, M.; Hamid, R.; Mutalib, A.; Rasul, M. Comparison of engineering skills with IR 4.0 skills. *Int. J. Online Biomed. Eng. Ijoe* **2019**, *15*, 15–28. [CrossRef]
20. Yavari, F.; Pilevari, N. Industry revolutions development from Industry 1.0 to Industry 5.0 in manufacturing. *J. Ind. Strateg. Manag.* **2020**, *5*, 44–63.
21. Atkeson, A.; Kehoe, P.J. *The Transition to a New Economy after the Second Industrial Revolution*; National Bureau of Economic Research Cambridge: Cambridge, MA, USA, 2001.
22. Salah, B. Real-Time Implementation of a Fully Automated Industrial System Based on IR 4.0 Concept. *Actuators* **2021**, *10*, 318.
23. Taalbi, J. Origins and pathways of innovation in the third industrial revolution. *Ind. Corp. Change* **2019**, *28*, 1125–1148.
24. Qureshi, A.H.; Alaloul, W.S.; Manzoor, B.; Musarat, M.A.; Saad, S.; Ammad, S. Implications of machine learning integrated technologies for construction progress detection under industry 4.0 (IR 4.0). In Proceedings of the 2020 Second International Sustainability and Resilience Conference: Technology and Innovation in Building Designs (51154), Sakheer, Bahrain, 11–12 November 2020; IEEE: New York, NY, USA, 2020; pp. 1–6.
25. Seltsikas, P.; Currie, W.L. Evaluating the application service provider (ASP) business model: The challenge of integration. In Proceedings of the 35th Annual Hawaii International Conference on System Sciences, Big Island, Hawaii, 10 January 2002; IEEE: New York, NY, USA, 2002; pp. 2801–2809.
26. Marques, I.A. Shaping the Future of Industry: How Can Vanguard Technology Improve Our Lives? 2021. Available online: https://run.unl.pt/handle/10362/138158 (accessed on 20 March 2023).
27. Alaloul, W.S.; Liew, M.; Zawawi, N.A.W.A.; Kennedy, I.B. Industrial Revolution 4.0 in the construction industry: Challenges and opportunities for stakeholders. *Ain Shams Eng. J.* **2020**, *11*, 225–230. [CrossRef]
28. Aghimien, D.; Aigbavboa, C.; Matabane, K. Impediments of the fourth industrial revolution in the South African construction industry. In *Collaboration and Integration in Construction, Engineering, Management and Technology*; Springer: Berlin/Heidelberg, Germany, 2021; pp. 223–227.
29. Carvalho, N.; Chaim, O.; Cazarini, E.; Gerolamo, M. Manufacturing in the fourth industrial revolution: A positive prospect in sustainable manufacturing. *Procedia Manuf.* **2018**, *21*, 671–678. [CrossRef]
30. Shahroom, A.A.; Hussin, N. Industrial revolution 4.0 and education. *Int. J. Acad. Res. Bus. Soc. Sci.* **2018**, *8*, 314–319. [CrossRef]
31. Erol, S.; Jäger, A.; Hold, P.; Ott, K.; Sihn, W. Tangible Industry 4.0: A scenario-based approach to learning for the future of production. *Procedia CiRp* **2016**, *54*, 13–18. [CrossRef]
32. De Lange, M.; De Waal, M. Owning the city: New media and citizen engagement in urban design. In *Urban Land Use*; Apple Academic Press: Cambridge, MA, USA, 2017; pp. 109–130.
33. Maresova, P.; Soukal, I.; Svobodova, L.; Hedvicakova, M.; Javanmardi, E.; Selamat, A.; Krejcar, O. Consequences of industry 4.0 in business and economics. *Economies* **2018**, *6*, 46. [CrossRef]
34. Bloem, J.; Van Doorn, M.; Duivestein, S.; Excoffier, D.; Maas, R.; Van Ommeren, E. The fourth industrial revolution. *Things Tighten* **2014**, *8*, 11–15.
35. Popkova, E.G.; Ragulina, Y.V.; Bogoviz, A.V. Fundamental differences of transition to industry 4.0 from previous industrial revolutions. In *Industry 4.0: Industrial Revolution of the 21st Century*; Springer: Berlin/Heidelberg, Germany, 2019; pp. 21–29.
36. Dallasega, P.; Rauch, E.; Linder, C. Industry 4.0 as an enabler of proximity for construction supply chains: A systematic literature review. *Comput. Ind.* **2018**, *99*, 205–225. [CrossRef]
37. Forcael, E.; Ferrari, I.; Opazo-Vega, A.; Pulido-Arcas, J.A. Construction 4.0: A literature review. *Sustainability* **2020**, *12*, 9755. [CrossRef]

38. Rastogi, V.; Srivastava, S.; Mishra, M.; Thukral, R. Predictive Maintenance for SME in Industry 4.0. In Proceedings of the 2020 Global Smart Industry Conference (GloSIC), Chelyabinsk, Russia, 17–19 November 2020; IEEE: New York, NY, USA, 2020; pp. 382–390.
39. Vu, T.L.A.; Le, T.Q. Development orientation for higher education training programme of mechanical engineering in industrial revolution 4.0: A perspective in vietnam. *Mech. Eng. Res. Dev* **2019**, *42*, 68–70.
40. Jung, E. 4th Industrial Revolution and Boundry: Challenges and Opportunities. In Proceedings of the 2018 IEEE International Electron Devices Meeting (IEDM), San Francisco, CA, USA, 1–5 December 2018; IEEE: New York, NY, USA, 2018; pp. 1.1.1–1.1.10.
41. Oesterreich, T.D.; Teuteberg, F. Understanding the implications of digitisation and automation in the context of Industry 4.0: A triangulation approach and elements of a research agenda for the construction industry. *Comput. Ind.* **2016**, *83*, 121–139. [CrossRef]
42. Bücker, I.; Hermann, M.; Pentek, T.; Otto, B. Towards a methodology for Industrie 4.0 transformation. In Proceedings of the International Conference on Business Information Systems, Hannover, Germany, 14–17 June 2016; Springer: Berlin/Heidelberg, Germany, 2016; pp. 209–221.
43. Liu, Y.; Ma, X.; Shu, L.; Hancke, G.P.; Abu-Mahfouz, A.M. From Industry 4.0 to Agriculture 4.0: Current status, enabling technologies, and research challenges. *J. Manuf. Syst.* **2020**, *17*, 4322–4334. [CrossRef]
44. Rymarczyk, J. Technologies, opportunities and challenges of the industrial revolution 4.0: Theoretical considerations. *Entrep. Bus. Econ. Rev.* **2020**, *8*, 185–198. [CrossRef]
45. Insights, D.J. The Fourth Industrial Revolution: At the Intersection of Readiness and Responsibility. 2020. Available online: https://www2.deloitte.com (accessed on 20 March 2023).
46. Moon, Y.; Seol, S.-S. Evaluation of the Theory of the 4 th Industrial Revolution. *Asian J. Innov. Policy* **2017**, *6*, 245–261.
47. Imran, M.; Salisu, I.; Aslam, H.D.; Iqbal, J.; Hameed, I. Resource and information access for SME sustainability in the era of IR 4.0: The mediating and moderating roles of innovation capability and management commitment. *Processes* **2019**, *7*, 211. [CrossRef]
48. Tien, N.H.; Nogalski, B. Developing high quality human resource to benefit from CP-TPP and IR 4.0. *Int. J. Res. Manag.* **2019**, *1*, 4–6. [CrossRef]
49. Potočan, V.; Mulej, M.; Nedelko, Z. Society 5.0: Balancing of Industry 4.0, economic advancement and social problems. *Kybernetes* **2020**, *50*, 794–811. [CrossRef]
50. Paschek, D.; Mocan, A.; Draghici, A. Industry 5.0—The expected impact of next industrial revolution. In Proceedings of the Thriving on Future Education, Industry, Business, and Society, Proceedings of the MakeLearn and TIIM International Conference, Piran, Slovenia, 15–17 May 2019; pp. 15–17.
51. Humayun, M. Industrial Revolution 5.0 and the Role of Cutting Edge Technologies. *Int. J. Adv. Comput. Sci. Appl.* **2021**, *12*, 0121276. [CrossRef]
52. Sarfraz, Z.; Sarfraz, A.; Iftikar, H.M.; Akhund, R. Is COVID-19 pushing us to the fifth industrial revolution (society 5.0)? *Pak. J. Med. Sci.* **2021**, *37*, 591. [CrossRef] [PubMed]
53. Tien, N.H.; Phong, V.M.T.; Thoi, B.; Duc, L.D.M. Developing High Quality Human Resource to Take Advantages from CPTPP and IR 4.0. *Int. J. Res. Financ. Manag.* **2019**, *2*, 67–69. [CrossRef]
54. Poszytek, P.J.S. The Landscape of Scientific Discussions on the Competencies 4.0 Concept in the Context of the 4th Industrial Revolution—A Bibliometric Review. *Sustainability* **2021**, *13*, 6709. [CrossRef]
55. Chen, S.; Xu, H. Meta-heuristic algorithm-based human resource information management system design and development for industrial revolution 5.0. *Soft Comput.* **2022**, *27*, 4093–4105. [CrossRef]
56. Leng, J.; Sha, W.; Wang, B.; Zheng, P.; Zhuang, C.; Liu, Q.; Wuest, T.; Mourtzis, D.; Wang, L. Industry 5.0: Prospect and retrospect. *J. Manuf. Syst.* **2022**, *65*, 279–295. [CrossRef]
57. Sindhwani, R.; Afridi, S.; Kumar, A.; Banaitis, A.; Luthra, S.; Singh, P.L. Can industry 5.0 revolutionize the wave of resilience and social value creation? A multi-criteria framework to analyze enablers. *Technol. Soc.* **2022**, *68*, 101887. [CrossRef]
58. Demir, K.A.; Döven, G.; Sezen, B. Industry 5.0 and human-robot co-working. *Procedia Comput. Sci.* **2019**, *158*, 688–695. [CrossRef]
59. George, A.S.; George, A.H. Industrial revolution 5.0: The transformation of the modern manufacturing process to enable man and machine to work hand in hand. *J. Seybold Rep. ISSN NO* **2020**, *1533*, 9211.
60. Golić, Z. Finance and artificial intelligence: The fifth industrial revolution and its impact on the financial sector. *Zb. Rad. Ekon. Fak. Istočnom Sarajev.* **2019**, *8*, 67–81. [CrossRef]
61. Lai, C.; Chundra, U.; Lee, M. Teaching and learning based on IR 4.0: Readiness of attitude among polytechnics lecturers. *J. Phys. Conf. Ser.* **2020**, *1529*, 032105.
62. Maddikunta, P.K.R.; Pham, Q.-V.; Prabadevi, B.; Deepa, N.; Dev, K.; Gadekallu, T.R.; Ruby, R.; Liyanage, M. Industry 5.0: A survey on enabling technologies and potential applications. *J. Ind. Inf. Integr.* **2022**, *26*, 100257. [CrossRef]
63. Marinelli, M. From Industry 4.0 to Construction 5.0: Exploring the Path towards Human–Robot Collaboration in Construction. *Systems* **2023**, *11*, 152. [CrossRef]
64. Yitmen, I.; Almusaed, A.; Alizadehsalehi, S. Investigating the Causal Relationships among Enablers of the Construction 5.0 Paradigm: Integration of Operator 5.0 and Society 5.0 with Human-Centricity, Sustainability, and Resilience. *Sustainability* **2023**, *15*, 9105. [CrossRef]
65. Hu, C.; Liu, P.; Yang, H.; Yin, S.; Ullah, K. A novel evolution model to investigate the collaborative innovation mechanism of green intelligent building materials enterprises for construction 5.0. *AIMS Math.* **2023**, *8*, 8117–8143. [CrossRef]

66. Pan, M.; Linner, T.; Pan, W.; Cheng, H.; Bock, T. A framework of indicators for assessing construction automation and robotics in the sustainability context. *J. Clean. Prod.* **2018**, *182*, 82–95. [CrossRef]
67. Ikudayisi, A.E.; Chan, A.P.; Darko, A.; Yomi, M. Integrated practices in the Architecture, Engineering, and Construction industry: Current scope and pathway towards Industry 5.0. *J. Build. Eng.* **2023**, *73*, 106788. [CrossRef]
68. Roach, D.J.; Hamel, C.M.; Dunn, C.K.; Johnson, M.V.; Kuang, X.; Qi, H.J. The m4 3D printer: A multi-material multi-method additive manufacturing platform for future 3D printed structures. *Addit. Manuf.* **2019**, *29*, 100819. [CrossRef]
69. Khalid, A.; Khan, Z.H.; Idrees, M.; Kirisci, P.; Ghrairi, Z.; Thoben, K.-D.; Pannek, J. Understanding vulnerabilities in cyber physical production systems. *Int. J. Comput. Integr. Manuf.* **2022**, *35*, 569–582. [CrossRef]
70. Chen, Y.; Huang, D.; Liu, Z.; Osmani, M.; Demian, P. Construction 4.0, Industry 4.0, and Building Information Modeling (BIM) for sustainable building development within the smart city. *Sustainability* **2022**, *14*, 10028. [CrossRef]
71. Baghalzadeh Shishehgarkhaneh, M.; Keivani, A.; Moehler, R.C.; Jelodari, N.; Roshdi Laleh, S. Internet of Things (IoT), Building Information Modeling (BIM), and Digital Twin (DT) in construction industry: A review, bibliometric, and network analysis. *Buildings* **2022**, *12*, 1503. [CrossRef]
72. Nihalani, S.; Joshi, U.; Meeruty, A. Smart materials for sustainable and smart infrastructure. *Mater. Sci. Forum* **2019**, *969*, 278–283.
73. Apicella, A.; Ianniello, D.; Aversa, R. Application of nanostructured smart materials in sustainable buildings. *Int. J. Sustain. Manuf.* **2010**, *2*, 66–79. [CrossRef]
74. Kaygusuz, A.; Keles, C.; Alagoz, B.B.; Karabiber, A. Renewable energy integration for smart sites. *Energy Build.* **2013**, *64*, 456–462. [CrossRef]
75. Reja, V.K.; Varghese, K. Impact of 5G technology on IoT applications in construction project management. *ISARC Proc. Int. Symp. Autom. Robot. Constr.* **2019**, *36*, 209–217.
76. Letourneau-Guillon, L.; Camirand, D.; Guilbert, F.; Forghani, R. Artificial intelligence applications for workflow, process optimization and predictive analytics. *Neuroimaging Clin.* **2020**, *30*, e1–e15. [CrossRef] [PubMed]
77. Akter, S.; Michael, K.; Uddin, M.R.; McCarthy, G.; Rahman, M. Transforming business using digital innovations: The application of AI, blockchain, cloud and data analytics. *Ann. Oper. Res.* **2022**, *308*, 7–39. [CrossRef]
78. Egege, C.O. Off-site modular construction as a method of improving construction quality and safety. *Int. J. Struct. Civ. Eng. Res.* **2018**, *7*, 259–268. [CrossRef]
79. Xu, X.; Lu, Y.; Vogel-Heuser, B.; Wang, L. Industry 4.0 and Industry 5.0—Inception, conception and perception. *J. Manuf. Syst.* **2021**, *61*, 530–535. [CrossRef]
80. Akundi, A.; Euresti, D.; Luna, S.; Ankobiah, W.; Lopes, A.; Edinbarough, I. State of Industry 5.0—Analysis and identification of current research trends. *Appl. Syst. Innov.* **2022**, *5*, 27. [CrossRef]
81. Skobelev, P.O.; Borovik, S.Y. On the way from Industry 4.0 to Industry 5.0: From digital manufacturing to digital society. *Ind. 4.0* **2017**, *2*, 307–311.
82. Javaid, M.; Haleem, A.; Singh, R.P.; Haq, M.I.U.; Raina, A.; Suman, R. Industry 5.0: Potential applications in COVID-19. *J. Ind. Integr. Manag.* **2020**, *5*, 507–530. [CrossRef]
83. Nahavandi, S. Industry 5.0—A human-centric solution. *Sustainability* **2019**, *11*, 4371. [CrossRef]
84. Gagnidze, I. Industry 4.0 and industry 5.0: Can clusters deal with the challenges?(A systemic approach). *Kybernetes* **2023**, *52*, 2270–2287. [CrossRef]
85. Golovianko, M.; Terziyan, V.; Branytskyi, V.; Malyk, D. Industry 4.0 vs. Industry 5.0: Co-existence, Transition, or a Hybrid. *Procedia Comput. Sci.* **2023**, *217*, 102–113. [CrossRef]
86. Aslam, F.; Aimin, W.; Li, M.; Ur Rehman, K. Innovation in the era of IoT and industry 5.0: Absolute innovation management (AIM) framework. *Information* **2020**, *11*, 124. [CrossRef]
87. Lallée, S.; Pattacini, U.; Lemaignan, S.; Lenz, A.; Melhuish, C.; Natale, L.; Skachek, S.; Hamann, K.; Steinwender, J.; Sisbot, E.A. Towards a platform-independent cooperative human robot interaction system: III an architecture for learning and executing actions and shared plans. *IEEE Trans. Auton. Ment. Dev.* **2012**, *4*, 239–253. [CrossRef]
88. Ugochukwu, N.A.; Goyal, S.; Arumugam, S. Blockchain-Based IoT-Enabled System for Secure and Efficient Logistics Management in the Era of IR 4.0. *J. Nanomater.* **2022**, *2022*, 7295395. [CrossRef]
89. Aheleroff, S.; Huang, H.; Xu, X.; Zhong, R.Y. Toward sustainability and resilience with Industry 4.0 and Industry 5.0. *Front. Manuf. Technol.* **2022**, *2*, 951643. [CrossRef]
90. Pratiwi, R.; Cakranegara, P.A.; Kamal, I.; Surahman, S. The Strategic Role of Islamic Law in Strengthening Entrepreneurial Character Education in The Era of The Industrial Revolution 5.0. *J. Manaj. Bisnis* **2022**, *9*, 234–241.
91. Adel, A. Future of industry 5.0 in society: Human-centric solutions, challenges and prospective research areas. *J. Cloud Comput.* **2022**, *11*, 40. [CrossRef]
92. Kasinathan, P.; Pugazhendhi, R.; Elavarasan, R.M.; Ramachandaramurthy, V.K.; Ramanathan, V.; Subramanian, S.; Kumar, S.; Nandhagopal, K.; Raghavan, R.R.V.; Rangasamy, S. Realization of Sustainable Development Goals with Disruptive Technologies by Integrating Industry 5.0, Society 5.0, Smart Cities and Villages. *Sustainability* **2022**, *14*, 15258. [CrossRef]
93. Müller, J.J.E.C. *Enabling Technologies for Industry 5.0*; European Commission: Brussels, Belgium, 2020; pp. 8–10.
94. Longo, F.; Padovano, A.; Umbrello, S.J.A.S. Value-oriented and ethical technology engineering in industry 5.0: A human-centric perspective for the design of the factory of the future. *Appl. Sci.* **2020**, *10*, 4182. [CrossRef]

95. Ghobakhloo, M.; Iranmanesh, M.; Mubarak, M.F.; Mubarik, M.; Rejeb, A.; Nilashi, M. Identifying industry 5.0 contributions to sustainable development: A strategy roadmap for delivering sustainability values. *Sustain. Prod. Consum.* **2022**, *33*, 716–737 [CrossRef]
96. Chen, X.; Eder, M.A.; Shihavuddin, A.; Zheng, D. A human-cyber-physical system toward intelligent wind turbine operation and maintenance. *Sustainability* **2021**, *13*, 561. [CrossRef]
97. Fazal, N.; Haleem, A.; Bahl, S.; Javaid, M.; Nandan, D. Digital Management Systems in Manufacturing Using Industry 5.0 Technologies. In *Advancement in Materials, Manufacturing and Energy Engineering, Vol. II*; Springer: Berlin/Heidelberg, Germany, 2022; pp. 221–234.
98. Hanif, M.I.; Iftikhar, L. Post COVID-19 industrial revolution 5.0. The dawn of cobot, chipbot and curbot. *Pak. J. Surg.* **2020**, *1*, 122–126. [CrossRef]
99. Saniuk, S.; Grabowska, S.; Grebski, W. Knowledge and Skills Development in the Context of the Fourth Industrial Revolution Technologies: Interviews of Experts from Pennsylvania State of the USA. *Energies* **2022**, *15*, 2677. [CrossRef]
100. Margherita, E.G.; Braccini, A.M. Socio-technical perspectives in the Fourth Industrial Revolution-Analysing the three main visions: Industry 4.0, the socially sustainable factory of Operator 4.0 and Industry 5.0. In Proceedings of the 7th International Workshop on Socio-Technical Perspective in IS Development (STPIS 2021), Trento, Italy, 14–15 October 2021.
101. Sakin, M.; Kiroglu, Y.C. 3D Printing of Buildings: Construction of the Sustainable Houses of the Future by BIM. *Energy Procedia* **2017**, *134*, 702–711. [CrossRef]
102. Es-sebyty, H.; Igouzal, M.; Ferretti, E. Improving stability of an ecological 3D-printed house-a case study in Italy. *J. Achiev. Mater. Manuf. Eng.* **2022**, *111*, 18–25.
103. Hu, Y.; Yang, S.; Yang, J.; Vincent, G.; Chen, X. Botanical gardens support urban sustainable development: A case study from the Shanghai Chenshan Botanical Garden. *Biodivers. Sci.* **2017**, *25*, 951. [CrossRef]

sustainability MDPI

Review

A Systematic Literature Review of Research on Social Procurement in the Construction and Infrastructure Sector: Barriers, Enablers, and Strategies

Catherine Xiaocui Lou [1,*], Riccardo Natoli [1], David Goodwin [2], Barbara Bok [3], Fang Zhao [4] and Peng Zhang [5]

1 Institute for Sustainable Industries & Liveable Cities, Victoria University, Melbourne, VIC 3000, Australia; riccardo.natoli@vu.edu.au
2 Nottingham University Business School, University of Nottingham Malaysia, Semenyih 43500, Malaysia; david.goodwin@nottingham.edu.my
3 Faculty of Health, Arts and Design, Swinburne University of Technology, Hawthorn, VIC 3122, Australia; bbok@swin.edu.au
4 Tobin College of Business, St. John's University, New York, NY 11432, USA; ceo@strategyandops.net
5 School of Engineering, Design and Built Environment, Western Sydney University, Penrith, NSW 2751, Australia; p.zhang@westernsydney.edu.au
* Correspondence: catherine.lou@vu.edu.au

Abstract: In Australia, a new feature of public policy is the requirement by governments that large-scale infrastructure projects integrate social procurement practices that alter the traditional focus on balancing price and quality. Social procurement has been gradually developing in practice, but the academic literature has not kept pace. Although past research has identified some of the barriers affecting social procurement implementation in the construction industry, the nature of the barriers impeding its proliferation has not to date been systematically reviewed. This paper undertakes a review of the social procurement literature published from January 2012 to 30 June 2022, with 49 papers chosen under selective criteria. This critical review employs the "Preferred Reporting Items for Systematic Reviews and Meta-Analyses" (PRISMA) technique to retrieve secondary data on social procurement from available peer-reviewed academic papers through three databases (Scopus, EBSCOhost, Web of Science). The literature analysis focuses on three themes: (1) barriers; (2) enablers; and (3) strategies to overcome the barriers. The paper finds that social procurement as a field of practice is evolving and expanding, but its role in contributing to social value creation remains an under-theorised concept. Recommendations for practice and future research are identified, including the need to measure the real-world impacts of policy.

Keywords: barriers; enablers; construction industry; infrastructure; social procurement

Citation: Lou, C.X.; Natoli, R.; Goodwin, D.; Bok, B.; Zhao, F.; Zhang, P. A Systematic Literature Review of Research on Social Procurement in the Construction and Infrastructure Sector: Barriers, Enablers, and Strategies. *Sustainability* **2023**, *15*, 12964. https://doi.org/10.3390/su151712964

Academic Editor: Antonio Caggiano

Received: 15 July 2023
Revised: 19 August 2023
Accepted: 22 August 2023
Published: 28 August 2023

1. Introduction

Traditionally, across a range of industries, procurement activities have focused on balancing price and quality. Increasingly, with the assistance of government initiatives, there has been an observable shift in focus in many large-scale procurement programs, drawing upon the social benefit potential of procurement. Social procurement differs from traditional procurement. It provides social benefits to local communities, in addition to the direct contribution of product and service purchasing activities [1]. Although government expenditure has long been recognised for its potential to deliver social impact via sustainable public procurement [2], in recent times, governments have used their considerable purchasing power to influence supply chains indirectly by mandating social outcome conditions in their contracts with suppliers [3].

Social procurement has been required by many government-initiated projects. For example, the State Government of Victoria in Australia announced a social procurement framework and associated policies in 2018, requesting all its departments and agencies to embed a prescribed social procurement framework and approach within their buying activities [4]. Similar initiatives have since been established in other Australian states, such as New South Wales [5], Western Australia [6], and Queensland [7]. Consequently, across the Australian continent, an increasing number of projects integrate social value creation elements into their processes. In the United Kingdom, Social Value Act 2012 requires all public bodies to consider how what they are proposing to buy might improve economic, social, and environmental well-being [8]. A policy procurement note states that all central government departments and agencies must evaluate social value with a "minimum overall weighting of 10%" for the total procurement [9].

Conversely, barriers have been identified in various projects by researchers such as Loosemore, et al. [10] and Loosemore, et al. [11]. For instance, industry practitioners and small and medium sized enterprises (SMEs) find it difficult to embed social procurement in practice, and have little knowledge of social procurement. Barriers to social procurement implementation in the construction industry have been discussed in prior research. Such studies explored this from the perspective of either tier-one construction contractors [12], social enterprises [13,14], Indigenous enterprises [15], migrants and refugees [16], ex-offenders [17,18], youth homeless [19], or from performance perspectives such as cross-sector collaboration [14].

However, such barriers are yet to be systematically reviewed for social procurement in the construction and infrastructure sectors. In fact, according to Troje and Kadefors [20], despite the prominence of social procurement from a policy perspective, there remains a fundamental lack of knowledge regarding the barriers that have impeded implementation of those policies and how they can be overcome. Furthermore, the manner in which social procurement is embedded into daily practices also remains something of a mystery, notwithstanding that social procurement implementation has been slowly developing in practice. Loosemore [13] stated the importance of more extensive research into social procurement barriers and the OECD [3] opined that the literature on the risks, barriers, and enablers of promoting the responsible conduct of non-governmental businesses throughout supply chains had received little attention.

This paper, therefore, addresses the following research questions:

RQ1: What are the documented barriers that impede the proliferation of social procurement?
RQ2: What are the countervailing enablers and strategies that can assist its implementation in practice?

A systematic review of the work undertaken in the field of social procurement to date can support academic and industry professionals to attain a more sophisticated understanding of the various barriers, enablers, and potential strategies that could emerge from future studies in the field. Such a review could also provide an overview of the development of social procurement and serve as a guide for practitioners and stakeholders (policymakers, tier-one construction companies, subcontractors, suppliers, social beneficiaries, etc.).

The systematic review undertaken in this study contributes to the extant literature in the following ways. First, it will provide a greater understanding of the enablers, barriers, and associated strategies of the implementation of social procurement. This knowledge gap is important to bridge as it has become increasingly evident that governments alone cannot be responsible for the social well-being of their citizens given the persistence of adverse social conditions faced by disadvantaged people, refugees, and growing inequality [21]. Second, a more advanced understanding of areas benefiting from social procurement could assist governments to better utilise their policy implementation to achieve greater social value creation. Third, this review can form a foundation of fundamental knowledge to assist future studies and provide helpful insights for practitioners and key stakeholders eager to better engage in social procurement.

To achieve this, the present literature review focuses on the following:

(1) Barriers and enablers in the implementation, management practices, and processes of social procurement in the construction and infrastructure sectors;

(2) Strategies that can be applied to overcome social procurement barriers to take advantage of enablers in the implementation and management of social procurement in the construction and infrastructure sector.

This paper has undertaken a systematic literature review for the period 1 January 2012 to 30 June 2022. The review addresses barriers in the social procurement process for the construction and infrastructure sector and also categorises the strategies via a thematic analysis.

This paper is structured as follows: Section 2 reviews social procurement definitions and their conceptualisation; Section 3 introduces the methodology applied in this study; Section 4 presents the findings from the systematic review; Section 5 sets out some limitations of this research, discusses the findings, investigates the implications from previous studies, makes recommendations for future research, and proposes recommendations that have the potential to positively impact social procurement implementation practices.

2. Social Procurement Concept

There are a variety of procurement approaches whose societal contributions are not fully encapsulated by established conceptions of either economic impact or social procurement. Typically, these procurement approaches go beyond the economic and the social and also include the environmental. Examples include public procurement, sustainable procurement, and green procurement, where the latter refers to the integration of environmental criteria (e.g., reduction in greenhouse gas emissions) into the public procurement of products and services [22].

Although strong links can be drawn between social procurement and broader goals of sustainable procurement, this research focuses on social impact related to the sustainability agenda rather than unpacking and outlining the entire field of sustainable procurement.

The focus of this paper is on procurement activities that promote social impact, specifically the creation of social value for communities. Additionally, the scope of this research is confined to social procurement in construction and infrastructure projects.

2.1. Definition of Social Procurement

Social procurement has been defined in many different ways. For example, Furneaux and Barraket [23] (p. 269) defined social procurement as: "… the acquisition of a range of assets and services, with the aim of intentionally creating social outcomes (both directly and indirectly)."

There is a tendency in the literature to define social procurement as encompassing all the dimensions of sustainable procurement. For instance, Wirahadikusumah, et al. [24] (p. 939) argued that sustainable procurement can be defined as "an effort of improvement to the traditional procurement by adding sustainability principles into consideration to procurement's important areas."

Willar, et al. [25] (p. 116) defined social procurement in the context of sustainable procurement construction projects as follows: "… sustainable procurement in government projects is understood as a process whereby the government, in the context of meeting the needs for construction works and services, assesses not only the project cost and capability aspects of service providers but also assesses social and economic aspects and the minimum damage to the environment".

Other researchers have provided explanations of what they believe constitutes social procurement while falling short of providing an explicit definition. These explanations are summarised in Table 1 below.

Table 1. Summary of the main focal points of social procurement.

Researcher	Main Focal Element
Hutchins and Sutherland [26]	Social procurement is a dimension of sustainable and responsible purchasing and procurement practices. It adds the social facets of sustainability that have often been overshadowed by environmental and economic dimensions
Furneaux and Barraket [23]	Develops a typology of social procurement implementation that utilises dimensions of direct/indirect perspectives to arrive at the creation of social outcomes with acquisitions
Wirahadikusumah, Abduh, Messah, and Aulia [24]	Improvement of traditional procurement by adding sustainability principles into the processes conducted within the procurement sphere. This reflects a more pronounced focus on sustainable development practices among firms undertaking social procurement
Willar, Waney, Pangemanan, and Mait [25]	Social procurement is one aspect of sustainable procurement; meeting the needs for construction works and services, not only the project cost and capability but also social, economic, and environmental aspects
Loosemore, Denny-Smith, Barraket, Keast, Chamberlain, Muir, Powell, Higgon, and Osbourne [27]	Social procurement policies are an emerging policy instrument being used by governments around the world to leverage infrastructure and construction spending to address intractable social problems in the communities they represent
Victorian Government [4]	Social procurement is when organisations use their buying power to generate social value above and beyond the value of the goods, services, or construction being procured. In the Victorian Government context, social value means the benefits that accrue to all Victorians when the social and sustainable outcomes in this framework are achieved

Part of the broader conceptualisation that lies at the heart of social procurement involves the notion of social value creation. As Loosemore, et al. [10] state, the creation of social value remains under-conceptualised, which has led to an ongoing debate in this area.

2.2. Social Procurement in Construction and Infrastructure Sectors

The construction and infrastructure sectors are seen as important facilitators of social procurement due to their size and potential money multiplier effects [28]. Loosemore, Alkilani, and Mathenge [10] posit that this is the reason why the construction industry is widely seen by governments as a focus for newly emerging social procurement policies. Furthermore, spending in these sectors is capable of being leveraged to provide employment and training opportunities for disadvantaged groups such as Indigenous people, those experiencing disabilities, migrants and refugees, women at risk, youth at risk, long-term unemployed, and ex-offenders [13,29].

3. Methods

This study systematically reviewed academic publications that address the topic of social procurement. Our review process is inspired by Preferred Reporting Items for Systematic Reviews and Meta-Analyses (PRISMA) [30] and follows the four steps (described next) for standalone systematic literature reviews [31]. The review covers the literature between 1 January 2012 and 30 June 2022.

3.1. Step 1: Planning

Planning involved understanding social procurement in the context of the construction and infrastructure sector and deciding on suitable databases to search. This required a series of database searches to assess the suitability of different databases and search keywords. For example, one determination was to establish a judicious selection of search terms. Figure 1 displays the results from three academic database searches using synonyms for "social procurement". The results suggested that the synonyms have quite dissimilar meanings in the literature. Based on results such as these, it was decided to use two search terms ("social procurement" and "sustainable procurement") in combination with two further terms (construction and infrastructure) to search three academic research databases (Scopus, EBSCOHost, Web of Science).

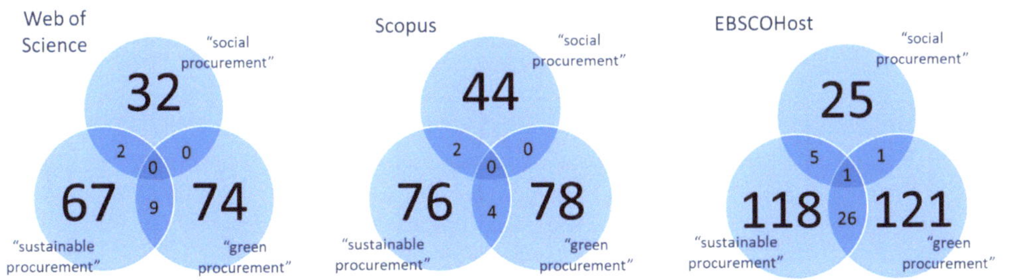

Figure 1. Example of experimental searches to understand the literature, returned from three database searches for three search terms.

3.2. Step 2: Search and Selection

The search and selection criteria deemed most suitable for this review were:

- Academic peer-reviewed journal articles written in English and published over the past 10 years (1 January 2012 to 30 June 2022) reflecting the growth that has occurred in social procurement research over the last decade;
- Research that presented findings relevant to barriers, enablers, strategies, and social value creation related to social procurement to match the focus of the research;
- Given the nascency of research on social procurement and a corresponding limitation on theory development, to address the research aims the authors expected to collect a wide range of materials from a wide range of academic literature with different quality rankings. Articles of sufficient quality as assessed against the minimum quality criteria derived from the Consolidated criteria for REporting Qualitative research (COREQ) checklist [32] that were sufficiently generic of suitable research reporting to be used to assess studies irrespective of their research methods (quantitative, qualitative, or mixed).

The search and selection process with results are presented in the PRISMA flow chart in Figure 2. The database searches identified 223 records. First, article screening consisted of two authors independently evaluating the suitability of each of the 103 identified articles by reading their titles, abstracts, keywords, contexts (e.g., industry), methodology, and any other clarification details in the full contents to decide on an article's inclusion or exclusion; second, resolving their selection differences through discussion with a third author. After the removal of duplicate records and checking for full article availability, 50 articles remained. One article was excluded for failing the quality test. Finally, 49 articles were selected.

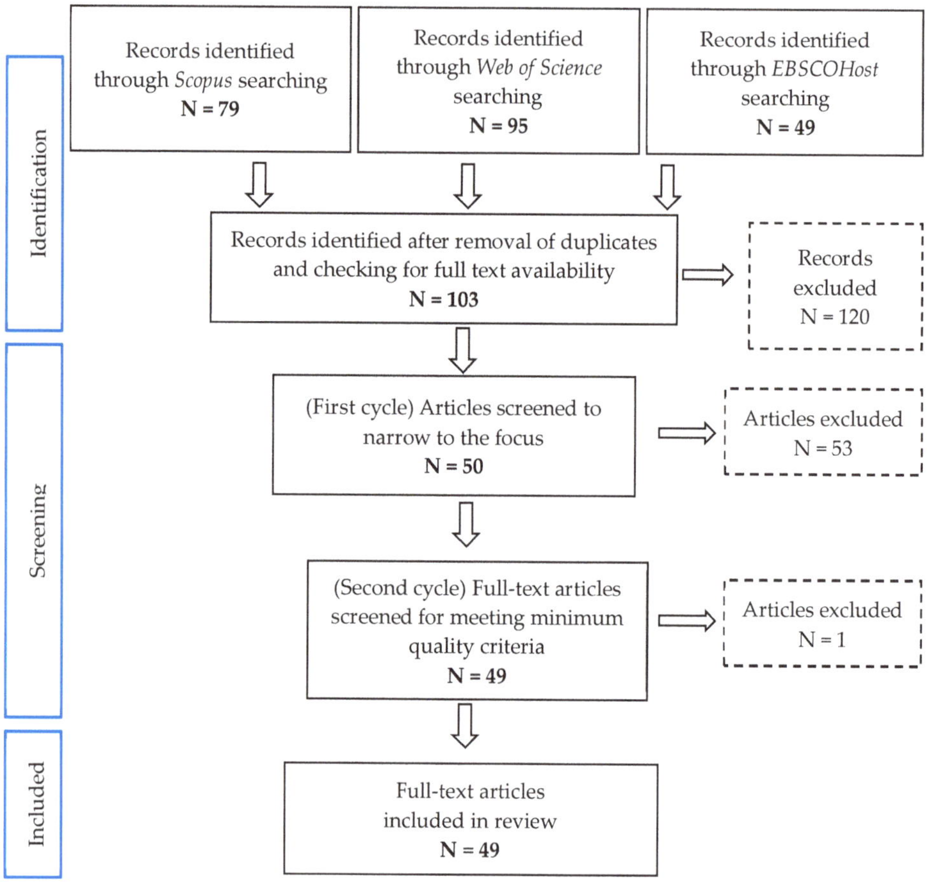

Figure 2. PRISMA flow diagram for this paper.

3.3. Step 3 Extraction

Data to answer the research questions in terms of barriers, enablers, and strategies were extracted into a single document for further synthesis in Step 4.

3.4. Step 4 Execution

Thematic analysis, an inductive analytic method, was selected as the method for categorising barriers, enablers, strategies, and social value creation into common themes by following Guest, et al. [33] (2012) and Saldana [34]. The thematic analysis process with an example is provided in Appendix A.

4. Findings

4.1. Overview

This section presents an overview of the 49 articles selected (see Table A2 in Appendix B) for this systematic literature review in terms of the: (i) number of articles published by year and by country between 2012 and 2022, (ii) journal distribution, (iii) leading authors of social procurement research, and (iv) most frequently used keywords.

4.1.1. Articles by Year and by Country

Figure 2 shows that no articles from 2012 to 2014 fell into our selection criteria and were thus not included. However, the number of articles that touch on our research questions started to grow between 2015 and 2019, although the growth was uneven. There has been a clear upward trend since 2020, considering that our search ended in July 2022 (shown in Figure 3). Figure 1 shows that more studies have been conducted in Australia (19) than anywhere else. The studies from Australia (19), the U.K. (6), and Sweden make up more than half of the studies published.

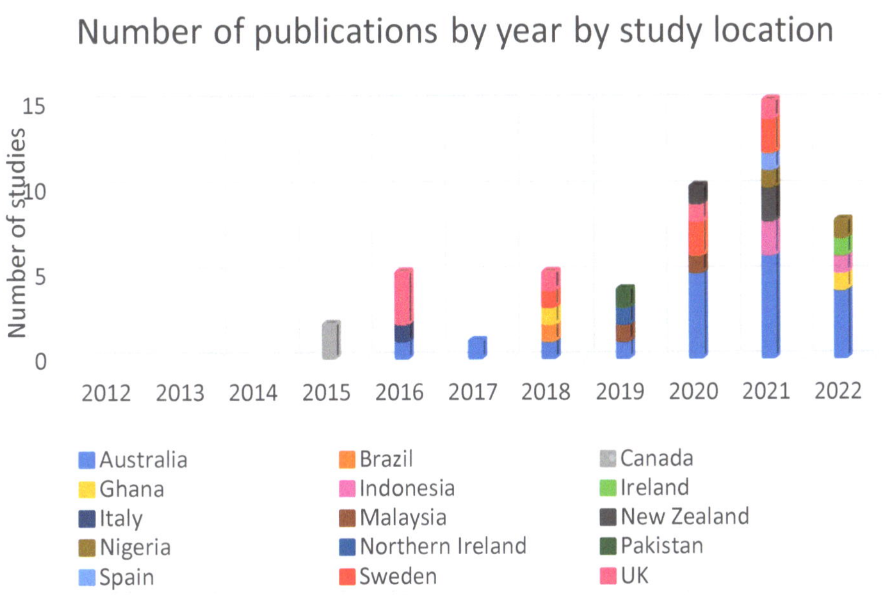

Figure 3. Social procurement publications by year and by country.

4.1.2. Journal Distribution

Our search found that the 49 articles included in our review were published in 25 different journals (see Appendix B). By subject areas, most articles were published in engineering (14), business, management, and accounting (12), social sciences (11), and environmental science (6), with a focus on the social aspects associated with social procurement and corporate social responsibility. The top five journals that published most of the articles selected for this review are Construction Management and Economics (7 articles), Engineering, Construction and Architectural Management (6), Sustainability (5), International Journal of Construction Management (4), and Built Environment Project and Asset Management (3).

4.1.3. Leading Authors

The most prolific authors in social procurement research among the 49 included articles and the country and years of publications are listed in Table 2. Only authors appearing as first authors at least twice are included in our analysis and reported in this table.

Table 2. Leading authors in social procurement research.

First Authors	Studied Country	Number of Articles as First Author	Years Published
Loosemore M.	Australia, U.K., global	15	2016–2022
Troje D.	Sweden	5	2018, 2020–2021
Barraket J.	Australia, U.K.	2	2018, 2020
Denny-Smith G.	Australia	2	2017, 2021
Ershadi M.	Australia	2	2021
Ogunsanya O.A.	Nigeria	2	2021–2022
Ruparathna R.	Canada	2	2015

Professor Martin Loosemore from Australia led 15 publications in social procurement-related topics from 2016 to 2021, crossing the topics of the employment experiences and capabilities empowerment of different disadvantaged or marginalised groups, the collaboration between institutions and social enterprises to achieve social innovation and the resultant social value and social impact created, and the professional practices and roles of social procurement champions in the construction industry.

Dr. Daniella Troje from Sweden looked into social procurement in construction industries from the employment requirement through an institutional perspective, contributing to the "policy-in-practice" literature and providing advice on policy implementation.

4.1.4. Most Frequently Used Keywords

Mapping keywords provides a way to visualise the field of social procurement research. Figure 4 presents a cloud map of the keywords and key phrases of the included articles. The figure highlights the main keywords and phrases in a broad context of social procurement.

Figure 4. Cloudmap of keywords and key phrases applied in the literature.

As presented in Figure 4, sustainability, employment, collaboration, social value, social innovation, corporate social responsibility, and social enterprises were the terms most of the research focused on. Social procurement was a key phrase used, with papers covering corporate social responsibility focusing on procurement processes in the construction sector.

From the cloud map summaries of the literature, social innovation and collaborations were of importance, along with employment. Of these, collaboration has been raised on numerous occasions in social procurement discussions.

4.1.5. Main Theories Applied to the Social Procurement Area

Although there is not a great deal of research on social procurement, most of the existing research focuses on problems/challenges occurring in practice and tends to be descriptive in nature as opposed to being oriented towards theoretical development [20]. Several researchers have pointed out there is a lack of conceptualisation and theoretical investigation in this area [13,35,36]. The theories applied in the social procurement literature within this systematic review are summarised in Table 3.

Table 3. Theories applied in social procurement research.

Theories Applied	Authors	Key Areas
Institutional Theory	Troje and Kadefors [20]	Employment requirement
New Institutional Theory	Troje and Andersson [37]	Social procurement practices and strategic and operative levels
	Loosemore, et al. [38]	Social procurement, social value, social impact, social outcomes, Institutions Isomorphism
	Loosemore, Keast, Barraket, Denny-Smith, and Alkilani [11]	Collaboration, intermediary, project management education, risk management, social value, social procurement, social innovation
Organisational Theory	Loosemore, et al. [39,40]	Social procurement champions in the construction and engineering industry, CSR
Practice Theory	Troje and Gluch [41] Troje [42]	Employment requirements, social sustainability, social value, Sweden
Job Performance Theory	Lam [43]	Construction development, holistic sustainability, design and construction, post-contract monitoring, KPIs, performance drivers
Social Value Theory/Value Theory	Denny-Smith, et al. [44]	construction employment, COVID-19, infrastructure investment, social procurement, social value
Resourced-Based View Theory	Ewuga and Adesi [45]	AEC, Republic of Ireland, strategy, sustainable procurement, suppliers' development, supply chain management
Principal-Agent Theory	Loosemore, Denny-Smith, Barraket, Keast, Chamberlain, Muir, Powell, Higgon, and Osborne [27]	Construction industry, collaboration, intermediaries, risk management, social procurement, corporate social responsibility
Social Exchange Theory	Loosemore, Bridgeman, and Keast [17]	Collaboration, construction, ex-offenders, social procurement, social value
Socio-Technical Transition Theory	Brooks and Rich [46]	London, sustainability, construction, socio-technical transitions, consumption, procurement
Field Theory	Barraket [47]	Intermediation, social procurement, field theory, social enterprise, social innovation
Ecological Modernisation Theory	Delmonico, et al. [48]	Sustainable operations, sustainable public procurement, sustainable supply chain, sustainable purchasing, Latin America

4.1.6. Methodologies Applied

The methodologies that have been applied in the reviewed papers include: quantitative methods (through surveys) and qualitative methods (e.g., semi-structured interviews, empirical fieldwork, focus groups, and case study). More details can be found in Appendix B.

4.2. Discussion of Themes

This study reviewed 49 papers. Three themes that emerged will be discussed in this section: (1) barriers; (2) enablers; and (3) applied strategies and recommendations for overcoming the barriers.

4.2.1. Main Barriers

From the 49 reviewed papers, the main barriers are summarised in Table 4 below. The main barriers can be categorised into nine groups: (i) knowledge, learning, tools, and awareness; (ii) organisational capacity and resources; (iii) policies and leadership; (iv) competitive forces/industry/organisational structure and culture; (v) procedures and practices; (vi) cost, administration, accounting, and funding; (vii) collaboration and engagement; (viii) marketing (communications, branding, products/services strategy); and (ix) resistance to change. These are presented in Table 4 below.

Table 4. A sorted list of main barriers identified for distinctive areas.

Main Barriers Identified	Included Articles	Description
Knowledge, learning, tools, and awareness	(14) [13,24,37,49–57]	Inadequate information, knowledge or awareness or the inadequate means to acquire the information and knowledge or raise awareness can be a barrier to efforts to implement social procurement. For example, lack of, inadequate, or ad hoc training programs, vague definitions and diversity of interpretations of key terms or the domain, lack of tools showing how to conduct sustainable procurement, or lack of platforms to exchange information and knowledge.
Organisational capacity and resources (including human resources development)	(13) [10,12,13,25,37,42,48,50,51,54, 56–58]	Constraints or issues with organisation's practices and human or other resources that affect their capacity to implement social procurement. For example, the complexity and uncertainty of role expectations and tasks, a lack of skilled labour or suitable candidates, lack of time to address sustainability issues, iterative role changes, recruitment difficulties, or unsustainable mandates.
Policies and leadership	(10) [10,13,42,48,50,52–54,57,58]	Constraints or issues related to policies or leadership. For example, vague or mismatched policies and policy goals, insufficient policies, regulations, or incentives, lack of leadership, lack of leadership motivation or demand from leadership for social procurement, or inflexible policies or policies that could change easily.
Competitive forces/industry/organisational structure and culture	(10) [10,12,13,25,48,51,52,55,57,58]	Issues related to competition in the industry. For example, increased competition, client silos, fragmented and transitory nature of the construction sector, lack of third-party pressures, barriers to entry to social enterprises, industry culture, transparency and governance factors of the industry, organisational short-term planning, and lack of effective strategy or partnership issues.
Procedures and practices	(9) [12,13,25,37,46,50,51,54,57]	Constraints, restraints, and issues related to procedures and policies. For example, lack of systematised practices or complicated procedures, lack of objective methods, standards, and certifications to vet bidders or evaluate bids or determine ethical credentials, restraints of existing procurement procedures and practices, difficulties with reporting and measuring social impact, lack of technical guidelines for implementation, or inadequate monitoring and control.

Table 4. *Cont.*

Main Barriers Identified	Included Articles	Description
Cost, administration, accounting, and funding	(9) [12,25,46,48,50,51,54,57,58]	Issues and constraints related to finances and capital and the administration of that. For example, lack of funding/capital and unwillingness to incur higher capital costs, push for lowest price, additional costs of tendering, administration, and compliance, additional unknown aspects of costs, risks, capabilities, and responsibilities, lack of financial support, poor cash flows, or underestimation of sustainability financing.
Collaboration and engagement	(6) [12,13,37,42,49,50]	Issues with collaboration between stakeholders and lack of engagement with social procurement. For example, lack of motivation or interest among interns, lack of engagement between social enterprises and construction companies, uncooperative attitudes and stakeholder fatigue, lack of teamwork, or low capacity of the supply chain to deliver on social outcomes.
Marketing (communications, branding, products/services strategy)	(5) [12,13,25,48,58]	Issues and constraints related to research and development (R&D), consumers, and markets. For example, insufficient R&D, lack of demand for sustainability products, small size and narrow scope of activities, market aspects, poor communication of value add, poor communication with construction firms.
Resistance to change	(5) [10,12,13,54,58]	Issues related to resistance to change. For example, resistance from industry incumbents to changing established relationships, displacement of existing informal recruitment networks and processes and lag in adoption of sustainable business concepts and practices.

There are also some other barriers identified by a few papers, such as industrial and employment relations, human resources management [56,58], supervision and teamwork [37,50], organisational strategic planning [13,46], research and development [54], consumer factors [53], financial management [58], and organisational development [13].

Different papers have deployed varying perspectives in their research and thus contributed different insights to understanding social procurement. For example, Loosemore, Alkilani, and Hammad [58] focused on barriers affecting Australian local disadvantaged job seekers. The researchers concluded that, for example, most stakeholders perceive social procurement in a negative light, as more of a risk than an opportunity; raise numerous cautionary concerns about the risks for creating harm by ineffective implementation; display a low level of engagement and a high level of suspicion that deters the collaborative effort needed to overcome the implementation barriers; there is a low level of engagement and a high level of suspicion by key stakeholders that also deters collaborative efforts; stakeholders perceive that the way the policies are being implemented as unjust and appearing to counter effective risk management; and finally, that education is needed for all stakeholders. In contrast, Delmonico, Jabbour, Pereira, Jabbour, Renwick, and Tavares Thome [48] explore barriers experienced by public authorities in the Brazilian public sector. Their study found that organisational cultural factors and perceptions of a gap between federal and state/local authorities can present significant barriers.

4.2.2. Enablers for Social Procurement Initiatives

From the 49 reviewed papers, the main enablers could be categorised into eight groups. First, three ecosystem-creating factors were identified by Barraket and Loosemore [14]: organisational, commercial, and institutional. Organisational factors include champions of social value creation, breadth and accessibility of organisational networks, and organisational purpose [14]. Commercial factors include competitive advantage, altruistic values linked to organisational history, and founders' passions [14], while institutional factors include organisational scale and structural position in the industry and new public governance trends, as well as increasing social and governmental expectations around collaboration [14].

Second, drivers of growth were categorised into six groups [13]: construction industry culture change, new social legislation and regulation, changing social expectations, potential impact of construction on society and environment, political trends, and changing public procurement priorities [13].

Third, social actors involved tended to vary, with project manager involvement not being necessary when the actors were aligned since this is where strategy and construction procurement primarily takes place [59].

A fourth enabler was the regulatory environment, which significantly influences sustainable procurement. This occurs via an adequate implementation of legislation such as the Infrastructure Concession Regulatory Act, 2005, adherence to the provisions of the Fiscal Responsibility Act, 2007, and government policies relating to public procurement as the key indicator [60].

In addition, organisational orientation, the fifth enabler, has a significant association with sustainable procurement, but only two variables have a strong association: the attitude of close competitors, and understanding that competitive advantage is enhanced through sustainability credentials [60].

Another (sixth) enabler, procurement method selection, has a significant influence on sustainable procurement, however the strength of the relationship is moderate when specific variables are considered, indicating that a one-size-fits-all approach will not be effective [60].

In terms of adoption of newer methodologies (seventh enabler), the adoption of the gateway process, competitive dialogue, cost-led procurement methods, and the two-stage open book model have a strong relationship with sustainable procurement, while e-procurement has a weak relationship [60]. Construction industry development directly influences sustainable procurement significantly [60].

Finally, stakeholder commitment, which refers to common shared beliefs in the organisations' goals and values for green procurement [61], was seen as the eighth enabler for social procurement. This is linked to project stakeholder technical competencies, which refers to project stakeholders having sufficient technical competencies to deliver on a green project [61]. Stakeholder ability to understand the bigger picture of green construction is also a factor and this can be bolstered by awareness creation and education in green practices before and during the project [61]. Knowledge sharing between the project stakeholders refer to the exchange of green practices throughout the organisations involved through, for example, training, meetings and benchmarking [61].

Two coercive factors (regulatory imperatives and client pressures), three mimetic factors (mimicking of competitors, cross-sector networks and alliances and supplier assessment programs) and no normative factors were found to explain social procurement, although alignment with the enterprise culture also appears to be central to social procurement policy implementation [38].

While procurement experts agree that understanding of sustainable procurement fundamentals, policies and strategies (e.g., leadership and roles and authorities), procurement organisation (e.g., procedures and systems), and sustainable procurement processes (procurement planning) affect the successful implementation of sustainable procurement of construction work, it is only sustainable procurement policies and strategies, and procurement organisations that have a statistically significant effect [62].

4.2.3. Implemented or Perceived Strategies to Enhance Social Procurement Practices

This study identified key strategies capable of being categorised into a series of themes, organised by stakeholder categories: policymakers; buyers; suppliers; collaborations and engagement; and general. For policymakers, the main themes are policies and legislation, monitoring, and auditing. The next stakeholder group, buyers, have themes comprising awareness, knowledge, learning, tools, and training; leadership; procedures and practices; and improving procurement method selection. For suppliers, key factors are knowledge and awareness and organisational capacity and resources. In the category of collaborations and engagement, there are themes of teamwork: helping local disadvantaged job seekers, integration and management, and sustainable infrastructure. General factors of importance to most stakeholders are cost and administration, measurements, and support for the change. Examples of key strategies are presented in Table 5.

There are also strategies to foster enablers. Bohari, Skitmore, Xia, Teo, and Khalil [61] recommended that all stakeholders have a sufficient level of knowledge of green practices so as to foster organisational capacity and resources. The policies, legislation, and leadership enablers can be fostered by: being cultivated as early as the beginning of the project; building commitment, which starts with creating awareness and nurturing a common understanding and interest between the stakeholders; developing a green orientation strategy, to be made available to all stakeholders, and articulated to other stakeholders involved, both internal and external, as early as possible; and effective communication, which is essential.

Table 5. Implemented or perceived strategies to enhance social procurement practices.

Stakeholders	Themes Identified	Example Strategies
Policymakers	Policies and legislation	- Sustain and support green policy tools by specific actions at national and local levels—Testa et al. Testa, Grappio, Gusmerotti, Iraldo, and Frey [55] - Develop a complete national framework of sustainable procurement implementation—Wirahadikusumah, Abduh, Messah, and Aulia [24] - Address in detail the micro tensions between the implementation of policy and practice. Before implementing social procurement policies, consider aligning the sector prerequisites, the local labour market conditions, and the prerequisites of the people the social policy is targeting—Troje [42] - Improve leadership barriers through legislation and updated policies. Consider a non-voluntary policy on green procurement—Ruparathna and Hewage [54] - Resolve policy conflicts and perverse incentives that undermine collaboration. Further, ensure that the risks and opportunities of social procurement are clear and shared and that policies reflect and acknowledge the sector's constraints and challenges. Finally, ensure that the parties responsible for implementation have the incentives, knowledge, resources, and time to collaborate—Loosemore, Denny-Smith, Barraket, Keast, Chamberlain, Muir, Powell, Higgon, and Osborne [27] - Provide a new construction procurement act with sustainability clauses that mandate not only design compliance but stipulates a statutory procurement requirement for contractors and suppliers—Ogunsanya, Aigbavboa, and Thwala [60]
	Monitoring	- Integrate environmental and social criteria into the supplier selection process. Monitor supplier environmental and social performance—Renukappa, Akintoye, Egbu, and Suresh [53] - Address information asymmetries and perverse incentives by government policymakers to monitor and measure the implementation—Loosemore, Denny-Smith, Barraket, Keast, Chamberlain, Muir, Powell, Higgon, and Osborne [27] - Establish construction industry development boards—Ogunsanya, Aigbavboa, and Thwala [60]
	Auditing	- Improve the sustainability auditing process—Ershadi, Jefferies, Davis, and Mojtahedi [50]

Table 5. *Cont.*

Stakeholders	Themes Identified	Example Strategies
	Awareness, knowledge, learning, tools, training	- Improve stakeholders' awareness of sustainability values and principles—Ershadi, Jefferies, Davis, and Mojtahedi [50] - Implement best practices through training and workshops in more tendering organisations and those with potential future suppliers—Testa, Grappio, Gusmerotti, Iraldo, and Frey [55] - Undertake training needs assessment in sustainability areas and plan effective training programs—Ershadi, Jefferies, Davis, and Mojtahedi [50]
	Leadership	- University procurement management needs to commit to sustainability and strongly encourage mid-level management to achieve sustainable goals—Zaidi, Mirza, Hou, and Ashraf [57] - At the corporate level, develop a sustainable procurement management (SPM) plan. Assign accountability to all levels of the organisation. Obtain senior management buy-in on the implementation plan—Ershadi, Jefferies, Davis, and Mojtahedi [50]
Buyers (Tier 1/2, etc.)	Procedures and practices	- Use tender templates to standardise the legal/normative part of tenders and the preparation of green tenders. Detail both the legal issues and general green technical specifications or contract performance clauses—Testa, Grappio, Gusmerotti, Iraldo, and Frey [55] - Provide more resources such as frameworks, tools, and databases so as to standardise the practices. Reinforce the practices with education, training, awareness programs, and research—Ruparathna and Hewage [54] - Maintain SPM metrics so the procurement team can track targets—Ershadi, Jefferies, Davis, and Mojtahedi [50] - Build maintenance mechanisms and formalise practices and routines. Plan for intangible knowledge retention, areas for knowledge exchange, and opportunities for continuous feedback—Troje and Andersson [37]
	Improving procurement method selection	- At the project level, ensure the right method is used. Choose the right and most appropriate procurement strategy. Adopt a gateway process, competitive dialogue, cost-led procurement method, the two-stage open book model, and e-procurement—Ogunsanya, Aigbavboa, and Thwala [60]

Table 5. *Cont.*

Stakeholders	Themes Identified	Example Strategies
Suppliers (SMEs), social enterprises, social benefit suppliers	Knowledge and awareness	- Implement best practices through training and workshops with potential future suppliers—Testa, Grappio, Gusmerotti, Iraldo, and Frey [55]
	Organisational capacity and resources	- Introduce organisational arrangement for career development, staff training, cost savings, and renewable energy—Wirahadikusumah, Abduh, Messah, and Aulia [24]
Collaborations and engagement	Teamwork	- Encourage stakeholders to engage in sustainable procurement. Establish team rules for all parties involved in the SPM implementation process. Encourage all parties to adhere to teamwork values—Ershadi, Jefferies, Davis, and Mojtahedi [50]
	Helping local disadvantaged job seekers	- Improve education in social procurement by addressing skills gaps and imparting transferable project skills and competencies. Involve intermediaries in facilitating collaboration and creating shared value. Address conflicting policy agendas. Improve education support at a lower level through work readiness. Increase government support for policy implementation. Develop policies that address supply and demand constraints—Loosemore, Keast, Barraket, Denny-Smith, and Alkilani [11]
	Integration and Management	- Employ a project management office (a centralised oversight structure) that: 1. Supports executives in making strategic partnership decisions that enable the achievement of business targets without compromising sustainability values, including undertaking a cost–benefit analysis of investments in sustainability and an analysis of the total cost of ownership. 2. Ensures the incorporation of sustainability requirements by adopting a collaborative (intra- and inter-organisational) implementation approach. 3. Supports the procurement team to evaluate and screen suppliers against sustainability criteria. 4. Systematises sustainability compliance and maintains the integration of sustainability controls from design through to delivery. 5. Undertakes post-delivery reviews to confirm that all sustainability tasks have been accomplished and SPM objectives have been met—Ershadi, et al. [63]
	Sustainable infrastructure	- Improve and strengthen cooperation among those with high-level and middle-level qualifications and their subcontractors/suppliers as well as civil society to create a sustainable infrastructure—Willar, Waney, Pangemanan, and Mait [25]

Table 5. *Cont.*

Stakeholders	Themes Identified	Example Strategies
General	Awareness and Education	- Establish criteria in the selection of services at the procurement phase of providers that emphasise green construction. Enhance their awareness, knowledge, and skills in the planning, implementation, and supervision phases of infrastructure projects. Ensure transfer of knowledge about sustainable principles among all project participants that are supported by well-trained and competent contractors regarding environmental, social, and economic viewpoints—Willar, Waney, Pangemanan, and Mait [25] - Strengthen institutional training, aligned with sustainability target-setting for public expenditures. Coordinate efforts among Federal, State and local (county) public institutions—Delmonico, Jabbour, Pereira, Jabbour, Renwick, and Tavares Thome [48]
	Cost and administration	- Agree on a common language between project investors/owners about SPM benefits based on a tangible cost–benefit analysis—Ershadi, Jefferies, Davis, and Mojtahedi [50]
	Measurements	- Adapt organisational performance targets to measure social value. Re-evaluate the organisation's financial modelling to accommodate costs on social value, viewing it as essentially a non-profit-making activity. Adopt a more person-centric approach to the design and implementation of solutions and a more holistic approach to the measurement of impacts and outcomes. Understand the project context and ensure bespoke solutions rather than use a standardised tool measured against project management metrics and targets—Murphy, and Eadie [51]
	Support on the change	- Appoint more than one social value champion to implement social procurement who each plays a multitude of roles at different organisational levels and across different organisational functions—Loosemore, Keast, and Barraket [39]

4.3. Validity of the Study

To safeguard the validity of the findings and conclusions of this systematic literature review, this study took several measures to reduce the risk of bias. This research focused particularly on sources of bias, selection bias, performance bias, and reporting bias. The assessment was largely based on the Cochrane bias risk assessment framework [64]. Table 6 outlines the process and measures of the bias risk assessment for the study.

Table 6. Risk bias assessment and measures for the literature review.

Bias	Measure	Purpose
Sources of bias	Used a comprehensive search strategy: multiple databases, relevant keywords, and controlled vocabulary	To capture a wide range of relevant literature
Selection bias	Applied consistently our clearly defined criteria for inclusion and exclusion to the literature reviewed	To minimise selection bias
Performance bias	Critically reviewing participants' responses and acknowledging the bias in this paper	To alert readers to take caution in considering research findings
Reporting bias	Used (PRISMA) technique to report methods and findings of our study Clearly stated the limitations of our literature review, including potential biases, and discussed how these limitations might affect the interpretation of the results and their implications	To ensure the transparency of our reporting To respond positively to the potential biases of the studies reviewed

5. Conclusions

Of the 49 included articles, 27 identified barriers in their research, and 18 identified enablers. Papers identifying both barriers and enablers totalled eight. Three articles [20,44,65] did not investigate barriers or enablers but included information about the social value created.

There are two tendencies in the research: one is to focus on employment as a specific delivered social value outcome, while the other is to look at a broad category of outcomes from social procurement (without focusing on any specific outcome). Focusing on employment as the social benefit means the results reported (e.g., barriers) are applicable to this type of social procurement, and this often gives the impression that creating employment is all that counts when discussing social procurement.

The current research has some limitations. For instance, in some of the studies reviewed in this research, it was observed that rather than obtain the views of a particular stakeholder group under study, researchers asked someone else to comment. This is particularly evident in the case of disadvantaged groups that are difficult to study directly; hence, it is much easier to ask employment providers to comment. There are additional examples of this, for example, surveying public universities and then purporting that the results represent the views of "the public sector". Researchers may well be correct in assuming that employment providers have a suitable overview of the problems faced by disadvantaged groups, and it should be noted that all the papers studied have been transparent about their methodology. Nonetheless, there are sensitive issues to be considered, including biases. These observations should be taken into account by those undertaking future research.

The high concentration of authors of the 49 articles seems to indicate that there are only a handful of researchers who are interested in social procurement within the construction industry. This concentration is another limitation of the current research.

Notwithstanding the limitations, a number of implications and areas for further exploration can be deduced from this study. Specifically, the role of social procurement in contributing to the creation of social value is an under-theorised concept, and more research into frameworks for measuring social outcomes is needed [66]. The conduct of future research into social value creation could help to articulate a social value chain by expanding on the building value chain, as presented by Groote and Lefever [67]. Investigations are needed to measure the real-world impacts of social procurement policy, including initiatives connected to construction projects, and to measure the gains delivered from social procurement [11,46,54,68]. There is scope for additional research to further investigate the issues associated with the integration of social procurement into different business models and different sizes of businesses [53].

All reviewed papers suggest that social procurement as a field of practice will evolve and expand, which will have significant implications. For instance, there could be greater workplace diversity in construction, more engagement of employees and work groups with communities, increased awareness and knowledge of the social value created by companies or specific projects and of what this means for people, companies and specific groups (e.g., [41,69]), and measurements showing increasing and different forms of social value created [41]. Moreover, more people may champion social value creation, and the value chains of companies that have adopted social procurement could be longer. This shows the potential influence of using social procurement on the wider policy area [20].

There is clear potential for future research to investigate changes in areas such as these and to develop theories elucidating how the evolution or expansion has occurred, the changes experienced by supply or value chain actors, affecting relationships, and the role of different policies and company cultures in the process of change (e.g., [44,45,70]). New barriers and enablers may appear and will need to be identified. The theoretical lenses used by the range of studies could be utilised to identify more details of practices.

Research on champions of social procurement or those in social procurement roles, to map the types of roles and developments in roles and practices, the effectiveness for social procurement of the influence they exert and understanding which of their roles are most appropriate in different situations, is recommended (e.g., [20,40,58]). More detailed mapping of the relationships between procurement managers and other decision makers is also needed [46].

Further research is required on the intermediaries that facilitate social procurement, their capabilities, and resource requirements (e.g., [69]). Additionally, research could investigate cases where successful implementation has occurred without the involvement of intermediaries.

It is also recommended that studies be conducted to assess the effectiveness of various procurement strategies, such as public–private partnerships, in actualising social policies into practice. A more comprehensive understanding of procurement's role in social value creation is necessary [42].

With respect to social procurement practices, the following recommendations arise from the current study. Specifically, the implementation of social procurement demands the establishment of new relationships, roles, knowledge, and skills, which poses risks of failure and increased complexity for the actors involved. This underscores the importance of intrapreneurship and creativity, as they are vital for developing new areas and roles that can account for a broader range of social value outcomes in the procurement process [29,46,66].

A common recommendation in the included articles is to use the results of their research to help different stakeholders to develop or improve their tools and practices and to act as driving forces to advance social procurement. This recommendation has been made, for example, in the context of construction subcontractors and indirectly related social enterprises, to address stigmas about ex-offenders as employees [18]; and in connection with Valencian public entities, to develop handbooks [71].

There have been calls for the development of training and educational courses for all stakeholders about the need to shift to sustainable procurement practices (e.g., [39,49,72]) and for the development of the skills and competencies needed to implement social procurement [11,29]. This is a clear need.

If the papers studied had incorporated a particular theoretical perspective or constructed typologies and models, these lenses or models could be utilised as normative standards or guideposts to steer institutions, businesses, and procurement personnel towards effective implementation of social procurement or realisation of social outcomes (e.g., [37,49]). However, they did not. Therefore, future studies should actively seek to apply a lens of theory to the analysis they undertake.

The efforts by governments and others to enhance the social well-being of disadvantaged groups, and address equity considerations, can be enhanced if the systems that surround the implementation of social procurement can be made more effective. This has a negative aspect—the reduction in barriers—and a positive aspect—the promotion of enablers. Social procurement is a promising new arena for the creation of social value.

Author Contributions: Conceptualisation, C.X.L.; Methodology, C.X.L., B.B., R.N., D.G. and F.Z. Abstract analysis, C.X.L., R.N. and B.B.; Figuration, B.B. and C.X.L.; Writing—original draft, C.X.L. and B.B.; Writing—review and editing, C.X.L., R.N., D.G., F.Z. and P.Z. All authors have read and agreed to the published version of the manuscript.

Funding: This study was funded by the Victorian Higher Education Strategic Investment Fund (2021–2022).

Institutional Review Board Statement: Not applicable.

Informed Consent Statement: Not applicable.

Data Availability Statement: Not applicable.

Acknowledgments: The authors are grateful to the guest editors and the anonymous reviewers for their useful suggestions, which have improved the quality of the manuscript.

Conflicts of Interest: The authors declare no conflict of interest.

Appendix A

Appendix A.1. Thematic Analysis Process with Example

Separate data on barriers and enablers were extracted from the included papers and saved as a Microsoft Word document. To categorise the barriers and enablers into common themes, the list of barriers and enablers was examined, and common themes were recorded according to which of the items (barriers or enablers) could be consolidated. The themes were strongly derived from the data and not made to fit into pre-existing theory. Table A1 presents an example to illustrate the process together with the following discussion. The first iteration of analysis involves immersion in the data by reading and rereading the descriptions of each barrier and generating descriptors or themes. From Table A1, in the first iteration, initial themes were assigned to barriers based on phrases in the included paper of what the barrier meant. Where such descriptions were missing, the meanings were interpreted from the contents of the paper overall or the name of the barrier. During the first iteration, the codes closely captured the meaning of each barrier. For example, the difference between the first two and third barriers in the table reflected the difference between lacking clear role descriptions (an HRM-related barrier) and the ad hoc way in which the role developed (a human resources development-related barrier). As more barriers were examined, so more codes were created. In the second iteration, codes were consolidated around common themes. For instance, since the role of unions is connected to the industrial relations function, which was the precursor function to employment relations, and since these two functions evolved into the contemporary organisational human relations management (HRM) function, the two codes were consolidated into the "Industrial and Employment Relations and Human Resources Management-related barriers". This was still taken to be sufficiently distinct from the role development code to warrant separate codes. It is nonetheless possible that other analysts would develop different codes and themes based on the extracted data.

Table A1. Example of thematic categorisation of barriers.

Iteration	Barrier Title and Some Relevant Descriptions of the Barrier	Initial Code	Iteration	Final Theme
1	Complexity and uncertainty in terms of role expectations and tasks [56] "has given rise to a new role", "this is not a well-defined professional role yet", "the role often was self-created and iteratively developed to align with both immediate and habitual needs", "the expectations on them had an in-built ambiguity", "they had to navigate between conflicting formal and informal roles and responsibilities", "the ERPs were detached from HR functions in the organization" [56]	Human resources management-related barrier	2	Industrial and employment relations and human resources management-related barriers
1	Unclear role boundaries and responsibilities, "a hybrid role with unclear boundaries and responsibilities were formed", "existence of both formal and informal roles suggests that the ERPs, despite the freedom to define their own role, do not yet have exclusive control and power of their work" [56]	Human resources management-related barrier	2	Industrial and employment relations and human resources management-related barriers
1	Iterative role development, "the work tasks of professionals are also in a continuous and iterative process that is simultaneously affected by the professionals themselves and/or formed through proxies such as various educational programs", "the roles and their included practices and tasks were developed in an iterative and ad hoc process shaped by emergent concerns, demands and incidents, like 'the refugee crisis'" [56]	Human resources development-related barrier	2	Human resources development-related barrier
1	Union opposition, "numerous new risks", "they do not believe that our people can do the job", "union opposition to social enterprises" [11]	Industrial relations-related barrier	2	Industrial and employment relations and human resources management-related barriers

Data categories by which the individual data items were synthesised, emerged mainly during the extraction process in an intuitive way and was assisted by the fact that high-level categories were already determined for this review (i.e., barriers, enablers, strategies, and social values created for specific stakeholders). This meant fewer groups of individual data items needed to be synthesised for the results reporting purposes.

Appendix B

The following Table A2 presents key information about the included papers in the systematic review process.

Table A2. Key information about the included papers.

Included Paper	Methodological Approach	Location	Data Collection and Analysis Methods	Sample Size	Theoretical Framework(s)/Key Literature	# Citations (Scopus)
Agbesi, Fugar, and Adjei-Kumi [70]	Quantitative variance	Ghana	Questionnaire survey, structural equation modelling	123	Sustainable procurement, diffusion of innovation theory, technology–organisation–environment framework	8
Agyekum et al. [49]	Quantitative	Ghana	Questionnaire survey, statistical analysis, semi-structured interviews	104	Sustainable public procurement	1
Allen [73]	Case study	New Zealand	Develops unique conceptual framework for analysis	1	Procurement policy, social procurement	4
Barraket [47]	Case method, longitudinal (2 waves)	Australia	Focus group, semi-structured interviews, narrative and thematic analysis	2 focus groups, 14 interviews	"New social procurement", field theory, intermediation, social innovation	19
Barraket and Loosemore [14]	Qualitative case study	Australia, U.K.	In-depth semi-structured interview, thematic analysis	3	Cross-sector collaboration, social enterprise	33
Bohari et al. [61]	Quantitative variance	Malaysia	Questionnaire survey, structural equation modelling	100	Stakeholder value and green procurement	27
Brooks and Rich [46]	Socio-technical transitions study	U.K.	Empirical fieldwork, questionnaires, interviews, document analysis, code mapping	70 questionnaires, 9 interviews	Socio-technical transition theory	28
Delmonico et al. [48]	Quantitative variance	Brazil	Survey questionnaire, statistical analysis	54	Sustainable operations management, industrial sustainability, sustainable public procurement	74
Denny-Smith and Loosemore [15]	Indigenous research, constructivism	Australia	Electronic questionnaire survey	33	Indigenous enterprises	11
Denny-Smith et al. [44]	Qualitative descriptive	Australia	Opinion survey, thematic analysis	107	Social procurement, employer of choice	18
Ershadi et al. [63]	Qualitative descriptive	Australia	Opinion survey, thematic analysis, expert consensus,	20	Project and sustainable procurement management	3

Table A2. *Cont.*

Included Paper	Methodological Approach	Location	Data Collection and Analysis Methods	Sample Size	Theoretical Framework(s)/Key Literature	# Citations (Scopus)
Ershadi et al. [50]	Qualitative descriptive case study	Australia	Interviews, thematic analysis Delphi validation	1 case study, 6 interviews	Sustainable procurement management	3
Ewuga and Adesi [45]	Quantitative analytical	Ireland	Online survey, descriptive statistical analysis, interviews	62	Sustainable supply chain management, suppliers' development, resource-based view	0
Fuentes-Bargues, et al. [71]	Qualitative descriptive	Spain	Content analysis	967 tenders	Social procurement	1
Hurt-Suwan and Mahler [66]	Qualitative experimental	New Zealand	Questionnaire, thematic analysis	8	Social procurement. precarious employment	2
Lam [28]	Qualitative	U.K.	Questionnaire survey, regression analysis	60	Holistic sustainability	2
Loosemore [13]	Qualitative descriptive	U.K.	Interviews, narrative analysis	12	Social procurement, social enterprise	93
Loosemore, Alkilani, and Mathenge [10]	Quantitative	Australia	Online survey, descriptive and inferential statistical analysis	70	Social procurement	20
Loosemore, Alkilani, and Hammad [16]	Qualitative descriptive	Australia	Online survey, descriptive and inferential statistical analysis	79	Employment-seeking experiences of refugees and migrants	5
Loosemore, Alkilani, and Hammad [58]	Qualitative descriptive	Australia	Online survey, nonparametric statistics	25	Refugee job-seeking experiences	1
Loosemore, Alkilani, and Murphy [38]	Social constructivist and interpretivist epistemology	Australia	Semi-structured interviews and thematic analysis	16	New institutional theory	7
Loosemore, Bridgeman, and Keast [17]	Social constructivist and interpretivist epistemology, case study	Global	Multi-method (semi-structured interviews, text, and artefact analysis, observation)	11	Cross-sector collaboration	9
Loosemore, Bridgeman, Russell, and Zaid Alkilani [19]	Social constructivist and interpretivist epistemology, case study	U.K.	Semi-structured interviews and thematic and documentary analysis	1 case study/10 interviews	Homelessness, capability approach	1

Table A2. *Cont.*

Included Paper	Methodological Approach	Location	Data Collection and Analysis Methods	Sample Size	Theoretical Framework(s)/Key Literature	# Citations (Scopus)
Loosemore, Daniele, and Lim [18]	Qualitative descriptive	Australia	Online survey, nonparametric statistics	94	Ex-offenders	2
Loosemore, Denny-Smith, Barraket, Keast, Chamberlain, Muir, Powell, Higgon, and Osborne [27]	Social constructivist and interpretivist epistemology, case study	Australia	Focus groups, thematic analysis	1 case study, 5 focus groups	Cross-sector collaboration, social procurement	6
Loosemore, Higgon, and Osborne [29]	Social constructivist and interpretivist epistemology, case study	Australia	Semi-structured interviews, observation, document analysis, thematic analysis	1 case study, 73 interviews	New public governance, cross-sector collaboration	2
Loosemore, Keast, and Barraket [39]	Qualitative descriptive	Australia	Semi-structured interviews	15	Social procurement	1
Loosemore, Keast, and Barraket [40]	Social constructivist and interpretivist epistemology	Australia	Semi-structured interviews, thematic analysis	15	Champions and organisational theory	0
Loosemore, Keast, Barraket, Denny-Smith, and Alkilani [11]	Social constructivist and interpretivist epistemology, case study	Australia	Focus groups, thematic analysis	1 case study, 5 focus groups, 35 stakeholder groups	Social procurement, cross-sector collaboration	0
Loosemore, Osborne, and Higgon [69]	Case study	Australia	Semi-structured interviews, daily diaries, observation, assessments, survey	34	Facilities management	2
Loosemore and Reid [12]	Qualitative descriptive	Australia	Interviews, narrative analysis	8	Social procurement	13
Messah, Wirahadikusumah, and Abduh [62]	Quantitative variance	Indonesia	Questionnaire survey, multivariate analysis	49	Sustainable procurement	1
Murphy and Eadie [51]	Quantitative descriptive	Northern Ireland	Survey	50	Service innovation, social innovation, socially responsible procurement	19
Ogunsanya, Aigbavboa, and Thwala [60]	Quantitative variance	Nigeria	Questionnaire survey, structural equation modelling	317	Sustainable procurement	0

Table A2. *Cont.*

Included Paper	Methodological Approach	Location	Data Collection and Analysis Methods	Sample Size	Theoretical Framework(s)/Key Literature	# Citations (Scopus)
Ogunsanya, Aigbavboa, Thwala, and Edwards [52]	Quantitative variance	Nigeria	Questionnaire survey, factor analysis	320	Sustainable procurement	22
Renukappa, et al. [53]	Mixed methods	U.K.	Questionnaire survey	53 questionnaires, 17 interviews	Sustainable procurement	17
Ruparathna and Hewage [54]	Mixed methods	Canada	Questionnaire survey, semi-structured interviews, content analysis	30 questionnaires, 9 interviews	Sustainable construction procurement	26
Ruparathna and Hewage [68]	Mixed methods	Canada	Questionnaire survey, document review, semi-structured interviews	9 interviews, 30 questionnaires	Sustainable procurement	99
Staples and Dalrymple [59]	Qualitative case study	Australia	Semi-structured interviews	5 case studies, 20 interviews	Public sector procurement	2
Testa, Grappio, Gusmerotti, Iraldo, and Frey [55]	Qualitative descriptive	Italy	Content analysis	164 tenders	Green public procurement	55
Troje [42]	Qualitative descriptive	Sweden	Semi-structured interviews, thematic analysis	28	Social procurement policies	1
Troje and Andersson [37]	Qualitative descriptive	Sweden	Semi-structured interviews, thematic analysis	46	Social procurement. Institutional work	9
Troje and Gluch [41]	Qualitative descriptive case studies	Sweden	Interviews, thematic analysis	3 cases, 23 interviews	Social procurement, practice theory	7
Troje and Gluch [56]	Qualitative descriptive	Sweden	Interviews, thematic analysis	21	Professional identity, role, and work practices	26
Troje and Kadefors [20]	Mixed methods	Sweden	Case study, interviews	3	Social procurement	16
Waris, Panigrahi, Mengal, Soomro, Mirjat, Ullah, Azlan, and Khan [65]	Qualitative case study	Malaysia	Multicriteria decision making, questionnaire survey	1 case study, 10 questionnaires	Sustainable procurement	43

Table A2. *Cont.*

Included Paper	Methodological Approach	Location	Data Collection and Analysis Methods	Sample Size	Theoretical Framework(s)/Key Literature	# Citations (Scopus)
Willar, et al. [25]	Quantitative descriptive	Indonesia	Questionnaire survey, statistical analysis	158	Sustainable construction	17
Wirahadikusumah, et al. [24]	Mixed methods	Indonesia	Questionnaires, interviews	20	Sustainability	5
Zaidi, Mirza, Hou, and Ashraf [57]	Quantitative intuitive modelling	Pakistan	Questionnaire survey, interpretive structural modelling	43	Sustainable public procurement	37

References

1. Bonwick, M.; Daniels, M. *Corporate Social Procurement in Australia: Business Creating Social Value*; Social Traders: Melbourne, Australia, 2014.
2. Hamilton, S.G. Public procurement—price-taker or market-shaper? *Crit. Perspect. Int. Bus.* **2022**, *18*, 574–615. [CrossRef]
3. OECD. *Integrating Responsible Business Conduct in Public Procurement*; OECD Publishing: Paris, France, 2020.
4. Victorian Government. *Victoria's Social Procurement Framework*; The State of Victoria: Melbourne, Austrilia, 2022.
5. Government of New South Wales. *Procurement Policy Framework*; Government of New South Wales: Sydney, Australia, 2022.
6. Government of Western Australian. *Western Australian Social Procurement Framework*; Government of Western Australian: Perth, Australia, 2022.
7. Queensland Government. *Consider Social Procurement*; Queensland Government: Brisbane, Australia, 2021.
8. Public Services (Social Value) Act 2012. 2012. Available online: https://www.legislation.gov.uk/ukpga/2012/3/enacted (accessed on 11 August 2023).
9. Procurement Policy Note 06/20—Taking Account of Social Value in the Award of Central Government Contracts. UK Cabinet Office, Department for Culture, Media and Sport, and Department for Digital, Culture, Media and Sport. 2020. Available online: https://www.gov.uk/government/publications/procurement-policy-note-0620-taking-account-of-social-value-in-the-award-of-central-government-contracts (accessed on 11 August 2023).
10. Loosemore, M.; Alkilani, S.; Mathenge, R. The risks of and barriers to social procurement in construction: A supply chain perspective. *Constr. Manag. Econ.* **2020**, *38*, 552–569. [CrossRef]
11. Loosemore, M.; Keast, R.; Barraket, J.; Denny-Smith, G.; Alkilani, S. The risks and opportunities of social procurement in construction projects: A cross-sector collaboration perspective. *Int. J. Manag. Proj. Bus.* **2022**, *15*, 793–815.
12. Loosemore, M.; Reid, S. The social procurement practices of tier-one construction contractors in Australia. *Constr. Manag. Econ.* **2019**, *37*, 183–200. [CrossRef]
13. Loosemore, M. Social procurement in UK construction projects. *Int. J. Proj. Manag.* **2016**, *34*, 133–144. [CrossRef]
14. Barraket, J.; Loosemore, M. Co-creating social value through cross-sector collaboration between social enterprises and the construction industry. *Constr. Manag. Econ.* **2018**, *36*, 394–408.
15. Denny-Smith, G.; Loosemore, M. Integrating Indigenous enterprises into the Australian construction industry. *Eng. Constr. Archit. Manag.* **2017**, *24*, 788–808.
16. Loosemore, M.; Alkilani, S.Z.; Hammad, A.W. The job-seeking experiences of migrants and refugees in the Australian construction industry. *Build. Res. Inf.* **2021**, *49*, 912–929.
17. Loosemore, M.; Bridgeman, J.; Keast, R. Reintegrating ex-offenders into work through construction: A case study of cross-sector collaboration in social procurement. *Build. Res. Inf.* **2020**, *48*, 731–746.
18. Loosemore, M.; Daniele, F.; Lim, B.T.H. Integrating ex-offenders into the Australian construction industry. *Constr. Manag. Econ.* **2020**, *38*, 877–893. [CrossRef]
19. Loosemore, M.; Bridgeman, J.; Russell, H.; Zaid Alkilani, S. Preventing youth homelessness through social procurement in construction: A capability empowerment approach. *Sustainability* **2021**, *13*, 3127.
20. Troje, D.; Kadefors, A. Employment requirements in Swedish construction procurement–institutional perspectives. *J. Facil. Manag.* **2018**, *16*, 284–298.
21. Barraket, J.; Keast, R.; Furneaux, C. *Social Procurement and New Public Governance*; Routledge: New York, NY, USA, 2016.
22. Cheng, W.; Appolloni, A.; D'Amato, A.; Zhu, Q. Green Public Procurement, missing concepts and future trends—A critical review. *J. Clean. Prod.* **2018**, *176*, 770–784.
23. Furneaux, C.; Barraket, J. Purchasing social good(s): A definition and typology of social procurement. *Public Money Manag.* **2014**, *34*, 265–272. [CrossRef]
24. Wirahadikusumah, R.; Abduh, M.; Messah, Y.; Aulia, M. Introducing sustainability principles into the procurement of construction works—case of Indonesian developers. *Int. J. Constr. Manag.* **2021**, *21*, 932–944. [CrossRef]
25. Willar, D.; Waney, E.V.Y.; Pangemanan, D.D.G.; Mait, R.E.G. Sustainable construction practices in the execution of infrastructure projects: The extent of implementation. *Smart Sustain. Built Environ.* **2021**, *10*, 106–124. [CrossRef]
26. Hutchins, M.J.; Sutherland, J.W. An exploration of measures of social sustainability and their application to supply chain decisions. *J. Clean. Prod.* **2008**, *16*, 1688–1698.
27. Loosemore, M.; Denny-Smith, G.; Barraket, J.; Keast, R.; Chamberlain, D.; Muir, K.; Powell, A.; Higgon, D.; Osborne, J. Optimising social procurement policy outcomes through cross-sector collaboration in the Australian construction industry. *Eng. Constr. Archit. Manag.* **2020**, *28*, 1908–1928. [CrossRef]
28. Lam, T.Y. Driving sustainable construction development through post-contract key performance indicators and drivers. *Smart Sustain. Built Environ.* **2022**, *11*, 483–499.
29. Loosemore, M.; Higgon, D.; Osborne, J. Managing new social procurement imperatives in the Australian construction industry. *Eng. Constr. Archit. Manag.* **2020**, *27*, 3075–3093.
30. Page, M.J.; McKenzie, J.E.; Bossuyt, P.M.; Boutron, L.; Hoffmann, T.C.; Mulrow, C.D. The PRISMA 2020 statement: An updated guideline for reporting systematic reviews. *BMJ* **2021**, *372*, n71. [CrossRef]
31. Okoli, C. A Guide to Conducting a Standalone Systematic Literature Review. *Commun. Assoc. Inf. Syst.* **2015**, *37*, 879–910.

32. Tong, A.; Sainsbury, P.; Craig, J. Consolidated criteria for reporting qualitative research (COREQ): A 32-item checklist for interviews and focus groups. *Int. J. Qual. Health Care* **2007**, *19*, 349–357. [CrossRef] [PubMed]
33. Guest, G.; MacQueen, K.M.; Namey, E.E. *Applied Thematic Analysis*; SAGE Publications, Inc.: Thousand Oaks, CA, USA, 2012.
34. Saldana, J. *The Coding Manual for Qualitative Researchers*, 2nd ed.; Sage: London, UK, 2013.
35. Walker, H.; Brammer, S. The relationship between sustainable procurement and e-procurement in the public sector. *Int. J. Prod Econ.* **2012**, *140*, 256–268.
36. Amann, M.; Roehrich, J.K.; Eßig, M.; Harland, C. Driving sustainable supply chain management in the public sector: The importance of public procurement in the European Union. *Supply Chain Manag. Int. J.* **2014**, *19*, 351–366.
37. Troje, D.; Andersson, T. As above, not so below: Developing social procurement practices on strategic and operative levels. *Equal Divers. Incl. Int. J.* **2021**, *40*, 242–258.
38. Loosemore, M.; Alkilani, S.Z.; Murphy, R. The institutional drivers of social procurement implementation in Australian construction projects. *Int. J. Proj. Manag.* **2021**, *39*, 750–761.
39. Loosemore, M.; Keast, R.; Barraket, J. How are social procurement actors shaping a new field of professional practice in the Australian construction industry? *Eng. Constr. Archit. Manag.* **2022**. ahead-of-print. [CrossRef]
40. Loosemore, M.; Keast, R.; Barraket, J. A typology of social procurement champions in the construction and engineering industry *Constr. Manag. Econ.* **2022**, *40*, 391–405.
41. Troje, D.; Gluch, P. Beyond policies and social washing: How social procurement unfolds in practice. *Sustainability* **2020**, *12*, 4956
42. Troje, D. Policy in practice: Social procurement policies in the Swedish construction sector. *Sustainability* **2021**, *13*, 7621. [CrossRef]
43. Lam, T.Y.M. A sustainable procurement approach for selection of construction consultants in property and facilities management *Facilities* **2020**, *38*, 98–113. [CrossRef]
44. Denny-Smith, G.; Sunindijo, R.Y.; Loosemore, M.; Williams, M.; Piggott, L. How construction employment can create social value and assist recovery from COVID-19. *Sustainability* **2021**, *13*, 988. [CrossRef]
45. Ewuga, D.; Adesi, M. Integrating sustainability practices into the Irish construction supply chain: Main contractors' perspective *Built Environ. Proj. Asset Manag.* **2023**, *13*, 105–122. [CrossRef]
46. Brooks, A.; Rich, H. Sustainable construction and socio-technical transitions in London's mega-projects. *Geogr. J.* **2016**, *182*, 395–405. [CrossRef]
47. Barraket, J. The role of intermediaries in social innovation: The case of social procurement in Australia. *J. Soc. Entrep.* **2020**, *11*, 194–214. [CrossRef]
48. Delmonico, D.; Jabbour, C.J.C.; Pereira, S.C.F.; Jabbour, A.B.L.d.S.; Renwick, D.W.S.; Tavares Thome, A.M. Unveiling barriers to sustainable public procurement in emerging economies: Evidence from a leading sustainable supply chain initiative in Latin America. *Resour. Conserv. Recycl.* **2018**, *134*, 70–79. [CrossRef]
49. Agyekum, A.K.; Fugar, F.D.K.; Agyekum, K.; Akomea-Frimpong, I.; Pittri, H. Barriers to stakeholder engagement in sustainable procurement of public works. *Eng. Constr. Archit. Manag.* **2022**. ahead-of-print. [CrossRef]
50. Ershadi, M.; Jefferies, M.; Davis, P.; Mojtahedi, M. Barriers to achieving sustainable construction project procurement in the private sector. *Clean. Eng. Technol.* **2021**, *3*, 100125. [CrossRef]
51. Murphy, M.; Eadie, R. Socially responsible procurement A service innovation for generating employment in construction. *Built Environ. Proj. Asset Manag.* **2019**, *9*, 138–152. [CrossRef]
52. Ogunsanya, O.A.; Aigbavboa, C.O.; Thwala, D.W.; Edwards, D.J. Barriers to sustainable procurement in the Nigerian construction industry: An exploratory factor analysis. *Int. J. Constr. Manag.* **2022**, *22*, 861–872. [CrossRef]
53. Renukappa, S.; Akintoye, A.; Egbu, C.; Suresh, S. Sustainable procurement strategies for competitive advantage: An empirical study. *Proc. Inst. Civ. Eng. Manag. Procure. Law* **2016**, *169*, 17–25. [CrossRef]
54. Ruparathna, R.; Hewage, K. Sustainable procurement in the Canadian construction industry: Challenges and benefits. *Can. J. Civ. Eng.* **2015**, *42*, 417–426. [CrossRef]
55. Testa, F.; Grappio, P.; Gusmerotti, N.M.; Iraldo, F.; Frey, M. Examining green public procurement using content analysis: Existing difficulties for procurers and useful recommendations. *Environ. Dev. Sustain.* **2016**, *18*, 197–219. [CrossRef]
56. Troje, D.; Gluch, P. Populating the social realm: New roles arising from social procurement. *Constr. Manag. Econ.* **2020**, *38*, 55–70. [CrossRef]
57. Zaidi, S.A.H.; Mirza, F.M.; Hou, F.; Ashraf, R.U. Addressing the sustainable development through sustainable procurement: What factors resist the implementation of sustainable procurement in Pakistan? *Socio-Econ. Plan. Sci.* **2019**, *68*, 100671. [CrossRef]
58. Loosemore, M.; Alkilani, S.Z.; Hammad, A.W.A. Barriers to employment for refugees seeking work in the Australian construction industry: An exploratory study. *Eng. Constr. Archit. Manag.* **2022**, *29*, 619–642. [CrossRef]
59. Staples, W.; Dalrymple, J. Construction procurement and State Government strategy: Aligned or disconnected? *Aust. J. Public Adm.* **2016**, *75*, 222–235. [CrossRef]
60. Ogunsanya, O.A.; Aigbavboa, C.O.; Thwala, D.W. Sustainable procurement model for publicly funded construction projects in developing nations–a structural equation modeling approach. *Int. J. Constr. Manag.* **2021**, *23*, 1–24. [CrossRef]
61. Bohari, A.A.M.; Skitmore, M.; Xia, B.; Teo, M.; Khalil, N. Key stakeholder values in encouraging green orientation of construction procurement. *J. Clean. Prod.* **2020**, *270*, 122246. [CrossRef]
62. Messah, Y.; Wirahadikusumah, R.; Abduh, M. Structural equation model (SEM) of the factors affecting sustainable procurement for construction work. *Int. J. Constr. Manag.* **2022**, *23*, 1–9. [CrossRef]

63. Ershadi, M.; Jefferies, M.; Davis, P.; Mojtahedi, M. Achieving sustainable procurement in construction projects: The pivotal role of a project management office. *Constr. Econ. Build.* **2021**, *21*, 45–64. [CrossRef]
64. Cumpston, M.; Li, T.; Page, M.J.; Chandler, J.; Welch, V.A.; Higgins, J.P.; Thomas, J. Updated guidance for trusted systematic reviews: A new edition of the Cochrane Handbook for Systematic Reviews of Interventions. *Cochrane Database Syst. Rev.* **2019**, *2019*, ED000142. [CrossRef]
65. Waris, M.; Panigrahi, S.; Mengal, A.; Soomro, M.I.; Mirjat, N.H.; Ullah, M.; Azlan, Z.S.; Khan, A. An application of analytic hierarchy process (AHP) for sustainable procurement of construction equipment: Multicriteria-based decision framework for Malaysia. *Math. Probl. Eng.* **2019**, *2019*, 6391431. [CrossRef]
66. Hurt-Suwan, C.J.P.; Mahler, M.L. Social procurement to reduce precarious employment for Māori and Pasifika workers in the construction industry. *Kōtuitui N. Z. J. Soc. Sci. Online* **2021**, *16*, 100–115. [CrossRef]
67. Groote, M.d.; Lefever, M. *Driving Transformational Change in the Construction Value Chain: Reaching the Untapped Potential*; Buildings Performance Institute Europe: Brussels, Belgium, 2016.
68. Ruparathna, R.; Hewage, K. Sustainable procurement in the Canadian construction industry: Current practices, drivers and opportunities. *J. Clean. Prod.* **2015**, *109*, 305–314. [CrossRef]
69. Loosemore, M.; Osborne, J.; Higgon, D. Affective, cognitive, behavioural and situational outcomes of social procurement: A case study of social value creation in a major facilities management firm. *Constr. Manag. Econ.* **2021**, *39*, 227–244. [CrossRef]
70. Agbesi, K.; Fugar, F.D.; Adjei-Kumi, T. Modelling the adoption of sustainable procurement in construction organisations. *Built Environ. Proj. Asset Manag.* **2018**, *8*, 461–476. [CrossRef]
71. Fuentes-Bargues, J.L.; Bastante-Ceca, M.J.; Ferrer-Gisbert, P.S.; González-Cruz, M.C. Analysis of the situation of social public procurement of works at the valencian region (Spain). *Sustainability* **2021**, *13*, 175. [CrossRef]
72. Natoli, R.; Lou, C.X.; Goodwin, D. Addressing Barriers to Social Procurement Implementation in the Construction and Transportation Industries: An Ecosystem Perspective. *Sustainability* **2023**, *15*, 11347. [CrossRef]
73. Allen, B. Broader outcomes in procurement policy–a case of New Zealand pragmatism. *J. Public Procure.* **2021**, *21*, 318–341. [CrossRef]

sustainability

MDPI

Article

Addressing Barriers to Social Procurement Implementation in the Construction and Transportation Industries: An Ecosystem Perspective

Riccardo Natoli [1,*], Catherine Xiaocui Lou [1] and David Goodwin [2]

[1] Institute for Sustainable Industries and Liveable Cities, Victoria University, Melbourne, VIC 3030, Australia; catherine.lou@vu.edu.au

[2] Nottingham University Business School, University of Nottingham Malaysia, Semenyih 43500, Malaysia; david.goodwin@nottingham.edu.my

* Correspondence: riccardo.natoli@vu.edu.au

Abstract: Although social procurement is viewed as an important part of social value creation, barriers to its implementation have resulted in a failure to realise the full societal benefits it was designed to achieve. As a key area of activity for government procurement projects, the construction and transportation industries have a big role to play in contributing positively to societal outcomes While prior studies have identified barriers from specific cohorts, no prior study has approached this from the perspective of the key stakeholders throughout the social procurement ecosystem within the construction and transport industries. To address this gap in social procurement research, interviews and a focus group totalling 42 participants were undertaken. Participants ranged from those implementing policy (government representatives), tendering for contracts (tier one companies) and providing specialised social procurement services (social enterprises), along with key intermediary support bodies. Results indicate that barriers exist throughout the entire social procurement ecosystem and highlight the need to develop an enhanced social procurement ecosystem capable of maximising the societal benefit that arises from social procurement. These findings provide a set of strategies for the key stakeholders in the ecosystem to consider adopting to improve social procurement outcomes.

Keywords: barriers; construction industry; transportation; social procurement; policy; practice; social enterprise; ecosystem

Citation: Natoli, R.; Lou, C.X.; Goodwin, D. Addressing Barriers to Social Procurement Implementation in the Construction and Transportation Industries: An Ecosystem Perspective. *Sustainability* **2023**, *15*, 11347. https://doi.org/10.3390/su151411347

Academic Editors: Zora Vrcelj and Malindu Sandanayake

Received: 17 June 2023
Revised: 10 July 2023
Accepted: 18 July 2023
Published: 21 July 2023

1. Introduction

Social procurement extends the idea of 'traditional' procurement by requiring that the supply chain delivers social benefit outcomes in addition to the goods and services being purchased [1,2]. Many governments practice social procurement (e.g., the United Kingdom [UK], Canada, South Africa, and the European Union [EU]). In Australia, the Victorian State Government announced its view that social procurement outcomes will accrue to all Victorians when the social and sustainable outcomes in the social procurement framework are achieved [3].

Definitions of social procurement focus on the intention of the buyer–supplier relationship to bring about additional value that would not be delivered by traditional procurement relationships. Organisations use their buying power to generate social value beyond the value of the goods, services, or construction being procured [3]. Governments can unleash significant untapped social value potential from their existing procurement spending by requiring construction firms to give back to the communities in which they build [4]. Specifically, governments use their position as volume buyers to influence their social procurement policies [5]. Prior studies confirm that the construction industry plays a vital role in the adoption of social procurement practices [6,7]. This potential is evidenced by research commissioned on social procurement in the West of Melbourne region. As per Figure 1 below, the report indicates that every AUD 100 million spent on construction with local businesses creates AUD 237 million of economic impact and 580 local jobs [8] while social procurement has the potential to contribute over 450 jobs for people in target cohorts [9].

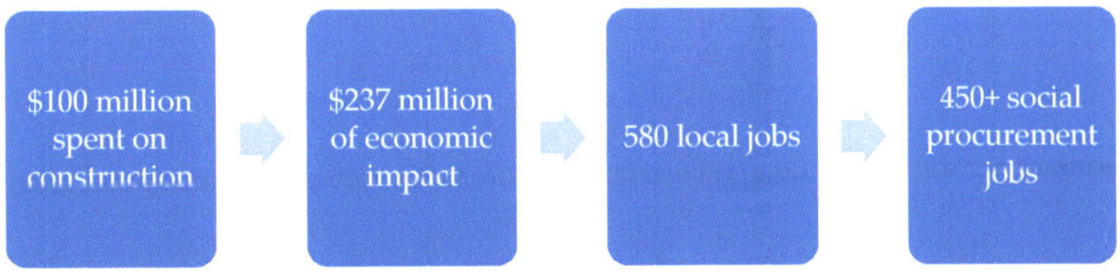

Figure 1. Construction flow-on effects to social procurement.

The motivation for this study is to address the paucity of empirical research examining social procurement issues throughout the whole supply chain [5] where poor social procurement implementation in construction could be due to constraints in their systems, structures and competencies [10]. The importance of more extensive research into social procurement barriers [4] had led some to opine that the literature on the barriers of promoting social procurement throughout supply chains had received little attention [11].

The focus on the whole supply chain refers to the business ecosystem which is a network of organisations—including suppliers, distributors, government agencies—involved in the delivery of a specific product or service [12]. Since each organisation affects the other, understanding their interactions within the ecosystem is important for its effectiveness. In the present research, the social procurement ecosystem refers to those who implement social procurement policy (government representatives), and those tendering for contracts (tier one contractors) (a tier one contractor is capable of delivering mega-projects over AUD 1 billion and has the ability to self-perform most of the required work on a project with its own employees), along with those providing specialised social procurement services (social enterprises) (the term social enterprises is used in this paper to denote both certified social enterprises and small and medium enterprises (SMEs) involved in the social procurement space) and key intermediary support bodies. The adoption of the ecosystem approach also responds to recent calls to include the perspective of policymakers in future social procurement research [13].

Prior studies assert that, unless the capacity of those within the social procurement ecosystem to deliver social value is considered, there is a danger that governments will fail to achieve their increasingly ambitious goals [14,15]. Problems with capacity have been linked to, among other things, barriers that key social procurement stakeholders face when trying to activate the social value that, theoretically, can arise from social procurement policies. It is no surprise therefore that, considered as a whole, social procurement projects have failed to realise the full societal benefits they were designed to achieve. Prior studies have identified barriers applicable to specific cohorts, but this paper seeks to contribute to the literature by answering the call to better understand the existing gamut of capacity and capability issues that are limiting the potential of social procurement [16].

The aim of this paper is to investigate perceived barriers to effectively implementing social procurement within construction and transportation infrastructure projects viewed from the perspective of key stakeholders in the social procurement ecosystem. Key stakeholders comprise: (i) social enterprises, (ii) tier one contractors, (iii) government representatives and (iv) key intermediary support bodies. The approach taken here, of including the perspectives of key stakeholders drawn from across the social procurement ecosystem, can be distinguished from the vast majority of published social procurement studies, which typically concentrate on one cohort. To achieve the main aim of this paper, the following research question is addressed:

RQ1: What do key stakeholders perceive to be the main barriers to the effective implementation of social procurement in the construction and transport industries?

By investigating this question, applying new institutional theory, support networks and organisational capability, this paper contributes to the emerging body of social procurement research by responding to the need for more construction and transportation research in this field. Answering this question is important as it provides a more comprehensive social procurement perspective, while also offering a more nuanced understanding of the challenges involved for each of the key stakeholders in the construction and transportation industries. This more nuanced understanding of areas benefitting from social procurement could assist governments to better utilise their policy implementation to achieve greater social value creation. To achieve this, interviews and a focus group comprising 42 participants were undertaken. Participants were obtained from the aforementioned four key stakeholder groupings who have responsibilities for implementing policy (government representatives), tendering for contracts (tier one contractors), providing specialised social procurement services (social enterprises) and providing support (key intermediary bodies).

The remainder of this paper is structured as follows. Section 2 provides an overview of the key literature on social procurement barriers. Section 3 outlines the research method adopted by this paper. Section 4 provides the results and discussion from the study while Section 5 draws links to the social procurement ecosystem. Section 6 concludes with some strategic recommendations arising from the research.

2. Literature Review

2.1. Social Procurement in Construction and Transportation Industries

As sustainability management increasingly makes its mark on the business landscape, social procurement, with its focus on attaining benefits beyond value for money, is being increasingly adopted by governments. There has been strong uptake in the UK, Canada, South Africa, the EU, as well as Australia. This has been supported by socially conscious private sector organisations keen to attain their corporate social responsibility objectives [2]. By changing their procurement policies, governments are recognising the role social procurement plays in contributing to sustainability outcomes and social value creation [17].

In 2018, the Victorian Government's Department of Treasury and Finance introduced a Social Procurement Framework (SPF). The SPF applies to all Victorian Government departments and agencies when they procure goods, services and construction at certain threshold conditions. The Victorian Government set up the SPF to add value to government purchases by: (i) creating jobs and skills-based training opportunities for local priority jobseekers; (ii) increasing business opportunities for social enterprises; and (iii) delivering social, economic and environmental benefits. The SPF includes procurement objectives for social impact and environmentally sustainable outcomes. It also targets outcomes for selected priority groups such as: Indigenous Victorians, Victorians with disability, disengaged youth, long-term unemployed people, migrants, refugees and asylum seekers, single parents and workers in transition [18].

The construction and transport industry is one of the main beneficiaries of government purchases via road, rail and infrastructure programs. These programs can act as a major catalyst to achieve social, economic and environmental benefits. Although the construction and transport industries offer enormous opportunities to implement social procurement, issues such as cross-sector collaboration with social enterprises [10] and notions of what an 'ideal' construction worker looks like [19] can be problematic. This is exacerbated by the fact that noncompliance under the SPF can lead to construction and transport infrastructure organisations being potentially struck off tender lists with government agencies. Thus, understanding the barriers that can hinder construction and transport industry participants from meeting their targets is increasingly important. The following subsection provides a brief overview of the main barriers to social procurement implementation in the construction and transport industries that have been identified in the literature.

2.2. Main Barriers to Social Procurement Implementation in Construction and Transportation

A review of the literature on social procurement in construction and transportation revealed six main barriers within these industries. The first main barrier concerned a lack of understanding and awareness of social procurement. An Indonesian study of owners of construction businesses mentions this barrier [1]. This finding reinforced an earlier study that interviewed eight tier one contractor senior managers and demonstrated that social procurement in construction is delivered mainly by existing industry incumbents who do not understand social procurement requirements [16]. This absence of people who understand social procurement hinders the monitoring of their implementation in practice as those with a lack of understanding demonstrate a tendency to view it as yet another compliance burden. The problem also manifests in inadequate training and a lack of platforms to exchange information and knowledge. This barrier was also identified in other construction social procurement studies [5,20–22].

A second main barrier is the perceived limited capacity of existing construction supply chain partners to deliver on social procurement clause requirements—particularly as many only work on projects for short periods of time [23]. Capacity issues can manifest via a lack of skilled labour [21,24]. Awareness of this barrier was reinforced by a study which examined social enterprise leaders in UK construction and found that there was a lack of supply of credible organisations capable of undertaking meaningful construction work [4]. Capacity gaps have also been identified in other social procurement studies [13,24–26].

A third main barrier is a lack of meaningful collaboration which has also been cited by social procurement experts and professionals as obstructing the integration of social enterprise organisations into supply chains [20]. This can occur via a lack of motivation and unco-operative attitudes [20,21]. For instance, there seems to be a perceived lack of trust in the ability of social enterprises to deliver work to the same standards as existing subcontractors [16]. This high level of suspicion leads to low social procurement engagement and resistance to change which adversely impacts the effective implementation of social procurement [14].

The fourth main barrier to social procurement implementation within the construction and transportation sector is the highly regulated nature of the industry and the difficulties which social service providers encounter in securing the necessary licences and certifications to even prequalify to tender on construction projects [4]. Other examples include a lack of technical guidelines regarding its implementation and systemised practices [25,26]. A recent study which undertook five focus group case studies disclosed that stakeholders perceived the way policies are being implemented to be unjust and appearing to be counter to effective risk management, leading to social procurement being viewed as more of a risk than an opportunity [24]. A lack of government support has been cited by other studies [5,21,27].

A fifth main barrier was the costs and administration effort associated with the implementation of social procurement. The pressure to pursue the lowest price has led to hesitancy to become involved in social procurement on a large scale [16,24]. A study of Indonesian construction practitioners showed that a number of specialised social procurement service providers were routinely more expensive to work with than nonproviders, resulting in those practitioners feeling hampered by the administrative burden associated with social procurement [26]. This was also identified in a study on sustainable procurement for public sector universities in Pakistan [28].

Finally, a sixth main barrier, organisational orientation, can act as an impediment to the effective implementation of social procurement. Specifically, the level of perceived pressure from competing construction and transportation industries impacts the extent of adoption of social procurement [4,16]. For instance, a study of construction industry professionals in Nigeria demonstrated that the attitudes of close competitors to social procurement and the need to retain a competitive advantage can act as barriers, or enablers, to social procurement [5]. Other studies cite similar effects [14,26,29].

Other barriers have been identified in the social procurement literature such as organisations lacking a long-term perspective, greater capital cost of research and development and resistance to change [21,24,27]. Although not main barriers, they comprise internal limitations among construction and transport organisations in the implementation of social procurement.

The studies reviewed above identified barriers from specific cohorts (e.g., tier one contractors, owners of construction businesses, social enterprise leaders, construction practitioners, construction industry professionals, etc.). By adopting an ecosystem perspective, this paper extends the literature and addresses the call for an integrated analysis of the main social procurement barriers in the construction and transport industries [5,11]. To achieve this, an analysis involving key stakeholder groups throughout the social procurement ecosystem is undertaken, focusing specifically on the context of construction and transportation infrastructure projects.

3. Research Method

To build upon the main barriers identified in the previous section, key stakeholders in the construction and transport infrastructure social procurement ecosystem were contacted. A qualitative analysis approach was employed to obtain an in-depth understanding of the perceived barriers to implementing social procurement in the construction and transport industries. The qualitative lens allowed for a more nuanced explanation of the research topic [30]. Following ethics approval, the qualitative data was collected through in-depth interviews and a focus group conducted online.

3.1. Data Collection

Purposive sampling was used to recruit interviewees [31]. To reflect the range of experiences of key stakeholders in the social procurement ecosystem, interview and focus group participants were selected from those exhibiting a range of expertise to capture the complexity that exists within the social procurement space. Specifically, a cross-representative selection of those with first-hand experience dealing with implementing policy (government representatives), tendering for contracts (tier one contractors) and those providing the specialised social procurement services (social enterprises and SMEs) were selected, along with representatives of key intermediary support bodies. As the State Government is the primary driving force behind social procurement in Victoria, tier one contractors had to have experience in large-scale construction and infrastructure projects. In keeping with the purposive sampling approach, communications with relevant people (e.g., procurement managers) were undertaken to identify the most knowledgeable person available to answer the interview questions. Key intermediary support bodies were identified primarily via established contacts from the research team. Their selection was based on their extensive experience with social procurement and acknowledged dealings with tier one contractors, social enterprises and state government. Government representatives were selected based on contact with the government department responsible for the social procurement policy. Finally, social enterprises had to have experience in the social procurement space; hence, they were either certified social enterprises or SMEs involved in social procurement. Their selection occurred via two main ways: (i) referrals from key intermediary bodies and (ii) responses to a call out via the internet for participants with social procurement experience within the construction and infrastructure industries. From there, a snowball strategy was employed.

Due to the COVID-19 pandemic, all interviews and the focus group dialogue were conducted via Zoom, following respondent consent. Emails were used to organise a day and time that reflected their preferred availability. Overall, in-depth interviews with 35 key stakeholders were conducted between March 2022 and October 2022. Interviews were stopped upon reaching data saturation [32]. Each interview lasted between 60 and 90 min and were audio recorded via Zoom. The focus group, conducted in May 2022, comprised service providers and consultants, along with major industry organisations with recent and current experience working on social procurement delivery to major projects in the western region of Melbourne (see Appendix A: Research Participants).

3.2. Qualitative Data Analysis

A total of 35 semi-structured interviews were conducted with key social procurement stakeholders. Broad individualisable questions were utilised along with spontaneous questions that followed up on unanticipated issues raised by respondents. In addition, a focus group of seven experienced industry representatives actively engaged in social procurement practice was formed. To ensure consistency and quality, two researchers conducted the interviews and focus group along with the transcriptions. Each transcript was then validated by a different member of the research team, which involved comparing the audio with the written transcript [33]. A content analysis of the transcripts was conducted for coding purposes which involved immersion and engagement with the interview data to identify themes [34]. The consistency and validity of the codes were checked through an intercoding technique [35], with the interview data analysed thematically. Figure 2 below presents the data analysis steps.

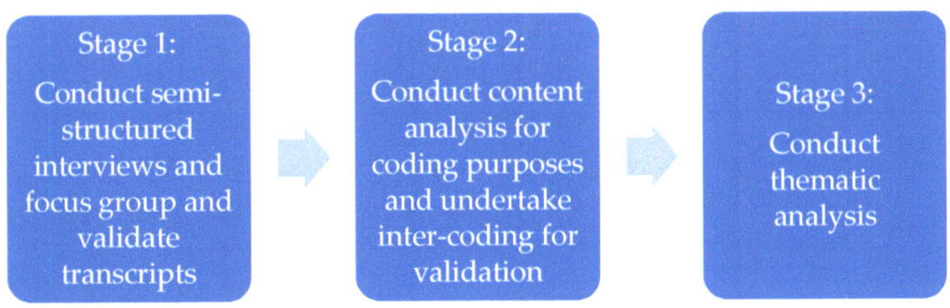

Figure 2. Data analysis steps.

4. Qualitative Results

The thematic analysis identified a total of three main themes and 14 subthemes that comprise key barriers to social procurement implementation in the construction and transportation industries. These are presented in Table 1 below while their discussion occurs in the associated subsection in parentheses.

Table 1. Key barriers to social procurement—a thematic presentation.

Area	Theme	Subthemes
Social Procurement Implementation in Construction and Transport	Supply Chain Process (Section 4.1)	Supply Chain Pressures (Section 4.1.1)
		Client Pressures (Section 4.1.2)
		Union Opposition (Section 4.1.3)
	Capacity and Capabilities of Social Enterprises (Section 4.2)	Capacity to Scale (Section 4.2.1)
		Non-Competitiveness of Social Enterprises (Section 4.2.2)
		Social Enterprise Capabilities (Section 4.2.3)
		Social Enterprise Resourcing (Section 4.2.4)
		Poor Cash Flows (Section 4.2.5)
	Support Networks (Section 4.3)	Lack of Support Networks (Section 4.3.1)
		Provision and Quality of Databases (Section 4.3.2)
		Certification and Accreditation Programs (Section 4.3.3)
		Social Procurement Education Support (Section 4.3.4)
		Supporting Priority Groups (Section 4.3.5)
		Tendering Process (Section 4.3.6)

4.1. Supply Chain Process

The research participants identified the supply chain process as a barrier to social procurement implementation in the construction and transport industries. Within this theme, the notion of pressure was a major issue—with both supply chain pressure and pressure from clients identified as barriers. In addition, trade union opposition was also identified as impacting the supply chain process and ultimately impacting the effectiveness of social procurement implementation. The first subtheme is reviewed below.

4.1.1. Supply Chain Pressures

A number of participants throughout the ecosystem identified supply chain pressures as a barrier to the effective implementation of social procurement. One main reason was that a number of existing suppliers were not capable of meeting social procurement requirements. Thus, when it came time to procure goods and services to meet social procurement requirements, tier one contractors had to use suppliers they had not worked with previously. This was viewed as a higher risk proposition.

> *It's very evident very early on that that risk is at the heart of most of these (supply chain) decisions, and that we're now creating greater risk, because we're asking them to use different suppliers that they're not necessarily used to. [Hence], there's probably a lot of suspicion around their (social enterprises and Indigenous businesses) ability to deliver … there was absolutely no desire to change supply chains, from what I could see.* [P2]

Thus, reluctance from tier one contractors to change suppliers was quite evident from both the social enterprise and intermediary perspectives as evidenced in the following quotes:

> *You'll sit in a room talking about social enterprise to subcontractors … and there's a little bit more resistance because there's: (a) why should I do this? (b) what's it going to cost? (c) who are you? (d) is the quality going to be as good?* [P3]

> *If you look at the major projects, it's the same social benefits suppliers [enterprises] that generally get engaged on every project. And of course, that changes over time because some are emerging but there's not a huge diversity of businesses being engaged.* [P30]

As tier one contractors required suppliers that were proficient with social procurement, many were having trouble identifying suitable supply partners. The resultant search was seen as an inefficient use of time which increased overall costs. The lack of maturity in the social enterprises landscape has meant that demand has outstripped supply—a notion that was highlighted by all key stakeholder groups.

> *If you look at the percentage of span that those projects need to direct to social enterprises, or Aboriginal owned businesses, I don't feel at the moment, there's enough of us to be able to meet that demand. So there's a supply side constraint that needs to be fixed.* [P12]

> *It's a little bit more challenging to find (social procurement) businesses as opposed to non-social procurement businesses. So that's probably one of the biggest barriers. There's not a pool of businesses out there per se that you can just call on when you need them and the challenge is, obviously, every major project in Victoria is fighting over that pool of talent as well.* [P33]

> *… sometimes it's about supply and demand. So are there enough social enterprises or Aboriginal businesses in the region or the area or suburb that a particular project is working in, that makes it more accessible to buy from those companies or to employ people?* [P36]

Supply chain pressures have been identified in prior studies as a barrier to social procurement implementation in construction and transportation [4,23]. To mitigate this known barrier, a potential area for action is to diversify the supply chain by looking beyond merely risk and price as the two differentiating factors. This requires a broader perspective of what value creation and social value is—the key outcomes of social procurement.

Furthermore, there is an opportunity for government to invest in the supply base to help make social procurement sustainable. This would assist in meeting the government's own social procurement targets and ultimately reap returns in terms of the economic benefits and employment.

4.1.2. Client Pressures

Some social enterprises spoke about the pressures clients placed on them regarding job performance due to differing priorities between the tender team and implementation team as well as the pressure to adequately meet their mandated targets set by government.

> *The tender team is different to the implementation team. Big Issue. Promise the world in the tender, then the implementation team comes in and they go, I can't do that.* [P2]

> *Did they actually do that (achieve their target)? And how did they do that? Or were they ... padding their outcomes, or were they using the same labour hire contracts that use the same ... people and move them around to different job sites and double counted them all at once?* [P4]

Moreover, other social enterprises opined that they were treated the same as traditional commercial operators even though what got them through the door in the first place was being a social enterprise supplier. This expectation meant that tier one contractors were not necessarily factoring social enterprise specific complexities when evaluating work progress/efficiency.

> *You're working with people who have got significant mental health issues. So you know, absenteeism can be quite high, you're working with people who have undergone significant trauma, which leads to mental health issues, anxiety, all sorts of stuff. So their productivity on certain days can be very low. You're trying to compensate for I mean, basically, commercial enterprises are all about productivity and efficiency and so on. Whereas a lot of these people are not able to operate at that level because of all the trauma and family violence, and whatever else is going on in their lives.* [P17]

Typically, the traditional position is that construction and infrastructure projects are delivered the way a client wants it [5]. As social procurement increasingly becomes a mandatory part of major project builds this mindset should shift, however as this theme demonstrates, the shift is yet to occur in a substantial practical manner.

4.1.3. Trade Union Opposition

Trade union opposition was also cited as a barrier to social procurement implementation in the construction and transport industries. The instances were specific to times where the social procurement contracted agreement was seen to be in conflict with trade union objectives.

> *I've had the union get involved and basically say that I can't have my people on their sites. ... they're not interested in supporting me and helping me to be on.* [P7]

> *They (social enterprises) find it hard to break in, or they run into issues with the union around their agreements. You know, it's a difficult space in some ways.* [P30]

This finding adds to the very limited body of knowledge about this barrier to social procurement implementation [24]. From a main theme perspective, the supply chain process is already experiencing difficulties in managing the transition to greater social procurement activities. Unless further government support is provided, such supply chain pressures will only be exacerbated by the influx of construction and transport infrastructure projects in the pipeline.

4.2. Capacity and Capabilities of Social Enterprises

A second main theme identified by research participants focused on the capacity and capabilities of social enterprises. Capacity gaps have also been identified in other social procurement studies [13,24–26]. Within this theme, the ability for social enterprises capacity to scale was seen as a major barrier to effective social procurement implementation, as were social enterprise's capabilities, resourcing, noncompetitiveness and poor cash flows. The first subtheme is reviewed below.

4.2.1. Capacity to Scale

Finding a social enterprise that can work at the scale required to service the social procurement project was seen as a major implementation barrier. One reason put forward by social enterprises was due to their relative infancy in the social procurement area.

> *... Tier one organisations want to procure their services but social enterprises can't deliver at that scale. ... a match of scale of demand and supply currently does not exist.* [P4]

> *The ability to scale is a definite barrier and is something that we're grappling with at the moment.* [P17]

> *It's interesting when you look at the social enterprise space across Australia, and certainly in Victoria, it's still relatively in its infancy ... we're not at scale yet. I think once we get to scale, that's when you'll start to see us (social enterprises) compete more effectively in the market, but we're still learning by doing at the moment.* [P12]

In fact, social enterprises and intermediaries identified the difficulty for social enterprises to access support to develop relevant skill sets to increase their capacity to scale.

> *Implementation is quite important and how we can scale that into the current business model. ... We have got so much work right now but we don't have the capacity of having the support (ongoing training and development) and employee workforce ... so there is a gap there that we're trying to fill and fix and grab and grow.* [P16]

> *Business and government can play a huge role in capacity building for social enterprises ... If it (building capacity) was built into policy to do pilot programmes with social enterprises to start small on a project and build that up over time ... a business or government organisation really partnering with them through the life of a project, the impact that that could have on social enterprise, and then this scalable impact being able to be generated through employment outcomes and environmental outcomes or whatever the impact model is of the social enterprise.* [P22]

The notion of government support to increase the capacity of social enterprises was recognised by government stakeholders who cited that the potential to assist with social benefit scaling is on the State Government agenda (e.g., government funding schemes).

> *Social enterprises to tap into grant money or funding for equipment that would help them scale up. ... ... the social benefit supply sector to scale up (is) definitely on our (state government) radar. How we grow the sector, how we grow each of those businesses I suppose is the challenge.* [P36]

The overall sense was that even though there were a lot of high-performing social enterprises in the ecosystem, support was needed to build their capacity to better respond to opportunities [21,24]. Given that players in the construction and transport space have greater scale expectations due to the size of projects undertaken in these sectors, government support was seen as critical to support organisations to scale up to meet demand.

4.2.2. Noncompetitiveness of Social Enterprises

The perceived main drivers of social enterprises' noncompetitiveness are high costs and low quality. Here, social enterprises felt that there was a disconnect between the value of the social procurement policy and the reality of how much things cost.

> *... when we're going in for tender, we lose all our tenders because we're seen as more expensive.* [P1]

> *... people automatically, not always, but they think that by engaging a social enterprise that the quality is not going to be as good.* [P3]

> *We're always balancing that commercial viability with our ability to deliver on our social purpose. So we're always having to negotiate how we cost in price in commercial contracts.* [P5]

> *I'm hearing a lot of excuses of quality or cost or time that I guess people are using as excuses. So I would say, that's probably the primary barrier.* [P10]

Tier one respondents seem to support this notion with price being an obvious factor when deciding upon the choice of supplier.

> *I wouldn't have a job if it (chosen social enterprise) was significantly more expensive, I'd get in trouble for that. So I wouldn't be able to implement it. If it was significantly more expensive I just would not be able to implement it. It just wouldn't happen.* [P26]

A key to overcome this is to push forward the entire value proposition that social enterprises offer into a final value calculation by offsetting some of the costs.

> *There is a cost associated with it (building social enterprise capacity). So, if government could help out with some of those costs, or making sure that when it is valued, that somehow it goes into your value calculation so then if your cost is slightly higher, then it's offset.* [P31]

This finding demonstrates the concern with the perceived lower quality of work performed by social enterprises. This goes beyond the very limited body of knowledge which had raised contrary insights about the nature and quality of the work undertaken [24].

4.2.3. Social Enterprise Capabilities

The perceived inability of social enterprises to develop and strengthen their capabilities is another social procurement implementation barrier which has been identified in the prior literature [4]. Intermediary bodies had been made aware of some uncertainty with social enterprise capabilities especially in the area of technical skills when working on construction sites.

> *I think the challenge (for social enterprises) is technical skills, particularly if you're working on site, you know, what are your safety systems? What are your environmental systems? That's what the big companies, that's where their red flags will be.* [P30]

Interestingly, some tier one contractors noted that, particularly in construction and transport, as there are a multitude of infrastructure projects, capabilities are the main restriction. This led to the following issue.

> *You really have to work with these (social) enterprises and your suppliers to either find matched capabilities at the outset, or create the jobs and work streams together through those relationships.* [P32]

This issue could be somewhat mitigated with greater engagement and support to build capability such as coaching and mentoring.

> *You need to engage with social enterprises to then build their capability and capacity. It goes far beyond just that usual supply chain partner that you engage with. ... from a capability perspective, you do need to help coach and mentor in those particular spaces, too. ... (currently) there's not many organisations that have the capability and capacity to do both large packages of work. ... they're probably the biggest bottlenecks.* [P31]

From a State Government perspective, it was felt that a better understanding of social enterprise supplier capabilities in the west was needed in the precontract stage to find matched capabilities to ensure potential social procurement targets can be met. Hence, to better match capabilities, the State Government were looking at ways to develop buyer capability.

Our (government department) team in particular are looking at how we develop government buyer capability to implement the framework ... (that's) a piece that has to happen to bring the market on the journey. [P36]

Although prior studies have identified capabilities as a barrier to effective social procurement implementation [36], the inclusion of the government perspective and their acknowledgement that further work in the precontract stage is required adds to the existing body of literature.

4.2.4. Social Enterprise Resourcing

Social enterprises, by virtue of their size, typically have limited resources to effectively implement social procurement. Thus, greater resources need to be provided to ensure their commercial viability. This may include access to funding via philanthropy groups or government funding schemes.

We're a non-profit, which has influence in the ways that we can access capital. Having access to that kind of capital through philanthropists for example, is really important for us. [P5]

I think a lot more grants to support social enterprises ... for start-up social enterprises, they're meaningful bundles of money that could actually go towards running costs, because it takes a good couple of years to really get traction and get a business going. [P13]

I think there's a lot of social enterprises that employ people from socially disadvantaged backgrounds. And a big barrier there that I've heard people talk about is this lack of appropriate funding, so the equivalent of the NDIS (National Disability Insurance Scheme) for people from disadvantaged backgrounds ... that's a big barrier. [P17]

This was a barrier also identified by tier one contractors and intermediaries who felt that assistance via the development of a social procurement resource toolkit for social enterprises would enhance their limited social procurement related resources

They (social enterprises) do need to be given the right tools to be able to play in this space, and it is quite a significant change to those smaller tier contractors, because they don't have the resources themselves. So that is a big barrier in the marketplace at the moment. ... I think there needs to be more financial resources and more toolbox resources to support these kinds of long term outcomes. [P23]

The major one is resourcing ... having a consistent workforce that can be engaged for long periods of time, and then providing a long term benefit to all these different contractors and government entities. [P25]

The notion of greater resources being made available via a social procurement resource toolkit adds to the very limited literature previously citing this as a means to overcome this barrier [27].

4.2.5. Poor Cash Flows

Access to cash flows, particularly for social enterprises, is vitally important to their ongoing operations. Thus, there was a call for the trading terms to be more favourable to social enterprises and to provide for prompt payment terms to enable them to obtain earlier access to funding to deliver on their social purpose.

Organisations who engage in social enterprise need to understand that trading terms are so important for cash flow for that enterprise that they need to address that whether it be from the government or from the actual tenderer. That is really important. I recently exited out of a contract because they wanted to pay me 45 days from end of the month. So how do I, as a social enterprise fund 75 days of cleaning. I'm not a bank. I'm not going to fund that. [P6]

I think the industry needs to do more to support new players, I think they need to pay better is a major thing. You know, access to cash flow, particularly for SMEs and social benefits suppliers is really fundamentally critical. . . . if you get paid 60 or 90 days, well, that's not good for your business, clearly. [P30]

Although this barrier has been identified in a previous study [24], the call for more favourable trading terms and prompt payments can help mitigate this barrier to arrive at a more effective social procurement implementation.

4.3. Support Networks

The third, and final, main theme identified by research participants as an impediment to social procurement implementation was support networks. The subthemes comprise a lack of support networks, provision and quality of databases, certification and accreditation programs, support for priority groups and the tendering process. The first subtheme is reviewed below.

4.3.1. Lack of Support Networks

A number of respondents pointed to a lack of an identifiable support network as impacting social procurement implementation.

I think the biggest challenge is connection, it is that network and knowing the right people having the right relationships. . . . it's not an easy space to crack into. [P30]

It's building those relationships, building those networks, working with subcontractors . . . main contractors (tier ones) don't do a lot of the work themselves so they bring it all in (via subcontractors). [P11]

Many were unsure of what was provided by the number of ancillary support service providers in the market, citing a general lack of support.

There seems to be a lot of support for ancillary services, but I'm not entirely sure what they deliver . . . doesn't actually provide any benefit or support for us. It's been no connections, no anything. [P13]

For a social enterprise trying to get in, it's about learning how to broker those relationships, learning where to go and how to navigate it, and how to write. [P10]

Another point of concern for social enterprises was the payments required to join certain support networks insofar as: (a) whether this acted as a source of exclusion for smaller organisations, and (b) whether it provided value for money.

The membership model, which I have always thought is fatally flawed, because it's the big businesses who are the members of those. [P30]

I just want to make sure that if we're going to utilise a service and pay . . . the services are not cheap so I'm going to make sure if I join a group that we're actually making benefit from it. [P28]

I don't think the intermediaries have been playing that well together . . . [we need to] develop a collective voice, because at the moment, you're just getting segment voices and no consistency in the way that they communicate with government . . . it's not a coherent conversation with government around social procurement, which I think is a real barrier. [P2]

Nonetheless, there was a sense that support networks that are linked throughout the ecosystem were a missing key element in the implementation of social procurement.

I think this ecosystem building is huge because all the ingredients are almost there. You have social suppliers, you can employ people, you have your tier ones who want to do this and provide that de-risking, by having that almost like an insurance to service provision or providing one of the service they need to provide. And it's just quite not linking up. But I feel like, even with government leadership, perhaps the procurement approach was able to encourage suppliers to connect up with tier ones to connect up with philanthropists and intermediaries and things like that ... (then) you could definitely come up with a solution that works and is able to meet everyone's needs and de-risk this whole thing for everyone involved. [P19]

Thus, there is a feeling that the ecosystem, which has multiple elements, is not being brought together appropriately. That is, the connections are not always clear and the vehicles to enable them to collaborate are not necessarily evident. Prior studies [21,24] have identified how improved support mechanisms are required to better manage risks associated with social procurement implementation.

4.3.2. Provision and Quality of Databases

Participants identified both the provision and quality of a database as an existing barrier to the effective implementation of social procurement in the construction and transport industries. The identification of both concerns constitutes a contribution to the literature. Regarding provision, a number of social enterprises cited the need for an accessible database to: (a) encourage them to use reliable suppliers; and (b) identify relevant organisations they should be in contact with. This was echoed by intermediary support bodies who saw that such tools could help save them time when accessing potential appropriate suppliers.

Maybe establish a resource database. I'm sick and tired of trying to look on Google every time I need to find something. I've got to go through 4000 different webpages until I find what I want. So a good resource database that is set on a benchmark, I think would be a fantastic opportunity for everyone. [P18]

Even your Tier one contractors, if they win a project ... they pretty much go through the Yellow Pages going, who's an Indigenous business, who's this who's that. They've got no idea who they're contacting, they waste a lot of time and that's when short cuts are made. [P11]

The inadequate quality of existing databases was also cited as a barrier to effective implementation. In fact, those that had access to a social procurement database bemoaned the quality of it. Specifically, they felt they still needed to do much work to understand the businesses listed in the database. Hence, both intermediary bodies and tier one contractors cited the need for improved descriptions of their capabilities and purpose to better understand the nature of their business.

If you go and ask industry around how beneficial all of those (existing social procurement databases) are, the answer won't be overly positive because you've still got to do all the work to understand the business. [P30]

The different regionalised social enterprises, who is around, what they do and what are their capabilities and where they've worked. So enabling tier ones to go, alright ... these are the people within my region, within my local network, I can come in and they are able to get started straight away. [P25]

There's (industry) platforms ... they will list at a basic level what the capabilities of an enterprise are, but they won't list its potential capabilities. [P29]

4.3.3. Certification and Accreditation Programs

To fulfil the social procurement criteria associated with major government infrastructure projects, tier one contractors are required to use certified social enterprises to meet their targets. Hence, social enterprises who do not meet the social procurement framework criteria for social enterprises, whether it be due to not being able to afford certification or not interested in attaining it, are excluded from the social procurement area of direct spend. Thus, those without accreditation tend only to be involved indirectly. Exclusionary criteria via the social procurement framework further water down the available supply base for social procurement. This constitutes a new barrier not previously identified in the existing social procurement literature.

> *There are a number of small to medium enterprises that do want to win government work, and they're obviously doing amazing things in the community. But for whatever reason they can't afford or don't want to be certified or, we have even anecdotal feedback of businesses that don't want to be classified.* [P36]

> *It can be a frustration because you've got businesses that are doing absolutely fantastic things and they don't get recognised because they're not labelled as a social enterprise, where you're potentially doing more impact than a social enterprise.* [P11]

On the other hand, the process of certification helps mitigate reputational risk for social enterprises. It also allows tier ones to avoid undertaking verification or further due diligence.

> *When our clients are looking for a bonafide social enterprise, they want to see that (accreditation) tick, because it just mitigates their risk. They know then that you've been through an independent process.* [P14]

> *Reputational risks if you're dealing with a social enterprise that turns out to not be doing what they say they do, and I think that's where some of the certifications are beneficial for that.* [P32]

Some suggest a government review of the social enterprise certification process may be needed.

> *It would be really beneficial to us, and it sounds like to others, if the way in which social enterprises were accredited was to be looked at by the government, it seems that the government sort of hitched their wagon to [X] who are considered the kind of oracle on all things social enterprise.* [P14]

4.3.4. Social Procurement Education Support

A lack of understanding of social procurement was seen as a major barrier to its implementation. Thus, support is needed to educate key stakeholders to improve their level of social procurement knowledge.

> *It's educating those people, getting them to understand how to do this, getting them to understand how to do a social procurement, not just you know, you just write this and it's all good. It's not cookie cutter.* [P2]

> *… educating the team in their workforce on what actually social procurement is (such as) aligning basic definitions, getting their heads around legislation, identifying internal stakeholders, you can champion social procurement and build a lot more engagement internally.* [P29]

Not understanding the purpose of social procurement leads to misjudgements of the work undertaken by key social procurement parties.

> *If they (X organisation) actually understood what social enterprises were about, most social enterprises are not here because of the dollars, we're not here because it's a profitable business that's going to make them millions of dollars … we're here to support individuals who need our support.* [P6]

> *Another barrier is a lack of knowledge amongst the government and corporates around this whole space and how to work with organisations like us (social enterprises).* [P17]

> *With big transport infrastructure projects, where it's led by government and the prerequisites are given by government, there needs to be more people on the ground that actually understand and can work with industries and with corporates.* [P27]

This lack of understanding from tier one contractors leads to confusion regarding the sense of purpose for social enterprises in the social procurement space. This has led to a belief that some tier one contractors operate in a contextual vacuum without understanding the true purpose of social procurement which is to benefit society first, with profit being secondary to that. Although this barrier has been cited in previous studies, understanding what that gap is is very important in social procurement implementation.

4.3.5. Supporting Priority Groups

A key to the social procurement program is the creation of employment amongst priority groups that will maximise positive employment outcomes. Since a number of organisations have little experience in this area, the lack of support for priority groups which has been identified in the prior literature as a barrier [10], continues to seemingly be excluded from part of the financial calculations.

> *People from socially disadvantaged backgrounds have so many barriers to employment, trauma, family violence, childcare, transport, language, and they are not easy to get into work without support.* [P17]

> *... they're (social enterprises) lacking resources and they don't know how to support people who are have complex disadvantage, which means there needs to be extra support to support these people if they want long term outcomes. And that's not factored into any of this, financially or otherwise.* [P23]

To mitigate this, some respondents felt there is a need to design adequate employment models to support priority groups with complex problems.

> *One of the biggest barriers is definitely the challenge of the employment aspect of social procurement ... to come up with employment models that actually support different cohorts is a very complex problem.* [P19]

Government financial support was seen as necessary to subsidise the substantial investment required to train individuals from priority groups and to provide the suite of support services (i.e., beyond employment) required to reintegrate those individuals into society with designated pathways.

> *There are people within the priority groups who will never ever be able to be employed directly by a contractor ... and they actually need to be employed by a social benefit supplier who has that wraparound support, that won't be on site, that will be off site, but are in the supply chain. And a lot of employers don't understand how to do that. ... if they (the government) want the retention of those people from those priority groups, money needs to be spent on the support of those priority groups, because it is unfair for the employer, and it's unfair for the person who's been employed ... (it's) a big barrier and it's a big gap in the market at the moment.* [P23]

> *It does take a substantial investment by the company and the people to train them (individuals from priority groups) up, which is not just a month or two, we're talking about six months to 12 months, even years depending on the job that they want to go into. So there's substantial investment on both sides.* [P25]

> *You've picked up a social benefit supplier who helped people get new employment (but) what happens to those people after? ... I feel that there's a gap there. If we want it to be sustainable, we need to actually care about the asylum seekers or the minority groups or the disabled people that we're so called indirectly employing. Do they have enough social workers to help them reintegrate into society? Do they have enough funding?* [P27]

4.3.6. Tendering Process

The social procurement tendering process under the social procurement framework was identified as a barrier by tier one contractors due to its cumbersome nature. This has also been identified in the prior literature [15,24]. Specifically, the number of forms to be completed and submitted was perceived as a time consuming obstacle constituting a major strain on existing resources.

The amount of documentation that we have to put together is unbelievable. Literally, we just submitted a tender for the state and there was over 22 different attachments, so 250 pages. . . . If these organisations (non-tier ones) aren't resourced up to have contract managers or tender writers, all that kind of stuff, it's going to make it very, very hard for them. [P25]

If you make something really regulatory it can become an administration burden. And then it takes the joy out of it. [P26]

A potential mitigating factor would be if a more streamlined approach to the tendering process could be adopted to make the engagement process simpler and more efficient for participants.

In addition to the results and discussion presented in this section, the following section draws links to the social procurement ecosystem to determine the implications associated with its effective implementation.

5. Discussion of the Ecosystem Implications for Social Procurement Implementation

The key stakeholders comprising the social procurement ecosystem are listed in Table 2 below along with the subthemes that are key priorities for them to facilitate effective social procurement implementation.

Table 2. Key priorities for effective social procurement implementation.

Theme	Subthemes	Social Enterprises	Tier One Contractors	Key Intermediaries	Government
Supply Chain Process	Supply chain pressures	X	X		
	Client pressures	X	X		
	Union opposition	X			
Capacity and Capabilities of Social Enterprises	Capacity to scale	X		X	X
	Non-Competitiveness of Social Enterprises	X			
	Social Enterprise Capabilities	X		X	X
	Social Enterprise Resourcing	X		X	X
	Poor Cash Flows	X	X		
Support Networks	Lack of Support Networks	X		X	
	Provision and Quality of Databases			X	X
	Certification and Accreditation Programs			X	X
	Social Procurement Education Support			X	X
	Supporting Priority Groups	X	X	X	X
	Tendering Process		X		X

With respect to the first theme explored in this paper (supply chain process), as demand for social procurement within state sponsored major construction and transport infrastructure projects continues to outstrip supply, such supply side constraints mean there is a need for government representatives to give further consideration to the ability of the construction and transport industries to facilitate the goals of social procurement. As demonstrated via the thematic analysis, supply side pressures are a major concern for subcontractors (social procurement provider services) and this has led some to game the system by rotating the same priority group individuals across several projects to meet targets. Such gaming also serves as a warning to procurement managers (tier one contractors) to adhere to the monitoring and compliance aspect of social procurement.

The issues with the supply side seem to be a product of the immature nature of the social procurement ecosystem. Greater levels of maturity will bring new levels of complexity. A move towards greater maturity may result in an ecosystem where temporary labour-hire solutions are less frequent, and the focus becomes building long-term employment outcomes for priority groups. These issues can be contextualised in the light of new institutional theory which suggests that informal institutions (existing norms and practices that influence patterns of behaviour) can help advance, or in this case undermine, formal institutions (social procurement regulations, laws and policies). Existing attitudes maintained by informal institutions can manifest in behavioural practices such as the resistance to hiring from priority groups, and not factoring in complexity when evaluating their work progress. The supply chain process outlined above also links to the next main thematic barrier to effectively implementing social procurement in construction and transport, which is ways of building capacity within the sector.

The main findings from the second theme focus mostly on the specialised social procurement services area of the ecosystem regarding its capacity and capabilities. Issues such as capacity to scale, improving competitiveness regarding price and quality, enhancing technical skill capabilities and having to overcome limited resources were dominant discussion points in the consideration of barriers. Organisational capability theory provides an avenue for potentially mitigating these deficiencies. An organisation has the ability to respond to internal and external change and utilise organisational resources for the purpose of achieving a particular end result [37]. Given the size and scope of the problem, such capability development would initially focus on the service providers themselves and the intermediary support bodies. Intermediation capability development on both the supply and demand sides of procurement relationships can help mitigate some of the capacity and capability barriers.

Another ecosystem stakeholder group (government representatives) views the growth of the specialised social procurement services area as one of their biggest challenges [38]. This takes on increased significance when one considers the massive size of projects undertaken in the construction and transport industries. Thus, provision of government organisational resources in the form of grant money or funding to improve capacity and capabilities is a key to reducing this barrier.

For the third theme, the main findings suggest that a lack of support networks has been instrumental in impeding social procurement implementation. Viewed in the light of network theory (e.g., [39]), the implications for the social procurement ecosystem lie in the enhancement of certain key mechanisms. Specifically, key mechanisms such as resource and information channels (e.g., assistance with tendering process, advice seeking, education support), affiliations (e.g., shared memberships, certification programs, database provision) and formal contractual relationships (e.g., strategic alliances, buyer–supplier contracts) can mitigate the associated barriers.

From an ecosystem perspective, intermediary support bodies and government have a major role to play. Currently, the right mechanisms are not in place to enable place-based solutions to play a big role. Place-based solutions could spark new processes for enhancing the ecosystem of social procurement. An enhanced ecosystem approach could support a work integration social enterprise that not only has solid links into industry, but equally importantly, links into communities and employment services. This could potentially include important services such as pre-employment support and brokerage funding.

Furthermore, there is scope for intermediary support bodies and government to improve the quality of both specialised social procurement providers and tier one access to online databases to facilitate an easier transition to social procurement. In addition, intermediary bodies can act as mentors to those stakeholders who seek advice and guidance in embedding social procurement in their operations and building internal organisational capability. This would lead to communities of practice [40].

6. Conclusions

The aim of this paper was to consult with key stakeholders in the social procurement ecosystem to identify the main barriers preventing effective social procurement implementation in the construction and transport industries. The adoption of the ecosystem approach responds to recent calls to include the perspective of policymakers in future social procurement research [13]. Of the 14 subthemes identified, provision and quality of an extensive social procurement database as well as certification and accreditation programs constituted newly identified barriers. In addition, four subthemes: trade union opposition, noncompetitiveness of social enterprises, social enterprise resourcing and poor cash flows added to the limited literature that have identified these as barriers [24,27].

This paper makes the following contributions. First, the findings highlight discrepancies between policy and practice. For instance, it was acknowledged by government representatives that a better understanding of social enterprise supplier capabilities in the precontract stage was required to ensure social procurement targets can be met. In addition, greater support was needed for priority groups such as designing more adequate employment models to address long-term employment considerations. Thus, this paper helps bridge the gap between policy and practice by providing practical perspectives on policy implementation issues.

Second, this paper contributes to social procurement research by highlighting the importance of local conditions. For instance, the barriers: provision and quality of database, and certification and accreditation programs provide an empirical example of what is important to practically implement from a situated practice perspective to achieve improved social procurement implementation.

The findings also lead to the following strategy recommendations for the effective implementation of social procurement in the construction and transport industries. To increase capacity to scale, the authors recommend: (i) embedding capacity building into policy via pilot programmes that partner with selected social benefit suppliers; and (ii) existing key intermediary bodies to develop a social procurement resource toolkit for social benefit suppliers to enhance their limited social procurement related resources.

To enhance social enterprise capabilities, a strategic recommendation is to improve the ability of tier one contractors to identify matched capabilities to better meet social procurement targets. To rectify the lack of support networks, a potential strategy is to strengthen the brokerage role. For instance, the introduction of government approved brokers who will be accountable for: (i) brokering social benefit suppliers at scale and quality to meet the needs of tier one contractors; and (ii) building and fostering partnerships.

For certification and accreditation programs, it is recommended that a review from Government, in collaboration with key intermediary bodies, of what constitutes a certified social enterprise certification should occur. Certification and accreditation programs currently overlook a number of SMEs that are already delivering social impact outcomes. Regarding social procurement databases, the Victorian State Government should consider the provision of a free high-quality database accessible to all stakeholders to help achieve social procurement targets.

Given the importance of contextual settings, future comparative research should evaluate other countries' social procurement ecosystems which could differ in breadth and maturity from Victoria's. In addition, future research should apply the learnings from this study to other sectors beyond the construction and transport industries, which will exhibit characteristics that necessitate different approaches to social procurement implementation.

Author Contributions: Conceptualization, R.N. and C.X.L.; Methodology, R.N., C.X.L. and D.G.; Formal analysis, R.N.; Data curation, R.N.; Writing—original draft, R.N.; Writing—review & editing, C.X.L. and D.G. All authors have read and agreed to the published version of the manuscript.

Funding: This study was funded by Victorian Higher Education Strategic Investment Fund (2021–2022).

Institutional Review Board Statement: The study was conducted in accordance with the Human Research Ethics Committee of VICTORIA UNIVERSITY (HRE21-151 approved on 18-11-2021) for studies involving humans.

Informed Consent Statement: Informed consent was obtained from all subjects involved in the study.

Data Availability Statement: The data presented in this study are not available as per the conditions agreed to in the ethics approval.

Acknowledgments: The authors are grateful to the guest editors and the anonymous reviewers for their useful suggestions, which have improved the quality of the manuscript. The authors also wish to thank B. Bok for assisting with the project.

Conflicts of Interest: The authors declare no conflict of interest.

Appendix A. Research Participants

Stakeholder Type	Participant Role	Pseudonym
Specialised SP Service Provider	Owner	P1
Specialised SP Service Provider	Leader	P2
Specialised SP Service Provider	Manager	P3
Specialised SP Service Provider	Director	P4
Specialised SP Service Provider	Manager	P5
Specialised SP Service Provider	Owner	P6
Specialised SP Service Provider	Owner	P7
Specialised SP Service Provider	Manager	P8
Specialised SP Service Provider	CEO	P9
Intermediary Support Body	Leader	P10
Intermediary Support Body	Business Engagement Leader	P11
Specialised SP Service Provider	CEO	P12
Specialised SP Service Provider	Owner	P13
Tier One Contractor	Manager	P14
Specialised SP Service Provider	General Manager	P15
Specialised SP Service Provider	CEO	P16
Specialised SP Service Provider	CEO	P17
Intermediary Support Body	Director	P18
Intermediary Support Body	Manager	P19
Intermediary Support Body	Project Manager	P20
Intermediary Support Body	Director	P21
Intermediary Support Body	Head	P22
Intermediary Support Body	CEO	P23
Intermediary Support Body	Director	P24
Tier One Contractor	Program Director	P25
Tier One Contractor	Social Procurement Manager	P26
Tier One Contractor	Manager	P27
Tier One Contractor	Executive Director	P28

Stakeholder Type	Participant Role	Pseudonym
Intermediary Support Body	Head of Partnerships	P29
Intermediary Support Body	Manager	P30
Tier One Contractor	Contracts Manager	P31
Tier One Contractor	Social Procurement Manager	P32
Tier One Contractor	Procurement Manager	P33
Tier One Contractor	Employment Facilitator	P34
Tier One Contractor	Employment Facilitator	P35
Government Representative	Specialist	P36
Government Representative	Manager	P37
Government Representative	Director	P38
Government Representative	Director	P39
Government Representative	Manager	P40
Tier One Contractor	Associate Director	P41
Tier One Contractor	Social Procurement Advisor	P42

References

1. Wirahadikusumah, R.; Abduh, M.; Messah, Y.; Aulia, M. Introducing sustainability principles into the procurement of construction works—Case of Indonesian developers. *Int. J. Constr. Manag.* **2019**, *21*, 932–944. [CrossRef]
2. Loosemore, M.; Alkilani, S.Z.; Murphy, R. The institutional drivers of social procurement implementation in Australian construction projects. *Int. J. Proj. Manag.* **2021**, *39*, 750–761. [CrossRef]
3. Victoria State Government 2022. Victoria's Social Procurement Framework. Available online: https://www.buyingfor.vic.gov.au/social-procurement-framework (accessed on 7 October 2022).
4. Loosemore, M. Social procurement in UK construction projects. *Int. J. Proj. Manag.* **2016**, *34*, 133–144. [CrossRef]
5. Ogunsanya, O.A.; Aigbavboa, C.O.; Thwala, D.W.; Edwards, D.J. Barriers to sustainable procurement in the Nigerian construction industry: An exploratory factor analysis. *Int. J. Constr. Manag.* **2019**, *22*, 861–872. [CrossRef]
6. Renda, A.; Pelkmans, J.; Egenhofer, C.; Schrefler, L.; Luchetta, G.; Selçuki, C.; Ballesteros, J.; Zirnhelt, A.C. *The Up-Take of Green Public Procurement in the EU27*; Study Prepared for DG Environment, European Commission, CEPS in Collaboration with the College of Europe: Brussels, Belgium, 2012.
7. Varnäs, A.; Balfors, B.; Faith-Ell, C. Environmental consideration in procurement of construction contracts: Current practice, problems and opportunities in green procurement in the Swedish construction industry. *J. Clean. Prod.* **2009**, *17*, 1214–1222. [CrossRef]
8. ArcBlue Consulting 2020a. *Social Procurement in the West: Transforming Spend into Community Prosperity (Part A Head-Line Report)*; ArcBlue: Melbourne, Australia, 2020.
9. ArcBlue Consulting 2020b. *Social Procurement in the West: Transforming Spend into Community Prosperity (Part B: Roadmap to Implementation)*; ArcBlue: Melbourne, Australia, 2020.
10. Barraket, J.; Loosemore, M. Co-creating social value through cross-sector collaboration between social enterprises and the construction industry. *Constr. Manag. Econ.* **2017**, *36*, 394–408. [CrossRef]
11. OECD. *Integrating Responsible Business Conduct in Public Procurement*; OECD Publishing: Paris, France, 2020. [CrossRef]
12. Hayes, A. What Is a Business Ecosystem and How Does It Work? *Investopedia*, 20 January 2021. Available online: https://www.investopedia.com/terms/b/business-ecosystem.asp (accessed on 9 May 2023).
13. Troje, D. Policy in Practice: Social Procurement Policies in the Swedish Construction Sector. *Sustainability* **2021**, *13*, 7621. [CrossRef]
14. Loosemore, M.; Denny-Smith, G.; Barraket, J.; Keast, R.; Chamberlain, D.; Muir, K.; Powell, A.; Higgon, D.; Osborne, J. Optimising social procurement policy outcomes through cross-sector collaboration in the Australian construction industry. *Eng. Constr. Archit. Manag.* **2020**, *28*, 1908–1928. [CrossRef]
15. Raiden, A.; Loosemore, M.; King, A.; Gorse, C. *Social Value in Construction*; Routledge: London, UK, 2018. [CrossRef]
16. Loosemore, M.; Reid, S. The social procurement practices of tier-one construction contractors in Australia. *Constr. Manag. Econ.* **2018**, *37*, 183–200. [CrossRef]
17. Burkett, I. *Social Procurement in Australia*; Foresters Community Finance: Brisbane, Australia, 2010.
18. Victoria State Government 2021. About Social Procurement—Suppliers. The Victorian Government. Available online: https://www.buyingfor.vic.gov.au/about-social-procurement (accessed on 23 May 2022).
19. Ormerod, M.; Newton, R. Construction as a career choice for young disabled people: Dispelling the myths. *Constr. Manag. Econ.* **2013**, *31*, 928–938. [CrossRef]

20. Agyekum, A.K.; Fugar, F.D.K.; Agyekum, K.; Akomea-Frimpong, I.; Pittri, H. Barriers to stakeholder engagement in sustainable procurement of public works. *Eng. Constr. Arch. Manag.* 2022, *ahead-of-print*. [CrossRef]
21. Ershadi, M.; Jefferies, M.; Davis, P.; Mojtahedi, M. Barriers to achieving sustainable construction project procurement in the private sector. *Clean. Eng. Technol.* **2021**, *3*, 100125. [CrossRef]
22. Murphy, M.; Eadie, R. Socially responsible procurement A service innovation for generating employment in construction. *Built Environ. Proj. Asset Manag.* **2019**, *9*, 138–152. [CrossRef]
23. Loosemore, M.; Higgon, D. *Social Enterprise in the Construction Industry: Building Better Communities*; Routledge: London, UK, 2015.
24. Loosemore, M.; Keast, R.; Barraket, J.; Denny-Smith, G.; Alkilani, S. The risks and opportunities of social procurement in construction projects: A cross-sector collaboration perspective. *Int. J. Manag. Proj. Bus.* **2022**, *15*, 793–815. [CrossRef]
25. Troje, D.; Andersson, T. As above, not so below: Developing social procurement practices on strategic and operative levels. *Equal. Divers. Inclusion Int. J.* **2020**, *40*, 242–258. [CrossRef]
26. Willar, D.; Waney, E.V.Y.; Pangemanan, D.D.G.; Mait, R.E.G. Sustainable construction practices in the execution of infrastructure projects: The extent of implementation. *Smart Sustain. Built Environ.* **2021**, *10*, 106–124. [CrossRef]
27. Ruparathna, R.; Hewage, K. Sustainable procurement in the Canadian construction industry: Challenges and benefits. *Can. J. Civ. Eng.* **2015**, *42*, 417–426. [CrossRef]
28. Zaidi, S.A.H.; Mirza, F.M.; Hou, F.; Ashraf, R.U. Addressing the sustainable development through sustainable procurement: What factors resist the implementation of sustainable procurement in Pakistan? *Socio-Econ. Plan. Sci.* **2019**, *68*, 100671. [CrossRef]
29. Delmonico, D.; Jabbour, C.J.C.; Farias Pereira, S.C.; Jabbour, A.B.L.d.S.; Renwick, D.W.S.; Tavares Thome, A.M. Unveiling barriers to sustainable public procurement in emerging economies: Evidence from a leading sustainable supply chain initiative in Latin America. *Resour. Conserv. Recycl.* **2018**, *134*, 70–79. [CrossRef]
30. Kimbu, A.N.; Adam, I.; Dayour, F.; de Jong, A. COVID-19-Induced Redundancy and Socio-Psychological Well-Being of Tourism Employees: Implications for Organizational Recovery in a Resource-Scarce Context. *J. Travel Res.* **2021**, *62*, 55–74. [CrossRef]
31. Ngoasong, M.Z.; Kimbu, A.N. Informal microfinance institutions and development-led tourism entrepreneurship. *Tour. Manag.* **2016**, *52*, 430–439. [CrossRef]
32. Dayour, F.; Park, S.; Kimbu, A. Backpackers' perceived risks towards smartphone usage and risk reduction strategies: A mixed methods study. *Tour. Manag.* **2019**, *72*, 52–68. [CrossRef]
33. Halcomb, E.J.; Davidson, P.M. Is verbatim transcription of interview data always necessary? *Appl. Nurs. Res.* **2006**, *19*, 38–42. [CrossRef] [PubMed]
34. Braun, V.; Clarke, V. *Successful Qualitative Research: A Practical Guide for Beginners*; Sage: London, UK, 2013.
35. Hsieh, H.-F.; Shannon, S.E. Three Approaches to Qualitative Content Analysis. *Qual. Health Res.* **2005**, *15*, 1277–1288. [CrossRef] [PubMed]
36. Loosemore, M.; Osborne, J.; Higgon, D. Affective, cognitive, behavioural and situational outcomes of social procurement: A case study of social value creation in a major facilities management firm. *Constr. Manag. Econ.* **2021**, *39*, 227–244. [CrossRef]
37. O'Regan, N.; Ghobadian, A. The importance of capabilities for strategic direction and performance. *Manag. Decis.* **2004**, *42*, 292–313. [CrossRef]
38. Theodorakopoulos, N.; Ram, M.; Beckinsale, M. Human resource development for inclusive procurement by intermediation: A situated learning theory application. *Int. J. Hum. Resour. Manag.* **2013**, *24*, 2321–2338. [CrossRef]
39. Lin, N. Building a network theory of social capital. In *Social Capital*; Routledge: Abington, UK, 2017; pp. 3–28.
40. Barraket, J. The Role of Intermediaries in Social Innovation: The Case of Social Procurement in Australia. *J. Soc. Entrep.* **2019**, *11*, 194–214. [CrossRef]

sustainability

Article

Research on Data-Driven Dynamic Decision-Making Mechanism of Mega Infrastructure Project Construction

Guangzhong Hu *, Yuming Liu, Kai Liu and Xiaoxu Yang

School of Economics and Management, Beijing Jiaotong University, Beijing 100044, China
* Correspondence: 21120635@bjtu.edu.cn; Tel.: +86-178-6580-0896

Abstract: The construction of mega infrastructure projects has the characteristics of repeatability, long duration, and high complexity. Therefore, it is particularly important to implement dynamic decision-making in projects. This study takes data-driven decision-making mechanisms as the entry point and constructs a dynamic decision-making system for mega infrastructure projects consisting of an information collection subsystem, an information processing and transformation subsystem, a human–computer collaborative decision-making subsystem and an evaluation and feedback subsystem. On this basis, we established a system dynamics model of dynamic decision-making for mega infrastructure projects. Vensim PLE 9.3.5 software was used to simulate and analyze the operation law of dynamic decision-making for mega infrastructure projects from a data-driven perspective, and the sensitivity of the application rate of information management technology, the application rate of data analysis methods, the participation rate of experts in decision-making, the historical case information on this project, and the information on similar projects on the effectiveness of program implementation were simulated and analyzed. The results of the study showed that all five key influencing factors have a positive impact on the effectiveness of program implementation. In addition, the application rate of information management technology and the application rate of information analysis methods have a higher sensitivity to the effectiveness of program implementation, the participation rate of experts in decision-making and historical case information on this project have average sensitivity to the effectiveness of program implementation, and information on similar projects has lower sensitivity to the effectiveness of program implementation. This study provides some ideas and suggestions to promote the effective use of information technology and digital technology by each participant in the construction of mega infrastructure projects while improving their dynamic decision-making efficiency, scientificity, and accuracy.

Keywords: mega infrastructure projects; data-driven; dynamic decision-making; system dynamics

Citation: Hu, G.; Liu, Y.; Liu, K.; Yang, X. Research on Data-Driven Dynamic Decision-Making Mechanism of Mega Infrastructure Project Construction. *Sustainability* **2023**, *15*, 9219. https://doi.org/10.3390/su15129219

Academic Editors: Zora Vrcelj and Malindu Sandanayake

Received: 30 March 2023
Revised: 25 May 2023
Accepted: 5 June 2023
Published: 7 June 2023

1. Introduction

The dynamic decision-making in this article refers to the decision-making in the process of mega infrastructure project construction.

The world has entered the "Trillion Era" of mega infrastructure projects [1,2]. According to a report on global consulting by McKinsey, the total investment in infrastructure will reach up to USD 57 trillion by 2030 [3]. It is estimated that the global annual investment in mega infrastructure projects will reach up to USD 6–9 trillion [1], accounting for 8% of the global gross domestic product (GDP). Mega infrastructure projects are large, complex engineering projects, often costing USD 1 billion or more, requiring years of development and construction, involving multiple public and private stakeholders, are transformative, and affect millions of people [1]. Major infrastructure projects are engineering projects of great significance to the economic and social development of a country under a certain background of the times [4,5]. Mega infrastructure projects are understood differently by scholars and practitioners in different cultural contexts, but generally, they refer to large-scale and complex architectural, engineering, and construction (AEC) projects in spaces with significant investment and broad social and economic impact [6,7]. Compared with general engineering projects, mega infrastructure projects are characterized by large scale, high complexity, and a wide range of significance [8]. These bring great challenges to decision-making and project management. There are many problems in the field of mega infrastructure project construction today. "Over-investment, over-schedule, and low investment profit" have become the international "iron law" for mega infrastructure projects [9,10]. The economic benefits of contracted works are mostly unstable and low The implementation of technological innovation in the field of mega infrastructure projects is slow and inactive, as evidenced by the statistics [11,12]. Flyvbjerg et al. investigated 258 large-scale projects in 20 countries across 5 continents and found that 90% of megaprojects were subjected to cost overruns and delays in schedules, resulting in being, on average over budget by 28% [13]. Traditional project management concepts and strategies have proven less efficient for modern mega infrastructure projects [13]. There is an urgent need for the research and development of cutting-edge technologies in mega infrastructure project fields and scientific and effective management methods [14]. The construction process of mega infrastructure projects often generates many problems involving environmental pollution and ecological damage, and a moment earlier to propose a decision plan can be a moment earlier to reduce pollution of the environment. Therefore, to improve the efficiency of decision-making is key. The implementation of dynamic decision-making is imperative [15]. Therefore, in order to cope with these challenges in the process of mega infrastructure projects, dynamic management has become an effective means in recent years. Dynamic decision-making, as the core of dynamic management, plays a crucial role in the efficient and high-quality operation of mega infrastructure projects [16]. Dynamic decision-making implies real-time, circular feedback, sustainability, and environmental adaptability [17]. The purpose of the decision-making mechanism studied in this paper is mainly three-fold: first, to introduce the information on similar projects to support dynamic decision-making in mega infrastructure projects and reduce the uncertainty of decision-making in the process of mega infrastructure project construction; second, to design a dynamic decision-making mechanism model with real-time monitoring, circular feedback, sustainability, and environmental adaptability to cope with the problem of high risk in the process of mega infrastructure project construction; and third, to cope with the problem of long processing cycles and inefficient decision-making during the construction of mega infrastructure projects, reduce the uncertainty of decision-making, and make full use of the massive data generated during mega infrastructure project construction.

2. Literature Review

2.1. The Basic Process of Dynamic Decision-Making in Mega Infrastructure Projects

Decisions are at the heart of engineering projects; better decisions will lead to better engineering design [18]. Effective decision-making program functions are important tasks for decision-making [19]. The formation of the decision-making program is an instant regulation and design based on the current needs, industry standards, geological survey accuracy, technology and equipment maturity, and enterprise capabilities [20]. In the whole life cycle of a project, the function of the decision-making will be continuously optimized and expanded with the deepening of people's understanding of the decision-making problem, environmental changes, and new functional requirements [21].

The decision-making process of mega infrastructure projects not only reflects path dependence but is also full of uncertainties and dynamic evolution, which makes the core decision-making process of mega infrastructure projects have various complex phenomena.

Various models of the decision-making process have been proposed, such as single-criterion models [22], multiple-criteria models [23], Sutherland's model [24], Holt's model [25], models based on operational studies [26], cybernetic decision models [27], fuzzy data models [28], etc., as shown in Table 1.

Table 1. Models of decision-making processes [29].

Single-criterion models:	**Multiple-criteria models:**
1. Discovering the difficulty. Applying to evaluate problem solution variants; 2. Identifying the problem; 3. Determining the criterion; 4. Setting a list of solutions; 5. Describing effects of the implementation of each solution; 6. Selecting the best solution; 7. Implementing the decision.	1. Identifying the problem; 2. Identifying decision criteria; 3. Assigning weights to criteria; 4. Elaborating alternative solutions; 5. Evaluating alternative solutions; 6. Selecting the best solution; 7. Implementing the chosen solution; 8. Evaluating the efficiency of the decision implemented.
Sutherland's model:	**Holt's model:**
1. The need to make a decision (goal); 2. Primary information (opinions, theories); 3. Empirical studies; 4. Building a model; 5. Generating solutions; 6. Selecting criteria for evaluation; 7. Evaluation of variants; 8. Selection of the solution; 9. Making a decision; 10. Implementation; 11. Feedback to correct the model.	1. Identification of the problem; 2. Analysis of the context; 3. Definition of the problem; 4. Elaboration of solutions; 5. Evaluation of variant solutions; 6. Selection of a solution; 7. Implementation; 8. Evaluation of effects.
Model based on operational studies:	**Cybernetic decision model:**
1. Building a model (describing the situation using mathematical language); 2. Solving the problem presented in the form of a mathematical model; 3. Verification of the model—possible corrections; 4. Monitoring—feedback and correction of the decision made.	1. Input—primary, raw information; 2. Transformation—a decision-making process; 3. Output—secondary information in the form of a decision.
Fuzzy data model:	
1. Data collection stage—input signals; 2. The fuzzification stage; 3. The stage of fuzzy inference;	4. The stage of defuzzification; 5. Making a decision.

The aforementioned models of the decision-making process can be broadly divided into two categories. The first category consists of models that utilize single-criterion and multi-criteria approaches that aim to evaluate several alternative solutions to determine the best solution by assessing the possible effects of their implementation. The second category consists of methods that build mathematical models that present the implementation of previous decisions through mathematical language. These modeling methods collect and analyze information about the current situation in order to determine the goal and the measures that will be applied to determine to what extent this goal will be achieved. The final stage provides feedback that allows the user to correct the model or decision if necessary. The models given in the table vary in their approach to problem-solving and subsequent procedures, but, in each case, the completed procedures lead to problem-solving. In different areas of business activity, different approaches may be applicable to different decision situations. An analysis of the literature related to decision-making revealed the lack of decision models adapted to mega infrastructure construction activities.

The decision-making model process for mega infrastructure projects is to analyze the environment and gather information, define the problem that causes difficulties, determine the evaluation criteria for the solution of the problem, develop different solutions, select a method, evaluate possible solutions, evaluate all alternatives and select one, make a decision, implement the decision, obtain feedback, and correct input data and basic assumptions [29].

2.2. Support Conditions for Dynamic Decision-Making of Mega Infrastructure Projects

Information is the basis of decision-making, and dynamic decision-making in mega infrastructure projects is inseparable from the collection of real-time information. The decision-making process also includes how to collect and analyze new information [30], and the empowerment of any decision-making program as a man-made system is set by the subject through the theoretical thinking at the virtual engineering level, preceding the entity. However, the functional value and role of the decision-making program cannot exist separately from the system entity and must ultimately be realized through the engineering entity [19]. The more complex the core decision-making problem, the more relations of "gene" and "bloodline" it has to the situation, and the more we need to look at the problem, think about the problem, and analyze the problem in the overall situation where the problem is located to find a decision-making solution to solve the problem [5]. This requires us to build a sound survey, a forecasting, monitoring, and inspection system, and an information processing and analysis platform in the process of mega infrastructure project construction [31] and to design a sound decision-making program formation mechanism Analyzing and solving such decision-making problems generally require cross-field, interdisciplinary, and cross-professional technologies, means, and methods, so it is necessary for decision-makers to build a holistic cognitive platform with complete knowledge and a good working mechanism [32]. The dynamic management of mega infrastructure projects is based on the information platform, with the information platform as the core [33].

Expert knowledge information is a source of learning, and information that is organized and processed to the right people will benefit project decision-making [34]. The organizational structure in the process of mega infrastructure projects is very complex. For example, a mega infrastructure project under construction in Southwestern China is a three-level management system; each level has a very complex organizational structure, mainly including the owner, construction units, design units, consulting units, etc., where the construction units have many companies. These units have experts in the field under the jurisdiction of the unit. These experts in the project decision-making process contribute their knowledge for different issues and provide strong support for decision-making [18]. Dynamic decision-making in the process of mega infrastructure project construction must not only be supported by factual information and data but also have the intelligence of decision-making experts. The participation of decision-making experts is indispensable in the stage of cause analysis, the decision-making program design, and the decision-making plan evaluation of mega infrastructure project construction decisions [5]. With a deep theoretical foundation and practical experience, experts have keen insight into the handling and response of emergencies, professional analysis and judgment, and accurate intelligence decision-making ability [35,36]. The intelligence of experts is essential for architectural and engineering organizations, as the characteristics of each project are dynamic and unique [35].

Although experts usually operate within a bounded range where they are knowledgeable and comfortable, they sometimes confidently give information outside their range of expertise [20]. They can also miss the bigger picture. Furthermore, the outcome of decision-making is affected by various factors, such as the professional background, knowledge, experience, personality, and emotions of the decision-maker [37], and has strong subjectivity and uncertainty. Therefore, it is necessary for human–computer collaboration to make decisions.

The core problem of dynamic decision-making in mega infrastructure projects is to propose the relevant decision-making program, and the process of the subject proposing the decision-making program is actually through the combination of theoretical thinking and engineering thinking on the basis of respecting the general law and reflecting the unique intention of the subject [29].

In summary, dynamic decision-making has a very far-reaching significance for the construction of mega infrastructure projects. The information collection team (fact data), information analysis team (tools and methods), and decision-making team (expert intelligence) of mega infrastructure project construction work together on the information management platform to jointly provide support for the dynamic decision-making of mega infrastructure projects, which will be a development trend and more conducive to giving full play to the substantive role of fact data and expert intelligence in the dynamic decision-making of mega infrastructure projects and realize the interactive interconnection and harmonization of various subjects. However, there is currently a lack of a systematic framework to determine the dynamic decision-making process architecture in the construction process of mega infrastructure projects. From the project survey, monitoring, inspection, data analysis, and expert discussion to the formation of the final decision-making plan, the decision-making path needs to be further designed and clarified. In this paper, we will use the system dynamics method to design and construct the dynamic decision-making mechanism path of mega infrastructure projects, take human–computer collaborative intelligent decision-making thinking as the core, describe the dynamic complexity of the system through causal feedback, and simulate the dynamic evolution process of the system using computer simulation technology.

3. Overview of System Dynamics

3.1. Concept of System Dynamics

System dynamics is a discipline of analysis and research based on information feedback and is comprehensive in its understanding and solution of problems [38,39]. System dynamics proposes that its behavior patterns and properties are determined by the internal dynamic structure and feedback mechanisms [40,41]. For the study of complex problems, system dynamics is solved using a qualitative combined with a quantitative approach; that is, the construction of the models is based on the theory of system dynamics and the use of computers to perform simulations and, thus, the study of the problem [42].

3.2. Composition of System Dynamics Model

3.2.1. Cause-and-Effect Diagram

Before simulating a system, the cause–effect relationship needs to be analyzed, which is a necessary condition for successful modeling, and the relationship between different factors is represented with the help of a cause–effect diagram, as shown in Figure 1. The cause–effect diagrams are used to represent the logical relationships between the different factors, i.e., to represent them qualitatively. Arrows are used to connect the different factors so that each factor forms a certain relationship and, thus, becomes a whole. If there is a positive sign at the arrow, it means that the variable at the end of the arrow increases, causing the variable at the arrow to increase; if the variable at the end of the arrow decreases, the variable at the arrow decreases, i.e., the variables at both ends increase or decrease in the same direction. If there is a negative sign at the arrow, it means that an increase in the variable at the end of the arrow will cause a decrease in the variable of the arrow, and a decrease in the variable at the end of the arrow will cause an increase in the variable of the arrow.

If arrows connect the factors to each other to form a closed path, then it is called a feedback loop, and there are positive feedback and negative feedback loops. If a loop has an even number of negative signs, it is a positive feedback loop and is indicated by "+" in the center of the loop; when a loop has an odd number of negative signs, it is a negative feedback loop and is indicated by "-" in the center of the loop.

As shown in Figure 1, Figure 1a indicates that when A increases, B increases; and when A decreases, B decreases. Figure 1b shows that when A increases, B decreases; and when A decreases, B increases. Figure 1c,d are causal loop diagrams, and it is difficult to determine the beginning and end of the loop. Figure 1c has two negative signs, which is a positive feedback loop and is expressed as A increases → B increases → C decreases → A increases or A decreases → B decreases → C increases → A decreases. Figure 1d has one negative sign, which is a negative feedback loop, and the causal relationship is expressed as A increases → B increases → C increases → A decreases or A decreases → B decreases →

C decreases → A increases. The arrows of the cause–effect diagram only reflect the logical relationship of different factors, and there is no quantitative relationship.

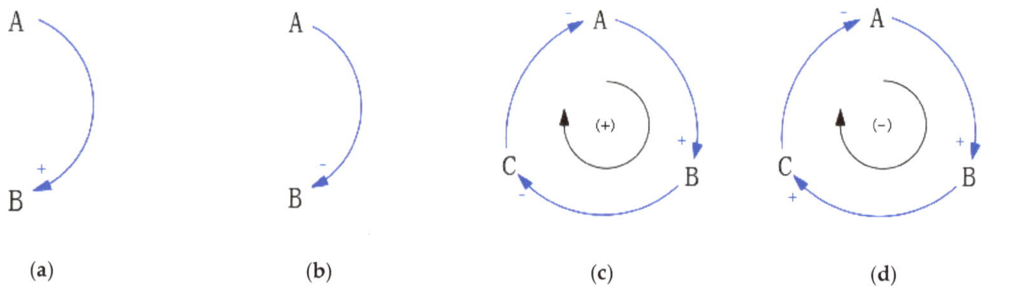

Figure 1. Example diagram of cause-and-effect relationship. (**a**) represents positive feedback, (**b**) represents negative feedback, (**c**) represents positive feedback loop, (**d**) represents negative feedback loop. (A,B,C represent different events.)

3.2.2. System Flow Diagram

The cause–effect diagram only reflects the increase or decrease among variables and cannot reflect the specific quantity of change, which is a qualitative description of different variables. Therefore, in order to quantitatively analyze the system and describe the whole change process, the first step is to transform the cause–effect diagram into a system flow diagram. A system flow diagram can quantitatively describe each variable in the system and make up for the lack of causality diagrams by assigning values to the variables and defining the variable relationships using formulas.

A stock–flow diagram captures the amount of accumulation resulting from a change in one variable leading to a change in another variable. Here, the main elements contained in the stock–flow diagram are described, as shown in Figure 2.

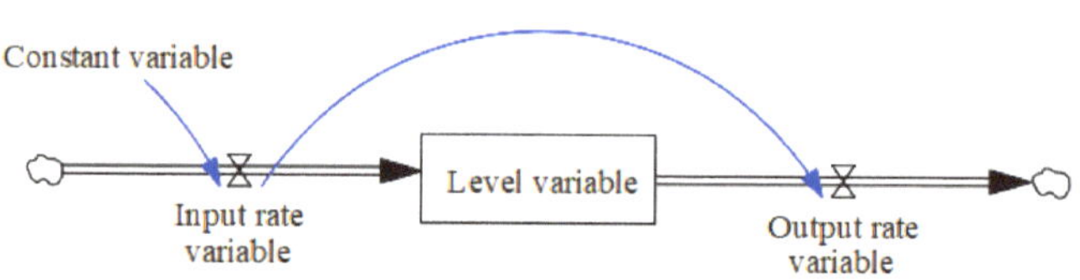

Figure 2. General form of system flow diagram.

(1) Level variable

The level variable is a variable that provides a description of the cumulative effect of the system. As time accumulates, this variable can demonstrate that the value at this moment is equal to the value at other previous moments plus the amount of effect accumulated during this time. The magnitude of the resulting value can reflect the actual state of the variable at a given moment in time. In the system flow diagram, the level variables are represented as rectangular boxes;

(2) Rate variable

The rate variable can reflect the rate of input or output of the state variable and is also a variable that can reflect the cumulative effect of the system. The rate of change of the system is also a variable that can reflect the rate of change of the system accumulation effect. In the system flow diagram, the rate variable is represented by a double arrow line with a funnel symbol together with the funnel symbol;

(3) Constant variable

Constants do not change as they accumulate over time. Constants can either point to rate variables or point to rate variables on top of auxiliary variables. However, no variable will point to a constant. There is no special notation for the representation of constants in the system flow diagram.

3.2.3. Equation

Before performing the simulation, equations are needed to establish the relationships between the variables, which are used to calculate the values of each variable. The equations to be used in this paper are:

$$Stock(t) = \int_{t_0}^{t} [Inflow(t) - Outflow(t)]dt + Stock(t_0) \quad (1)$$

$Stock(t)$ is the number of stocks at time t. $Inflow(t)$ is the inflow volume. $Outflow(t)$ is the outflow volume. $Stock(t_0)$ is the volume of stock at the initial moment. System dynamics expresses time as a continuous quantity. The equation is represented in the software Vensim PLE 9.3.5 by INTEG = (x, initial), with initial being the initial value.

3.3. The Modeling Process of System Dynamics

System dynamics can be analyzed based on actual problems, modeled based on the analysis results, and then simulated using software to analyze the obtained simulation results and finally provide relevant suggestions. The modeling process is shown in Figure 3.

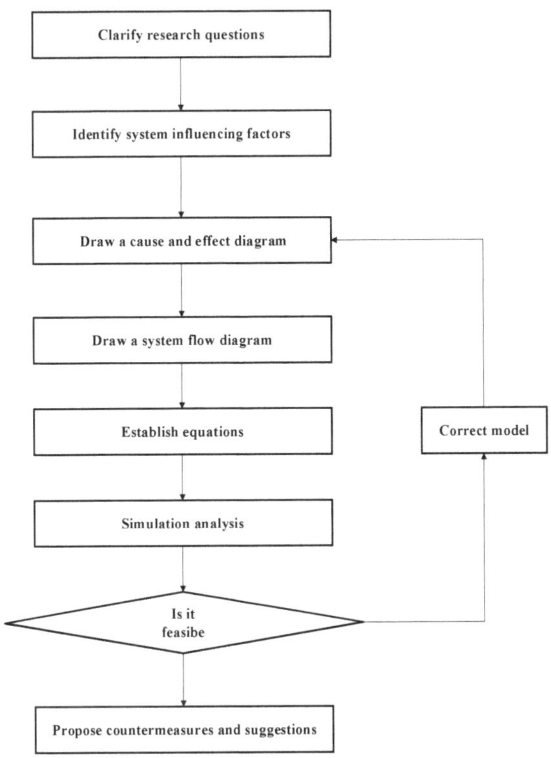

Figure 3. Modeling process of system dynamics.

The first step is to analyze and understand the problem in all aspects with the theory and method of system dynamics and to identify the problem to be studied. In the second step, the influencing factors of the system are clarified, and the relationship between the factors is analyzed. In the third step, a cause–effect diagram is drawn based on the analysis results. In the fourth step, on the basis of the cause–effect diagram, the system flow diagram is drawn. In the fifth step, the appropriate function is selected to establish the equation In the sixth step, on top of the system dynamics theory, the software is used to simulate the model and analyze the results. If the simulation results are found to be contrary to the reality, it is necessary to return to make modifications to the model. In the seventh step sensitivity analysis is performed on key elements to discover their influence laws.

4. System Construction

In the traditional dynamic decision-making system for the construction of mega infrastructure projects, it is difficult to complete the collation and analysis of massive data, while the dynamic decision-making information system for the construction of mega infrastructure projects in the information environment can break through the limitations of time and space, break through the constraints of funds and equipment, and carry out comprehensive and accurate data collation and analysis. The construction of mega infrastructure projects should follow the trends, seize the advantages, and use information management technology to analyze construction information data. In addition, the experience and intelligence of relevant industry experts are indispensable as leading roles in the dynamic decision-making process of mega infrastructure projects. Humans lead dynamic decision-making for mega infrastructure projects with the assistance of computers. Before using the system dynamics method to build the model, it is first necessary to establish a dynamic decision-making system for mega infrastructure project construction. According to the development and characteristics of mega infrastructure project construction, the dynamic decision-making system of engineering construction is divided into four subsystems: a decision support subsystem, an information processing and transformation subsystem, a human–computer collaborative decision-making subsystem, and a subsystem for evaluating the effect of the program. The four subsystems are described below, as shown in Figure 4.

4.1. *Information Collection Subsystem*

The collection of project implementation information is the basis for dynamic decision-making in mega infrastructure projects, and the comprehensiveness and accuracy of the collected information directly determine the smoothness of the project implementation information data analysis. Increasingly often, cutting-edge information technologies are a source of information and datasets supporting the process of arriving at a decision [35,43]. The collection of information on mega infrastructure project construction is inseparable from advanced space–sky–earth integrated information collection technology and equipment. In recent years, with the promotion of mega infrastructure projects and the rapid development of information technology, cutting-edge technologies, such as the ubiquitous Internet of Things [44], mobile Internet [45], GIS [46], and satellite remote sensing [47,48], have effectively supported the collection of information for mega infrastructure projects. Especially in the context of big data, when analyzing project information and making decisions, it is necessary to combine the information resources of other project cases [20,49] and policies, regulations, and industry codes. Therefore, by analyzing the source of project information and the mechanism of information acquisition, the project information collection can be divided into two aspects: static information data and dynamic information data. Static information data mainly includes the collection of historical case information, project plans and strategies, policies, regulations, and industry codes. Dynamic information data mainly includes project advance survey information, project monitoring, and inspection information [34,49]. Figure 4 shows the analysis of real-time dynamic information data to find problems combined with static information data to analyze the causes of the problems to design a decision-making program. For example, regarding investment information indicators, monthly investment information on the construction process of mega infrastructure projects is monitored and counted, and then the quarterly investment information on the construction process of mega infrastructure projects, annual investment information, and cumulative investment information from the start of construction are counted.

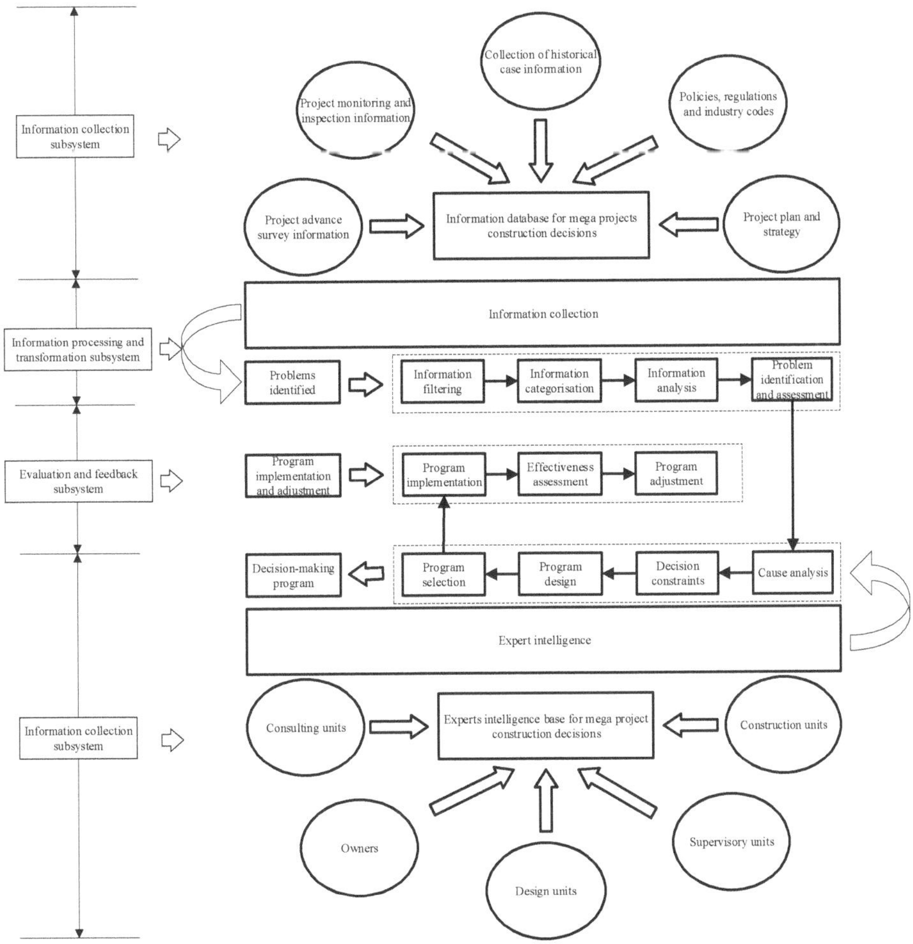

Figure 4. Decision-making and dynamic operation mechanism of mega infrastructure projects based on human–computer collaborative thinking.

4.2. Information Processing and Transformation Subsystem

The project information processing and transformation subsystem is an important part of the dynamic decision-making of mega infrastructure projects, which is led by human intelligence and assisted by computers. Using decision-making information preprocessing methods, decision-making information analysis methods, decision-making information evaluation methods, logical thinking, creative thinking, and accumulated knowledge and experience, the filtering and categorization of information data is achieved. In addition, statistical analysis methods are used to transform information into analysis results [50], from which problems can be identified and evaluated. In view of the complexity and long-term nature of the construction of mega infrastructure projects, the collected information data often have significant characteristics, such as being massive, multi-source, heterogeneous, and interdisciplinary. Moreover, it is increasingly difficult to process and analyze the information data of project implementation. The integrated application of "tool methods" is the "core" of mega infrastructure project decision-making. Therefore, project decision-making relies on the integration and innovation of new data processing analysis tools and methods. According to the existing research results, the information processing and transformation subsystem of mega infrastructure projects can be divided into four parts: information filtering, information categorization, information analysis, and problem identification and assessment. This is shown in Figure 4.

4.3. Human–Computer Collaborative Decision-Making Subsystem

Human–computer collaboration refers to the process of processing, transforming, and analyzing the massive monitoring and inspection information in the process of mega infrastructure projects by computer networks using modeling methods, from which problems can be identified. Then, experts from each unit can evaluate and make decisions. "Factual data" and "tool methods" have laid the foundation for the dynamic decision-making of mega infrastructure projects. However, each link of dynamic decision-making still relies on expert intelligence. The scientific nature of dynamic decision-making cannot be separated from the support of expert intelligence. Generally speaking, expert intelligence is mainly used in information analysis and evaluation, decision-making program formulation, and selection [51]. Therefore, under human–computer collaborative thinking, the real-time participation and collaboration of expert intelligence is particularly important Scientific and efficient interdisciplinary expert selection and opinion integration are conducive to maximizing the positive role of expert intelligence in mega infrastructure project decision-making. The specific operating mode is shown in Figure 4.

The dynamic decision-making mentioned in this paper mainly refers to the dynamic decision-making carried out in response to the problems encountered in the construction process of mega infrastructure projects. Through consulting experts in the related fields and the relevant literature, it was determined that the decision-making expert database in this paper mainly refers to the decision-making expert team composed of owners, design units, supervisory units, construction units, consulting units, etc., as shown in Figure 4. These experts contribute intelligence and knowledge in the dynamic decision-making process.

4.4. Evaluation and Feedback Subsystem

In the process of mega infrastructure project construction, the successful implementation of a decision-making program often requires trial and error. The evaluation and feedback of the implementation effect of the decision-making program are important parts of the dynamic decision-making process of mega infrastructure projects. Both the Holt's decision model [25] and the multi-criteria decision models [23] mentioned the evaluation and adjustment of the effect of the implemented decision program. Although the theoretical decision-making program combines "factual data", "tool methods", and "expert intelligence", the construction of mega infrastructure projects is dynamically changing, and various environmental factors are intertwined and complex. There is a certain uncertainty in the implementation of the decision-making program. Timely evaluation and adjustment are required. This process of the effectiveness of assessment and program adjustment is also repetitive and dynamic. This is shown in Figure 4.

5. Model Construction and Simulation

5.1. System Objective

The simulation research of the dynamic decision-making operation mechanism of mega infrastructure projects based on human–computer collaborative thinking starts from the establishment of the system objective. This paper mainly studies the dynamic decision-making service mechanism of mega infrastructure project construction led by an operation mechanism under the thinking of human–computer collaboration. Therefore, the goal of the system dynamics model is to comprehensively grasp the dynamic decision-making process of mega infrastructure projects, accurately identify the key elements that affect its operation process, explore the interaction between the key elements, and provide guidance for the owners of mega infrastructure projects to carry out the design and selection of decision-making programs and continuously optimize the operation process of dynamic decision-making of mega infrastructure projects.

5.2. System Boundary Determination

The dynamic decision-making system of mega infrastructure projects is a system with a complex structure and many influencing factors, which is affected by various constraints of the project, information management (collection, processing, and analysis) technology, expert intelligence, and other factors. Its internal structure is complex, and there is multi-level causal feedback among various factors, which is manifested as a nonlinear system with multiple complex feedback loops [52]. Due to the complexity of this multi-loop feedback and non-linear analysis, this feature fully meets the characteristics of system dynamics modeling and simulation. Therefore, this paper uses the system dynamics method to analyze the relationship between the factors of the dynamic decision-making system of mega infrastructure projects.

Under the influence of various factors in the dynamic decision-making of mega infrastructure projects, the system boundary was first defined. Combined with the work of the relevant departments of mega infrastructure projects, this paper uses the Delphi method, questionnaire survey method, field investigation of mega infrastructure projects, and consultation with experts in the industry to analyze the system boundary. The operational process, influencing factors, and participating subjects of the dynamic decision-making of mega infrastructure projects were delineated into the system. We invited experts from universities involved in the scientific research of mega infrastructure projects and industry experts involved in the construction of a mega infrastructure project in Southwest China—a total of eight people—to set up an expert group to define the boundaries of the dynamic decision-making system for the construction of mega infrastructure projects. Eight experts put forward their personal opinions; after three lots of feedback, we understood that the system mainly involves an information collection system, an information processing and transformation platform, a decision support information base, and a decision expert team.

5.3. Cause-and-Effect Diagram of Dynamic Decision-Making System for Mega Infrastructure Projects

In the analysis of the dynamic decision-making information system for mega infrastructure projects, a change in each factor has an impact on the results of the analysis. There is also interaction and mutual influence among various factors. In this paper, the system dynamics software Vensim PLE was used to establish a cause-and-effect diagram of the influencing factors, as shown in Figure 5.

As can be seen from Figure 5, there are two positive feedback loops in the dynamic decision-making operation of mega infrastructure projects, as shown in Table 2.

(1) Loop 1 represents the most basic process of the dynamic decision-making and operation of mega infrastructure projects. Information collection focuses on comprehensiveness and accuracy. By virtue of using the existing scientific and technological level and staffing situation, the implementation of the project is monitored and inspected to collect information. In addition, information is uploaded to the information management platform in a timely manner. The information management platform is used to process and analyze information in conjunction with historical cases of this project and similar projects' cases to improve the efficiency of mega infrastructure project decision-making. After that, the problem identification and assessment are carried out. The problem identification is carried out by the decision-making experts according to the project plan and strategy, policies, regulations, and industry codes. The problem assessment is carried out according to the relevant assessment methods and rules. The decision-making experts carry out the cause analysis by combining the historical cases of this project and the historical cases of

similar projects. The decision-making experts combine the problem assessment results, the causes of the problems, and the decision-making constraints, using certain decision-making models and methods to design programs, among which the owners select the best program Finally, the construction unit implements the decision-making program and the owners evaluate the effectiveness of program implementation;

(2) Loop 2 represents the process of updating the historical case database when the dynamic decision-making of mega infrastructure projects runs. Firstly, the reasons for the problem are identified by searching and matching from the historical case database of this project. It provides reference ideas for the decision-making program. Once the decision-making programs are completed, the optimal program needs to be selected and implemented and its effectiveness evaluated. Finally, the decision-making process is formed into a case to update the historical case database of the project, and it further improves and enriches the historical case database of the project.

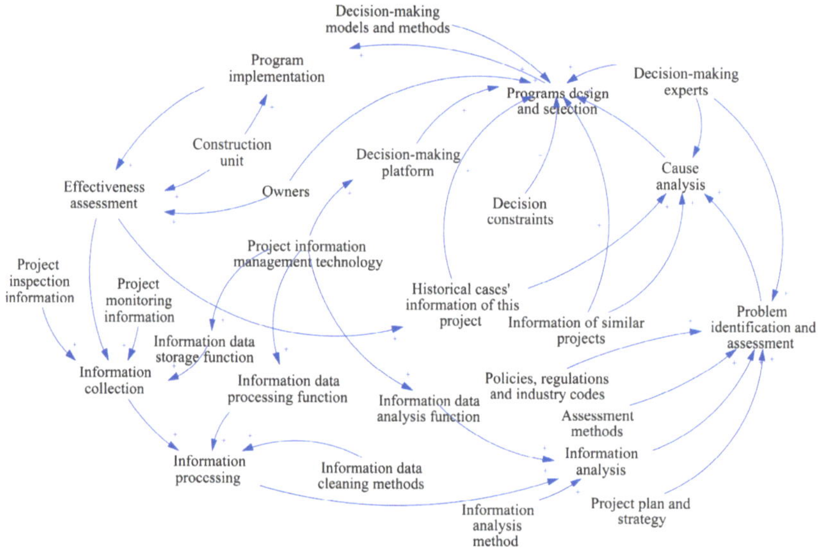

Figure 5. The cause-and-effect diagram of the dynamic decision-making mechanism for mega infrastructure projects based on human–computer collaborative thinking.

5.4. Simulation Flow Diagram of Dynamic Decision-Making System for Mega Infrastructure Projects

Although the logical relationship and feedback loops among the elements are clarified through the cause-and-effect diagram, the essence and structural relationship of the system elements cannot be fully revealed [53]. The system flow diagram has the ability to reflect the interaction form of various variables in the system. A dynamic system model with a feedback structure can be established after quantifying each feedback loop. Human–computer collaborative intelligent decision-making thinking has a guiding effect on the dynamic decision-making mechanism of mega infrastructure projects. The change of thinking leads to the application of related technologies and measures. The presentation of the influence of human–computer collaborative thinking is mainly based on the performance of relevant measures. Therefore, this paper comprehensively considers the reality of the dynamic decision-making practice of mega infrastructure projects and the scientific nature of data and constructs a system flow diagram of the dynamic decision-making mechanism of mega infrastructure projects based on human–computer collaborative decision-making thinking according to the cause-and-effect diagram, as shown in Figure 6.

Table 2. Dynamic decision-making operation feedback loop of mega infrastructure projects.

Loop Number	Type of Feedback	Loop Details
Loop 1	Positive feedback	Information collection → Information processing → Information analysis → Problem identification and assessment → Causes analysis → Solution design and selection → Decision-making solution implementation → Implementation effect evaluation → Information collection
Loop 2	Positive feedback	Historical case information on this project → Analysis of causes → Solution design and selection → Decision-making solution implementation → Evaluation of implementation results →Historical case information on this project

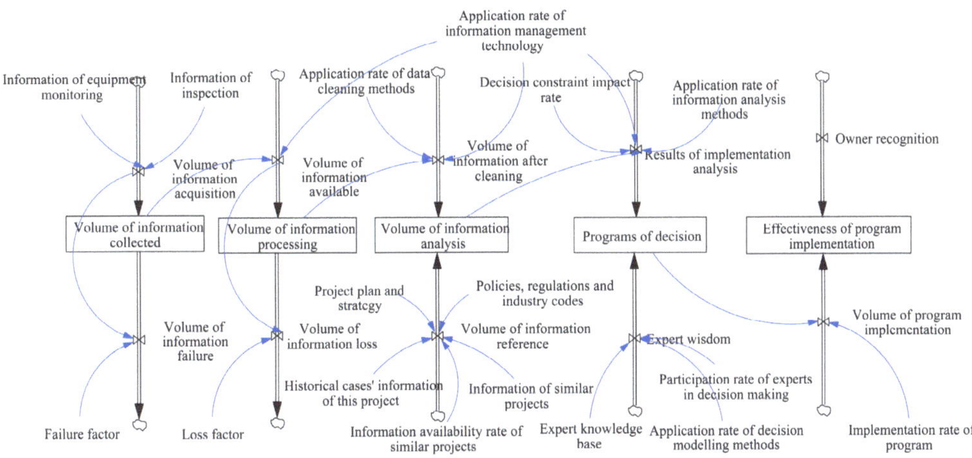

Figure 6. Flowchart of dynamic decision-making mechanism for mega infrastructure projects based on human–computer collaborative thinking.

According to Figure 6, the system is mainly composed of 32 variables, including 5 level variables, 10 rate variables, and 17 constants, as shown in Table 3.

Establishing system variable equations and setting parameters are digital representations oriented to the interaction and the way they act among the elements of the system [58]. The dynamic decision-making mechanism of mega infrastructure projects based on human–computer collaborative thinking is characterized by complexity and dynamism. It has certain theorization, broadness, and geographical differences in terms of quantifying the influencing factors. It is difficult to obtain specific data that can express the relationship and influence among the elements through questionnaires, research interviews, or historical data in a single way in concrete implementation. In addition, the process of simulation is not about how realistic it is but about its usefulness and the extent to which it reveals how things change. Although the parameters in the model, in many cases, lack accurate data, the system dynamics model reveals changes in the evolutionary trend of the whole system and does not require precise results [40]. The correctness of the structure of the system dynamics model is more important than the choice of the parametric values. Therefore, in this paper, the initial assignment of parameters was performed through the review of historical mega infrastructure project construction-related materials and interviews with experts in the field, who have participated in the construction of mega infrastructure projects. For example, the initial values of level variables and rate variables were determined by consulting 56 experts within the construction industry, and some constants were derived on the basis of reference to the existing research results and expert opinions. Based on the initial parameter settings, the final model was determined after several times of debugging and verification. Information on the consulting experts is shown in Table 4. The specific equation design and parameter descriptions are shown in Table 5.

Table 3. Names and types of variables.

Variable Type	Name of Variables	Reference
Level variables (5)	Volume of information collected (y_1), volume of information processing (y_2), volume of information analysis (y_3), programs of decision-making (y_4), effectiveness of program implementation (y_5)	[29,54,55]
Rate variables (10)	Volume of information acquisition (x_1), volume of information failure (x_2), volume of information available (x_3), volume of information loss (x_4), volume of information after cleaning (x_5), volume of information reference (x_6), results of implementation analysis (x_7), expert intelligence (x_8), owner recognition (x_9), volume of program implementation (x_{10})	[29,56,57]
Constants (17)	Information on equipment monitoring (c_1), information on inspection (c_2), failure factor (c_3), application rate of information management technology (c_4), application rate of data cleaning methods (c_5), loss factor (c_6), project plan and strategy (c_7), policies, regulations, and industry codes (c_8), historical case information on this project (c_9), information on similar projects (c_{10}), information availability rate of similar projects (c_{11}), decision constraint impact rate (c_{12}), application rate of information analysis methods (c_{13}), expert knowledge base (c_{14}), participation rate of experts in decision-making (c_{15}), application rate of decision modeling methods (c_{16}), implementation rate of program (c_{17})	[11,20,57]

5.5. Model Simulation

The purpose of validity testing is to check the validity of the model results and verify whether the information and behavior associated with the model reflect the characteristics and change patterns of the actual system. Therefore, Vensim PLE 9.3.5 software was needed to check the dynamic decision system model of the mega infrastructure projects to ensure the proper operation of the system model simulation. After constructing the system dynamics model, it was necessary to perform simulation tests according to the pre-set equations and initial values of the parameters. We used Vensim PLE software to test the simulation effect of the system dynamics model and then analyzed the validity and rationality of the simulation results. On this basis, we selected the main variables to complete the sensitivity analysis operation to understand its effect on the dynamic decision-making operation mechanism of mega infrastructure projects.

5.5.1. Model Simulation Analysis

Validity testing was required to verify that the information and behavior associated with the model reflect the characteristics and patterns of change in the actual system. This paper uses Vensim PLE software to test the validity of the model of the dynamic decision-making system operation mechanism of mega infrastructure projects. With reference to the existing related research, the simulation time was limited to 12 months, and the time step was 1 month. The five main variables of the volume of information collected, the volume of information processing, the volume of information analysis, the programs of decision-making, and the effectiveness of program implementation were selected for monitoring. The change patterns of the above variables in the system were observed, and the simulation results of the main variables of the model under the established parameters are shown in Figure 7.

Table 4. Information on consulting experts.

	Variable	Category	Number	Percentage (%)
Project information	Project type	Railway	30	53.57
		Road	9	15.71
		Public building	11	19.11
		Hydropower	4	6.79
		Others	2	3.57
	Project duration (years)	1~3	10	18.21
		4~6	15	27.32
		7~9	17	30.36
		10~12	13	23.04
		>12	1	2.14
Information on respondents	Types of respondent firms	Owner	13	23.75
		Designer	11	19.64
		Contractor	15	26.79
		Supervisor	7	11.96
		Government	4	6.43
		Supplier	6	10.71
	Gender	Men	46	82.14
		Women	10	17.86
	Education background	College or below	7	12.32
		Bachelor	0	0.00
		Master	32	57.50
		Doctor	14	25.18
	Work experience (years)	<5	3	5.18
		5~10	4	7.14
		11~15	17	30.71
		16~20	23	40.89
		>20	8	14.11
	Professional qualification	Project manager	4	6.43
		Department manager	11	19.64
		Project engineer	9	15.71
		Professional technician	23	40.89
		Others	10	17.86

(1) The volume of information collected is increasing.

During the construction of a mega infrastructure project, the operation of the project is monitored by various monitoring equipment at all times, and there will also be supervisory units and owners visiting the site regularly to check the construction situation. All these means are carried out to collect information. On the one hand, the information can be collected by a variety of space–sky–earth integrated intelligent sensing technology and equipment to monitor the implementation process of mega infrastructure projects in real time. On the other hand, the information can be collected through the construction unit personnel self-inspection and the inspection of supervisory units and owners. The construction situation of mega infrastructure projects is complex and changeable, which will make some of the acquired information data lose its own value as time goes by and then lead to data failure. However, in this information and digitalization environment, efficient data acquisition technology and multi-source data collection channels can still make the amount of data collected by the project management information system increase continuously;

(2) The volume of information processing and analysis are growing in tandem.

The growth rate of information processing volume is slightly flat at first compared with the later stages due to the low popularity of information processing technology and tools and the impact of the data loss factor. It also restricted by the speed of information analysis. With the increasing maturity of information processing technology and the deepening and improvement of the scope of project information monitoring and inspection, the speed of information processing is improved. The historical case information on this project and the information on similar projects provide useful supplements for information analysis. Information analysis experts transform existing information resources by applying professional information analysis methods and the powerful information data analysis function of the information management platform, and the quality and quantity of information analysis results are continuously improved;

(3) The effectiveness of program implementation is positively correlated with the programs of decision-making, and both show an increasing trend.

Table 5. The description of equation design and parameter.

Variables	Equations and Initial Values	Description
Volume of information collected (y_1),	INTEG ($x_1 - x_2$, 5000)	Using integral functions
Volume of information processing (y_2)	INTEG ($x_3 - x_4$, 0)	
Volume of information analysis (y_3)	INTEG ($x_5 + x_6$, 0)	
Programs of decision – making (y_4)	INTEG ($x_7 + x_8$, 0)	
Effectiveness of program implementation (y_5)	INTEG ($x_9 \times x_{10}$, 0)	
Volume of information acquisition (x_1)	$c_2 + c_1$	Using linear correlation functions
Volume of information failure (x_2)	$x_1 \times c_3$	
Volume of information available (x_3)	$y_1 \times c_4$	
Volume of information loss (x_4)	$x_3 \times c_6$	
Volume of information after cleaning (x_5)	$y_2 \times c_4 \times c_5$	
Volume of information reference (x_6)	$c_7 + c_8 + c_9 + c_{10} \times c_{11}$	
Results of implementation analysis (x_7)	$y_3 \times c_{13} \times c_{12} \times c_4$	
Expert intelligence (x_8)	$c_{14} \times c_{15} \times c_{16}$	
Owner recognition (x_9)	0.6	Consulting with experts in the field of mega infrastructure project construction
Volume of program implementation (x_{10})	$y_4 \times c_{17}$	
Information on inspection (c_2)	1000	Based on existing research and experience in the construction of mega infrastructure projects
Information on equipment monitoring (c_1)	4000	
Failure factor (c_3)	0.03	
Loss factor (c_6)	0.02	
Application rate of information management technology (c_4)	0.5	
Application rate of data cleaning methods (c_5)	0.8	
Project plan and strategy (c_7)	1000	
Policies, regulations, and industry codes (c_8)	3000	
Historical case information on this project (c_9)	7000	
Information on similar projects (c_{10})	15,000	
Information availability rate of similar projects (c_{11})	0.05	
Application rate of information analysis methods (c_{13})	0.6	
Decision constraint impact rate (c_{12})	0.2	
Expert knowledge base (c_{14})	10,000	
Participation rate of experts in decision – making (c_{15})	0.7	
Application rate of decision modeling methods (c_{16})	0.6	
Implementation rate of program (c_{17})	0.8	

Programs of decision-making are influenced by factors such as the participation rate of experts in decision-making, the expert knowledge base, the application rate of decision modeling methods, the decision constraint impact rate, etc. The increase in the information analysis volume of mega infrastructure projects continuously promotes the output of implementation analysis results. These analysis results combined with expert intelligence jointly promote the output of decision-making programs. The programs of decision-making discussed and reasoned by relevant experts are approved by owners and then passed through docking to each unit of the project to implement. Constrained by the quality and quantity of the pre-decision programs, the volume of program implementation is low, which leads to less effective program implementation. With the continuous output of high-quality programs of decision-making, the effectiveness of program implementation is improving continuously. The improvement from the implementation of the program further enhances the influence of the dynamic decision-making system. Therefore, the effectiveness of program implementation is weak in the early stage and significantly enhanced in the middle and late stages.

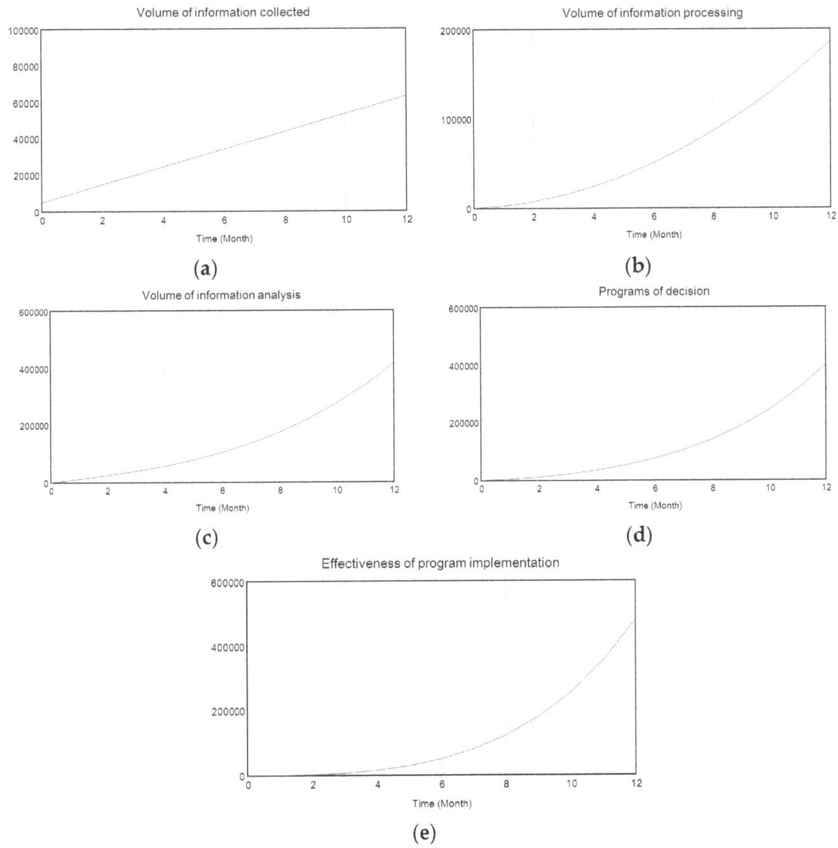

Figure 7. Simulation results of main variables of the system model. (**a**) represents Volume of information collected, (**b**) represents Volume of information processing, (**c**) represents Volume of information analysis, (**d**) represents Programs of decision, (**e**) represents Effectiveness of program implementation.

Thus, it can be seen that the performance of the volume of information collected, the volume of information processing, the volume of information analysis, the programs of decision-making, and the effectiveness of the program implementation is consistent with the reality of the dynamic decision-making operation of mega infrastructure projects, indicating that the model can accurately reflect the operation system of the dynamic decision-making mechanism of mega infrastructure projects based on human–computer collaborative thinking. It has a certain significance for the design of dynamic decision-making programs for mega infrastructure projects.

5.5.2. Sensitivity Analysis

Sensitivity analysis was performed by changing the parameter values of important variables to observe the resulting dynamic changes in the simulation results, thus analyzing the impact of the adjusted variables on the system and the degree of influence. On the basis of the analysis of the dynamic decision-making mechanism of the project, combined with the constructed cause–effect diagram and system flow diagram, five elements were selected for sensitivity analysis. They were the application rate of information management technology, the application rate of data analysis methods, the participation rate of experts in decision-making, the historical case information on this project, and the information on similar projects. In this paper, four subsystems were constructed, and five key factors are the key representative elements of the four subsystems, among which the application rate of information management technology has a key influence on all five subsystems. The data analysis method is the key element of the data processing and transformation subsystem Expert wisdom, historical case information on this project, and the information on similar projects are the keys of the human–computer collaborative decision-making subsystem The implementation effect of the engineering decision program is closely related to the information analysis and result production links, and the above five elements, as important factors supporting the output of information products and decision results, can significantly show their effect on the role of the system model.

(1) The application rate of information management technology

The parameter values of the application rate of the information management platform were set to 0.5, 0.1, and 0.9 to obtain three simulation curves, as shown in Figure 8. By comparing and analyzing the simulation results, it is concluded that the change of application rate of information management technology has an obvious positive impact on the implementation effect of the engineering decision-making program. The process of mega infrastructure project construction is filled with a huge amount of complex data, which needs to be collected, processed, and analyzed using information management technology. High-quality data are the source and basis of outputting high-quality decision-making programs, which can influence the effectiveness of program implementation. Therefore, mega infrastructure project construction should actively introduce information technology and digital technology to enhance operation efficiency and quality and strive to improve the impact of decision-making programs;

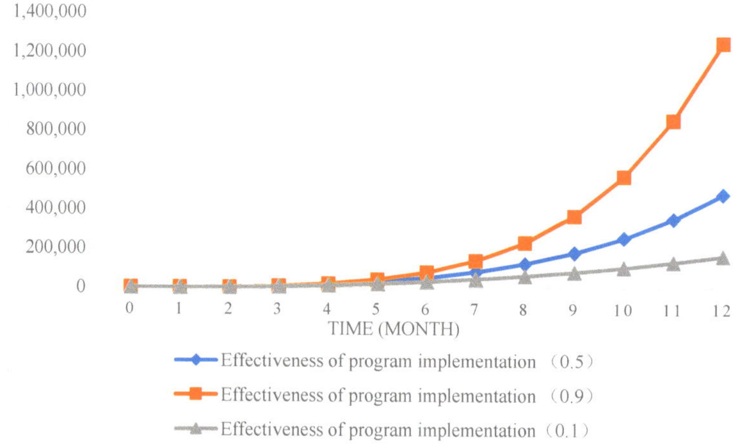

Figure 8. Changes in the sensitivity of the application rate of information management technology on the effectiveness of program implementation.

(2) Application rate of information analysis methods

The parameter values of the application rate of information analysis methods were set to 0.6, 0.2, and 0.9, and three simulation curves were obtained, as shown in Figure 9. By comparing and analyzing the simulation results, it is concluded that the application of information analysis methods can significantly improve the implementation effect of decision results. A variety of information analysis methods, such as information measurement analysis and content analysis, can improve the efficiency of information data result transformation under combined use with the information management platform and provide support for the engineering decision-making program design. Therefore, the construction of mega infrastructure projects should pay attention to the application of information analysis methods and combine the scientific and reasonable use of information analysis methods with the needs of mega infrastructure projects to promote the output of decision-making results;

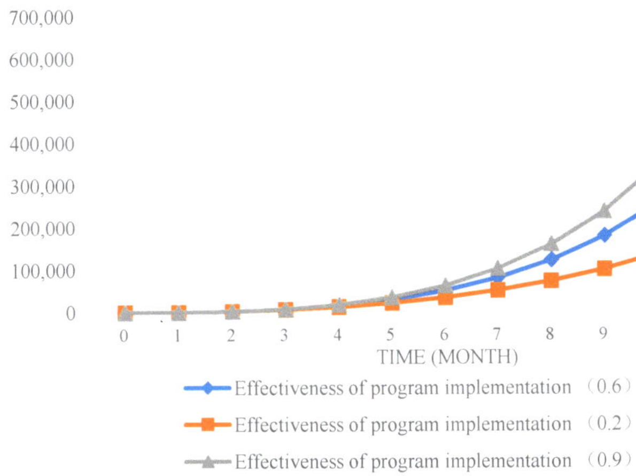

Figure 9. Changes in the sensitivity of the application rate of information analysis methods on the effectiveness of program implementation.

(3) Participation rate of experts in decision-making

The parameter values of the experts' participation rate in decision-making were set to 0.7, 0.2, and 1, and three simulation curves were obtained, as shown in Figure 10. By comparing and analyzing the simulation results, it is concluded that the experts' participation rate in decision-making and the effectiveness of program implementation are positively correlated. In addition, the implementation effect of the dynamic decision-making program of mega infrastructure projects needs the support of a construction expert team. The project information analysis results need to be sublimated with the participation of the expert team and processed into a decision-making program that can be delivered to the owners through the experts' empirical reasoning and deliberative thinking. Therefore, the construction of mega infrastructure projects should pay attention to the selection and application of an expert team and gather experts with both theoretical foundation and practical experience in the construction of mega infrastructure projects based on research projects. Through the synergy of the information management system and expert intelligence, the intelligent decision-making of mega infrastructure projects is carried out to enhance the effectiveness of decision-making program implementation;

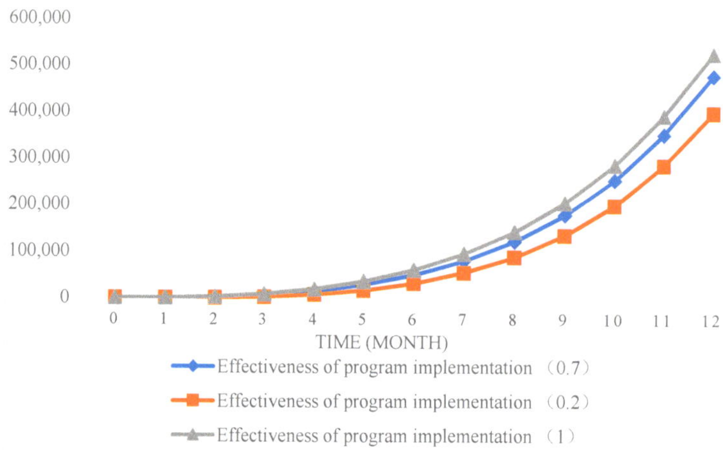

Figure 10. Sensitivity change of participation rate of experts in decision-making on the effectiveness of program implementation.

(4) Historical case information on this project

The parameter values of historical case information on this project were set to 7000, 2000, and 10,000, and three simulation curves were obtained, as shown in Figure 11. By comparing and analyzing the simulation results, it is concluded that the effectiveness of mega project construction decision-making program implementation is influenced by the historical case information on this project. The more available information obtained from it, the more significant the influence is. As they are decisions in the same project, the decision environment, influencing factors, and constraints have a high degree of fit. Therefore, the historical case information on this project has a direct and close connection with the new problems and decisions. Therefore, the usability is high, which can give reliable information support to the decision experts. Therefore, it can serve as a useful supplement to decision information analysis, shorten the information analysis time, and accelerate the output of results. It has a more obvious impact on project decision-making in the later information analysis and result production stages;

(5) Information from similar projects

The parameter values of information on similar projects were set to 15,000, 2000, and 30,000, and three simulation curves were obtained, as shown in Figure 12. Through the comparative analysis of the simulation results, it is concluded that the effectiveness of program implementation is related to the information on similar projects. In addition, the information on similar projects can promote the effectiveness of program implementation. Through the curve changes caused by the numerical changes, we found that the available information provided by the similar project cases is significantly reduced due to the influence of the information availability rate of similar projects. The information provided by the similar project cases can have an impact on the effectiveness of program implementation to a certain extent, but it does not cause too much fluctuation. The information on similar projects is composed of various types of construction project information, which can meet the cross-discipline information resource demand of mega infrastructure project construction decision-making by breaking the information barrier and alleviating the problem of an "information silo". Therefore, regardless of the amount of information provided, the information on similar projects will have a positive impact on the effectiveness of program implementation.

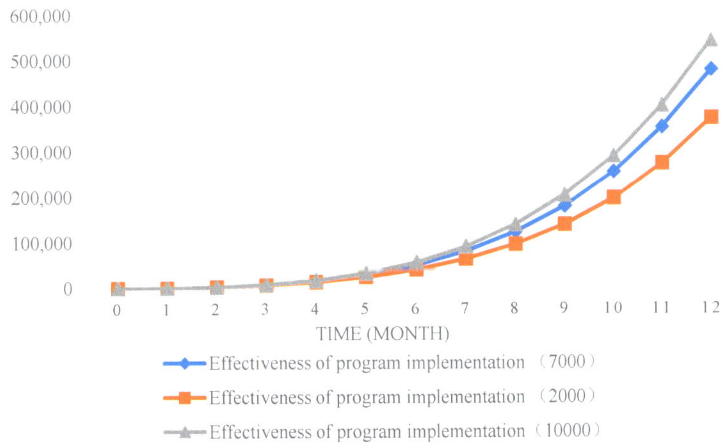

Figure 11. Sensitivity change of the historical case information on this project on the effectiveness of program implementation.

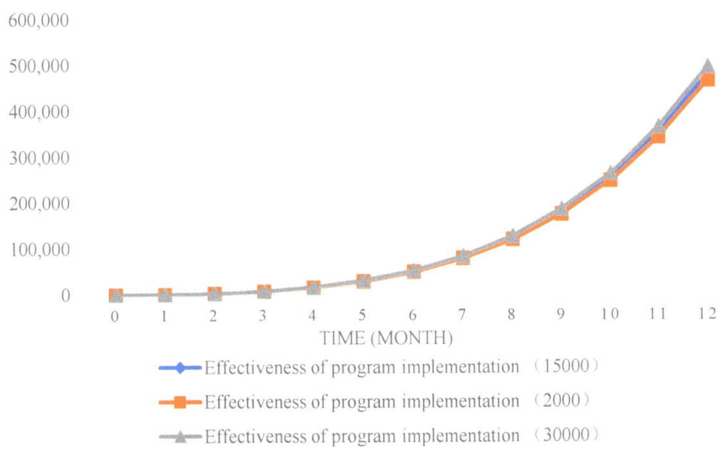

Figure 12. Sensitivity change of the information on similar projects on the effectiveness of program implementation.

6. Discussion

6.1. About Model Construction

Szafranko, E [28] considered decision-making to be a process that includes activities such as expressing decision-making needs, collecting and processing data to support decision-making, measuring results, and finally, evaluating the implementation of the chosen program and the extent to which it meets the evaluation criteria at the beginning of the process. The dynamic decision-making model of mega infrastructure projects constructed in this paper basically agrees with the views of the scholar, but there are still different views in some aspects. For example, this paper believes that the first step of dynamic decision-making in major projects should be collecting information and discovering problems through the analysis of actual information, rather than expressing decision-making needs first and then collecting and processing data. This is related to the characteristics of mega infrastructure projects; the major strategic position of mega infrastructure projects and their complexity determines that they must be monitored and inspected in real time from which problems are identified, and then decisions are made. In addition, this paper not only specifies the content of information collection but also further refines the information data processing process into four processes: information filtering, information categorization, information analysis, and problem identification and assessment.

6.2. About Key Influencing Factors

The results of the sensitivity analysis of the five key influencing factors show that the application rate of information management technology and the application rate of information analysis methods have higher sensitivity to the effectiveness of program implementation, the participation rate of experts in decision-making, and the historical case information on this project have average sensitivity to the effectiveness of program implementation, and the information on similar projects has lower sensitivity to the effectiveness of program implementation. Therefore, more attention should be paid to the improvement of the application rate of information management technology and the application rate of information analysis methods in the process of mega infrastructure project decision-making This is consistent with the idea mentioned by Sheng, Z. in his book [5], which is verified in this paper using model simulation and sensitivity analysis.

7. Conclusions

This paper researches the dynamic decision-making mechanism in the construction process of mega infrastructure projects based on the perspective of human–computer collaboration and intelligent decision-making thinking and constructs the dynamic decision-making mechanism of mega infrastructure projects based on intelligent decision-making thinking in three aspects: the organization mechanism, the operation mechanism, and the guarantee mechanism. On this basis, a simulation and sensitivity analysis were carried out by establishing a system dynamics model. The results show that the performance of key variables, such as the volume of information collected, the volume of information processing, the volume of information analysis, the programs of decision-making, and the effectiveness of program implementation is in line with the reality of the mega infrastructure project dynamic decision-making operation. In addition, the five elements of the application rate of information management technology, the application rate of information analysis methods, the participation rate of experts in decision-making, the historical case information on this project, and information from similar projects all have a positive impact on improving the effectiveness of the program implementation of mega infrastructure projects. The mechanism construction and simulation analysis provide useful reference for optimizing the design process of the decision-making program of mega infrastructure projects.

Based on the inheritance of the overall idea of traditional engineering decision-making, this study integrates the technical means of information and digitalization into the framework of the decision-making mechanism for the construction of mega infrastructure projects. It also introduces dynamic decision-making theory into the field of mega infrastructure project construction. This study clarifies the system of influencing factors and their interaction influence paths that affect the dynamic decision-making of mega infrastructure project construction from a data-driven perspective and establishes a theoretical model of the dynamic decision-making mechanism of mega infrastructure project construction based on the data. It further improves the framework of the theoretical system of dynamic decision-making for the construction of mega infrastructure projects.

Finally, there are two main limitations of the dynamic decision mechanism model for mega infrastructure project construction constructed in this paper. One is that it still needs to rely on expert knowledge and experience for brand-new and never-before-seen problems that arise during the construction of mega infrastructure projects. Second, this study has currently constructed a theoretical level decision-making mechanism based on the existing literature and actual research on mega infrastructure projects, but if it is to be implemented in the construction process of large infrastructure projects, it needs to be combined with the actual situation of the project's organizational structure and information collection technology. For example, the organizational structure of a mega infrastructure project in Southwest China is a three-level management mechanism, and the problems arising in the process of project implementation are subject to hierarchical decision-making, and the weight of decision experts is also considered in the decision-making process. These limitations will be further studied in depth in our subsequent research.

Author Contributions: Conceptualization, G.H. and Y.L.; methodology, G.H.; software, G.H. and K.L.; investigation, X.Y.; writing—original draft preparation, G.H.; writing—review and editing, Y.L. and K.L.; visualization, G.H.; supervision, Y.L. and X.Y. All authors have read and agreed to the published version of the manuscript.

Funding: This research was supported by the State Key Laboratory for Track Technology of High-speed Railway: [Grant Number 2021YJ111], China Academy of Railway Sciences.

Institutional Review Board Statement: Not applicable.

Informed Consent Statement: Not applicable.

Data Availability Statement: Not applicable.

Conflicts of Interest: The authors declare no conflict of interest.

References

1. Flyvbjerg, B. What you Should Know about Megaprojects and Why: An Overview. *Proj. Manag. J.* **2014**, *45*, 6–19. [CrossRef]
2. Tamosaitiene, J.; Sarvari, H.; Chan, D.W.M.; Cristofaro, M. Assessing the Barriers and Risks to Private Sector Participation in Infrastructure Construction Projects in Developing Countries of Middle East. *Sustainability* **2021**, *13*, 153. [CrossRef]
3. Lin, Y.-S.; Chou, J.-S.; Chiou, C.-H. Reengineered Governance Process for Assessing Core Public Infrastructure Projects. *Sustainability* **2021**, *13*, 9669. [CrossRef]
4. Swyngedouw, E.; Moulaert, F.; Rodriguez, A. Neoliberal urbanization in Europe: Large-scale urban development projects and the New Urban Policy. *Antipode* **2002**, *34*, 542–577. [CrossRef]
5. Sheng, Z. *Fundamental Theories of Mega Infrastructure Construction Management: Theoretical Considerations from Chinese Practices*; Springer: Cham, Switzerland, 2018. [CrossRef]
6. Hu, Y.; Chan, A.P.C.; Le, Y.; Jin, R.-z. From Construction Megaproject Management to Complex Project Management: Bibliographic Analysis. *J. Manag. Eng.* **2015**, *31*, 04014052. [CrossRef]
7. Li, Y.; Lu, Y.; Taylor, J.E.; Han, Y. Bibliographic and comparative analyses to explore emerging classic texts in megaproject management. *Int. J. Proj. Manag.* **2018**, *36*, 342–361. [CrossRef]
8. Wang, H.; Miao, L. *China Goes Global: The Impact of Chinese Overseas Investment on Its Business Enterprises*; Palgrave Macmillan: London, UK, 2016. [CrossRef]
9. Liu, M.; Li, Y.; Gao, X.; Le, Y.; Hu, Y. Grasping institutional complexity in infrastructure mega-projects through the multi-level governance system: A case study of the Hong Kong-Zhuhai-Macao Bridge construction. *Front. Eng. Manag.* **2018**, *5*, 52–63. [CrossRef]
10. Li, Y.; Han, Y.; Luo, M.; Zhang, Y. Impact of Megaproject Governance on Project Performance: Dynamic Governance of the Nanning Transportation Hub in China. *J. Manag. Eng.* **2019**, *35*, 05019002. [CrossRef]
11. Petrov, I.; Hakimov, A. Digital technologies in construction monitoring and construction control. *IOP Conf. Ser. Mater. Sci. Eng.* **2019**, *497*, 012016. [CrossRef]
12. Coyle, D. Shaping successful mega-project investments. *Oxf. Rev. Econ. Policy* **2022**, *38*, 224–236. [CrossRef]
13. Flyvbjerg, B.; Holm, M.K.S.; Buhl, S.L. What causes cost overrun in transport infrastructure projects? *Transp. Rev.* **2004**, *24*, 3–18. [CrossRef]
14. Luo, L.; Yang, Y.; Zheng, J.; Xie, J. Measuring Project Governance of Mega Infrastructure in China: A Scale Development Study. *Sustainability* **2022**, *14*, 593. [CrossRef]
15. Zhao, Z.-Y.; Zuo, J.; Zillante, G. Transformation of water resource management: A case study of the South-to-North Water Diversion project. *J. Clean. Prod.* **2017**, *163*, 136–145. [CrossRef]
16. Guo, J.; Xu, J.; He, Z. Study on location selection of client-supplied goods and materials support center for the Sichuan-Tibet railway based on dynamic intuitionistic fuzzy multi-attribute decision-making. *J. Intell. Fuzzy Syst.* **2021**, *40*, 5669–5679. [CrossRef]
17. Liu, C.; Lei, Z.; Morley, D.; AbouRizk, S.M. Dynamic, Data-Driven Decision-Support Approach for Construction Equipment Acquisition and Disposal. *J. Comput. Civ. Eng.* **2020**, *34*, 04019053. [CrossRef]
18. Salet, W.; Bertolini, L.; Giezen, M. Complexity and Uncertainty: Problem or Asset in Decision Making of Mega Infrastructure Projects? *Int. J. Urban Reg. Res.* **2013**, *37*, 1984–2000. [CrossRef]

19. Giezen, M.; Salet, W.; Bertolini, L. Adding value to the decision-making process of mega projects: Fostering strategic ambiguity redundancy, and resilience. *Transp. Policy* **2015**, *44*, 169–178. [CrossRef]
20. Brown, C.B.; Elms, D.G. Engineering decisions: Information, knowledge and understanding. *Struct. Saf.* **2015**, *52*, 66–77 [CrossRef]
21. Dikmen, I.; Atasoy, G.; Erol, H.; Kaya, H.D.; Birgonul, M.T. A decision-support tool for risk and complexity assessment and visualization in construction projects. *Comput. Ind.* **2022**, *141*, 103694. [CrossRef]
22. Wassenaar, H.J.; Chen, W. An approach to decision-based design with discrete choice analysis for demand modeling. *J. Mech. Des* **2003**, *125*, 490–497. [CrossRef]
23. Chen, Y.; Li, K.W.; Xu, H.; Liu, S. A DEA-TOPSIS method for multiple criteria decision analysis in emergency management. *J Syst. Sci. Syst. Eng.* **2009**, *18*, 489–507. [CrossRef]
24. Hutchins, M.J.; Sutherland, J.W. An exploration of measures of social sustainability and their application to supply chain decisions *J. Clean. Prod.* **2008**, *16*, 1688–1698. [CrossRef]
25. Clithero, J.A. Improving out-of-sample predictions using response times and a model of the decision process. *J. Econ. Behav Organ.* **2018**, *148*, 344–375. [CrossRef]
26. Vukomanovic, M.; Radujkovic, M. The Balanced Scorecard and Efqm Working Together in a Performance Management Framework in Construction Industry. *J. Civ. Eng. Manag.* **2013**, *19*, 683–695. [CrossRef]
27. Mohammadi, F.; Sadi, M.K.; Nateghi, F.; Abdullah, A.; Skitmore, M. A hybrid quality function deployment and cybernetic analytic network process model for project manager selection. *J. Civ. Eng. Manag.* **2014**, *20*, 795–809. [CrossRef]
28. Zavadskas, E.K.; Antucheviciene, J.; Turskis, Z.; Adeli, H. Hybrid multiple-criteria decision-making methods: A review of applications in engineering. *Sci. Iran.* **2016**, *23*, 1–20.
29. Szafranko, E.; Harasymiuk, J. Modelling of Decision Processes in Construction Activity. *Appl. Sci.* **2022**, *12*, 3797. [CrossRef]
30. Björnsson, I.; Larsson Ivanov, O.; Honfi, D.; Leander, J. Decision support framework for bridge condition assessments. *Struct. Saf.* **2019**, *81*, 101874. [CrossRef]
31. Stefanic, M.; Stankovski, V. A review of technologies and applications for smart construction. *Proc. Inst. Civ. Eng. Civ. Eng.* **2019**, *172*, 83–87. [CrossRef]
32. Woitsch, R.; Sumereder, A.; Falcioni, D. Model-based data integration along the product & service life cycle supported by digital twinning. *Comput. Ind.* **2022**, *140*, 103648. [CrossRef]
33. Lu, C.; Liu, J.; Liu, Y.; Liu, Y. Intelligent construction technology of railway engineering in China. *Front. Eng. Manag.* **2019**, *6*, 503–516. [CrossRef]
34. Matthews, J.; Love, P.E.D.; Porter, S.R.; Fang, W. Smart data and business analytics: A theoretical framework for managing rework risks in mega-projects. *Int. J. Inf. Manag.* **2022**, *65*, 102495. [CrossRef]
35. Cheng, M.-Y.; Roy, A.F.V. Evolutionary fuzzy decision model for construction management using support vector machine. *Expert Syst. Appl.* **2010**, *37*, 6061–6069. [CrossRef]
36. Yoon, J.H.; Pishdad-Bozorgi, P. Game Theory-Based Framework for Analyzing the Collaborative Dynamic of Tacit Knowledge Sharing and the Choice of Procurement and Contract Types in Mega Construction Projects. *Buildings* **2022**, *12*, 305. [CrossRef]
37. Leung, M.-y.; Yu, J.; Chan, Y.S. Focus Group Study to Explore Critical Factors of Public Engagement Process for Mega Development Projects. *J. Constr. Eng. Manag.* **2014**, *140*, 04013061. [CrossRef]
38. Schwaninger, M.; Hamann, T.K. Theory-Building with System Dynamics: Principles and Practices. In *Computer Aided Systems Theory—Eurocast 2005*; Diaz, R.M., Pichler, F., Arencibia, A.Q., Eds.; Springer: Berlin/Heidelberg, Germany, 2005; Volume 3643, pp. 56–62.
39. Forrester, J.W. System dynamics—A personal view of the first fifty years. *Syst. Dyn. Rev.* **2007**, *23*, 345–358. [CrossRef]
40. Thompson, B.P.; Bank, L.C. Use of system dynamics as a decision-making tool in building design and operation. *Build. Environ.* **2010**, *45*, 1006–1015. [CrossRef]
41. Vantsevich, V.V. Road and off-road vehicle system dynamics. Understanding the future from the past. *Veh. Syst. Dyn.* **2015**, *53*, 137–153. [CrossRef]
42. Elsawah, S.; Pierce, S.A.; Hamilton, S.H.; van Delden, H.; Haase, D.; Elmandi, A.; Jakeman, A.J. An overview of the system dynamics process for integrated modelling of socio-ecological systems: Lessons on good modelling practice from five case studies. *Environ. Model. Softw.* **2017**, *93*, 127–145. [CrossRef]
43. Szafranko, E.; Srokosz, P. Applicability of the theory of similarity in an evaluation of building development variants. *Autom. Constr.* **2019**, *104*, 322–330. [CrossRef]
44. Shen, W. Application of BIM and Internet of Things Technology in Material Management of Construction Projects. *Adv. Mater. Sci. Eng.* **2022**, *2022*, 5381252. [CrossRef]
45. Shi, Q.; Ding, X.; Zuo, J.; Zillante, G. Mobile Internet based construction supply chain management: A critical review. *Autom. Constr.* **2016**, *72*, 143–154. [CrossRef]
46. Song, Y.; Wang, X.; Tan, Y.; Wu, P.; Sutrisna, M.; Cheng, J.C.P.; Hampson, K. Trends and Opportunities of BIM-GIS Integration in the Architecture, Engineering and Construction Industry: A Review from a Spatio-Temporal Statistical Perspective. *ISPRS Int. J. Geo-Inf.* **2017**, *6*, 397. [CrossRef]
47. Donaldson, D.; Storeygard, A. The View from Above: Applications of Satellite Data in Economics. *J. Econ. Perspect.* **2016**, *30*, 171–198. [CrossRef]

48. Donaldson, D. Railroads of the Raj: Estimating the Impact of Transportation Infrastructure. *Am. Econ. Rev.* **2018**, *108*, 899–934. [CrossRef]
49. Chang, A.S.-T.; Tsai, Y.-W. Engineering Information Classification System. *J. Constr. Eng. Manag.* **2003**, *129*, 454–460. [CrossRef]
50. Wang, Q.; Tan, Y.; Mei, Z. Computational Methods of Acquisition and Processing of 3D Point Cloud Data for Construction Applications. *Arch. Comput. Methods Eng.* **2020**, *27*, 479–499. [CrossRef]
51. Brito, J.A.M.d. Judgement in geotechnical engineering practice. *Soils Rocks* **2021**, *44*, e2021063821. [CrossRef]
52. Liu, H.; Yu, Y.; Sun, Y.; Yan, X. A system dynamic approach for simulation of a knowledge transfer model of heterogeneous senders in mega project innovation. *Eng. Constr. Archit. Manag.* **2020**, *28*, 681–705. [CrossRef]
53. Xu, N.; Liu, Q.; Ma, L.; Deng, Y.; Chang, H.; Ni, G.; Zhou, Z.; Zhao, D. A Hybrid Approach for Dynamic Simulation of Safety Risks in Mega Construction Projects. *Adv. Civ. Eng.* **2020**, *2020*, 9603401. [CrossRef]
54. Lin, S.; Jia, Y.; Xia, S. Research and Analysis on the Top Design of Smart Railway. *J. Phys. Conf. Ser.* **2019**, *1187*, 052053. [CrossRef]
55. Liu, C.; Chen, X. Data-driven design paradigm in engineering problems. *Proc. Inst. Mech. Eng. Part G J. Aerosp. Eng.* **2016**, *231*, 1522–1534. [CrossRef]
56. Tchana, Y.; Ducellier, G.; Remy, S. Designing a unique Digital Twin for linear infrastructures lifecycle management. *Procedia CIRP* **2019**, *84*, 545–549. [CrossRef]
57. Krishnamenon, M.; Tuladhar, R.; Azghadi, M.R.; Loughran, J.G.; Pandey, G. Digital Twins and their significance in Engineering Asset Management. In Proceedings of the 2021 International Conference on Maintenance and Intelligent Asset Management (ICMIAM), Ballarat, Australia, 12–15 December 2021; pp. 1–6.
58. Tang, Z.W.; Ng, S.T.; Skitmore, M. Influence of procurement systems to the success of sustainable buildings. *J. Clean. Prod.* **2019**, *218*, 1007–1030. [CrossRef]

sustainability

MDPI

Article

Life Cycle Assessment for Geopolymer Concrete Bricks Using Brown Coal Fly Ash

Jingxuan Zhang [1], Sarah Fernando [1,2], David W. Law [1], Chamila Gunasekara [1], Sujeeva Setunge [1], Malindu Sandanayake [3] and Guomin Zhang [1,*]

[1] School of Engineering, RMIT University, Melbourne, VIC 3000, Australia; david.law@rmit.edu.au (D.W.L.); chamila.gunasekara@rmit.edu.au (C.G.); sujeeva.setunge@rmit.edu.au (S.S.)

[2] Faculty of Engineering, University of Peradeniya, Peradeniya 20400, Sri Lanka

[3] College of Sports, Health and Engineering, Victoria University, Footscray, VIC 3011, Australia; malindu.sandanayake@vu.edu.au

* Correspondence: kevin.zhang@rmit.edu.au

Abstract: Traditionally, the construction industry has predominantly used Portland cement (PC) to manufacture bricks, as it is one of the most-commonly available building materials. However, the employment of waste industrial material for brick production can lead to a significant improvement in terms of sustainability within the construction sector. Geopolymer bricks made from brown coal fly ash, a promising industrial waste by-product, serve as a potential alternative. Conducting a life cycle assessment (LCA), this study thoroughly evaluated the entire manufacturing process's environmental impact, from source material acquisition and transportation to brick manufacturing, distribution, usage, and end-of-life, for brown coal bricks as compared to PC bricks. The LCA of the brown coal bricks revealed that their primary environmental impacts stemmed from the raw material manufacturing and usage, while exhibiting substantial reductions in ozone depletion, water depletion, and metal depletion. These findings highlighted the environmental advantages of the brown coal bricks and their potential to revolutionize sustainable construction practices.

Keywords: life cycle assessment; brown coal fly ash; impact assessment; cost–benefit analysis

Citation: Zhang, J.; Fernando, S.; Law, D.W.; Gunasekara, C.; Setunge, S.; Sandanayake, M.; Zhang, G. Life Cycle Assessment for Geopolymer Concrete Bricks Using Brown Coal Fly Ash. *Sustainability* **2023**, *15*, 7718. https://doi.org/10.3390/su15097718

Academic Editor: You-Fu Yang

Received: 23 March 2023
Revised: 21 April 2023
Accepted: 23 April 2023
Published: 8 May 2023

1. Introduction

The gravity of climate change and other pressing environmental issues necessitates the prioritization of sustainable solutions. Portland cement (PC) concrete, the most-extensively used building construction material, contributes to an alarming 5–8% of global anthropogenic carbon dioxide emissions [1,2]. The building construction sector has shifted towards sustainable building design, recognizing the significance of the complete project life cycle, from raw material extraction to the disposal stage. The adoption of green strategies throughout a construction project's entire life cycle plays a pivotal role in achieving a building's sustainability. A sustainable construction process should incur minimal environmental impact, not only during the manufacturing and operational stages, but throughout the entire project's life cycle. Roughly 80% of greenhouse gas emissions and energy consumption in the construction sector can be attributed to the operational stages of buildings. In several studies, the implementation of novel technologies, policies, and mitigation measures/technologies has been conducted to reduce GHG emissions during the operational phase [3–5]. Nevertheless, it is equally imperative to curb and lessen the environmental impacts during the initial manufacturing and construction stages.

Brick is a prevalent construction material, and Asia contributes to over two-thirds of global brick production [6]. In Australia, the annual brick production stands at a staggering 1.6 billion [7]. Conventional bricks are typically manufactured from clay with high-temperature kiln firing or PC. Traditional masonry clay bricks are non-eco-friendly due to the considerable energy consumption [8]. The production of PC consumes a large quantity of raw materials and energy [9,10], which considerably impacts the environment. The production process also emits substantial amounts of greenhouse gases (GHGs) into the atmosphere, exacerbating environmental issues. To alleviate the challenges associated with cement and concrete production, several studies have explored the use of sustainable binders sourced from waste materials. Geopolymer, an alternative to PC binders, holds tremendous promise in the construction sector. Geopolymers are derived by alkali activation of aluminosilicate materials, which could be natural or synthetic materials or industrial by-products [11,12]. To evaluate the environmental impacts of geopolymer binders, life cycle assessment (LCA) is the most-appropriate tool available. It provides consensus framework, terminology, and methodological phases [13,14]. LCA facilitates a comprehensive, quantitative, and interpretive analysis of the environmental impact of a specific service, process, or product over its entire life cycle.

Researchers are exploring innovative ways of producing bricks in a more environmentally friendly manner by repolymerizing alternative waste by-products [15,16]. Brown coal fly ash geopolymer concrete masonry bricks' application presents a sustainable and cost-effective solution for the construction industry. According to estimates, global brown coal production reached 286 Gt in 2016, with Australia being the fourth-largest contributor at 76.5 Gt [17]. In Victoria, Australia, all the brown coal reserves are situated in the La Trobe Valley, primarily at two power plants, Loy Yang and Yallourn, where the brown coal ash is produced from lignite and sub-bituminous coals found in two separate seams [18]. Currently, there are no commercial applications for brown coal ash in the construction industry, and most of it ends up in landfills, leading to environmental contamination. Therefore, utilizing brown coal ash for geopolymer-based bricks would not only minimize the impact of this waste, but also eliminate the need for PC.

The quantification of sustainability factors based on real-life data is crucial to gain awareness in the different environmental impact categories. Early-stage LCA studies are essential to convey the knowledge required to reduce environmental impacts by including the building material manufacturing phases for the entire LCA study. Moreover, to date, most geopolymer brick LCA studies have focused only on the early stages of manufacturing, i.e., cradle-to-gate [19–22], and a limited number of impact categories [23–25]. This study undertook an exhaustive investigation of the LCA of the utilization of waste brown coal ash from the two power plants in the La Trobe Valley, Victoria, in the manufacture of geopolymer bricks, including the twelve major impact categories for the "cradle-to-grave" phases. The study covered twelve major impact categories for the cradle-to-grave phases of the brick's life cycle, offering a detailed analysis of the essential factors that arise from manufacturing brown coal ash geopolymer bricks and the variations due to differences in ash composition. Moreover, this study quantified the environmental benefits of using brown coal ash from landfill sites based on relevant impact categories to enhance the overall understanding of the environmental impact.

This study's objective was fourfold: (1) two distinct types of brown coal fly ash geopolymer bricks' environmental impacts were evaluated and compared with conventional PC bricks; (2) hotspot environmental impact factors were identified and avenues for improvements throughout the entire life cycle suggested; (3) a clear economic analysis of utilizing brown coal fly ash for the production of geopolymer bricks is provided; (4) the impacts based on a comprehensive range of impact categories and real-life data were classified.

Current research on LCA assessment of brown coal ash geopolymer bricks does not include a benefit analysis on the use of performance indicator methods. In order to identify opportunities for the improvement of the environmental and economic performance of brown coal bricks in their production, this study quantified the environmental impacts of two brown coal geopolymer bricks during their "cradle-to-grave" life cycle. Their environmental performance was also compared with that of conventional PC concrete bricks. The study lays the groundwork for forthcoming research methodologies aimed at maximizing the eco-sustainability of geopolymer bricks made from brown coal fly ash.

2. Research Methodology

2.1. Research Framework

Utilizing life cycle assessment (LCA), an in-depth study was conducted to scrutinize the environmental impacts and advantages associated with the production of geopolymer concrete bricks composed of brown coal ash sourced from varying power station locations.

Life cycle assessment (LCA) is a systematic approach that utilizes a methodology to gauge the ecological efficacy of products and processes throughout their life cycle, enabling the identification of areas that require refinement [26]. This method has been adopted by researchers, including [27–29]. The LCA procedure described in ISO 14040 [13] was followed to ensure the application of rigorous standards. The ultimate objective of LCA is to quantify and appraise the environmental impact performance of products/processes, facilitating informed decision-making [30]. This framework comprises four distinct, yet interconnected components: goal and scope definition, inventory analysis, impact assessment, and interpretation.

To conduct the impact assessment, the SimaPro (Version 8.2.0) LCA software was employed, which provides a comprehensive and precise analysis. For the purposes of impact assessment, the ReCiPe Mid-Point (Europe H) method (an exceptional feature of SimaPro) was chosen as the best-fitting tool. This sophisticated approach generated a total of eighteen midpoint impact categories, providing a thorough and detailed evaluation [31]

2.2. Goal and Scope Definition

The goal of this study was to analyze the environmental impacts and benefits associated with geopolymer concrete bricks manufactured from brown coal ash obtained from two La Trobe Valley power stations in Victoria.

This study also analyzed the economic benefits based on the fly ash's end-of-life storage location. The economic assessment quantified the benefits by linking the environmental impact categories in the comparison unit.

The functional unit selected allowed the normalization of the impacts for the different compressive strengths of the bricks for comparative and contribution analysis in the LCA. A performance indicator approach (unit of functional performance) was adopted for the analysis; thus, this avoided the distinction between the material scale and the structural scale [32].

The functional unit for the process was selected based on the functional performance unit reported by Damineli et al. [32]. This approach compared the environmental impacts associated with variable concrete types and performances (compressive strength) in the LCA study. Hence, a factor termed impact intensity (Equation (1)) was adopted to enable a comparison of the three types of bricks.

$$\text{Impact intensity (ix)} = x/cs \tag{1}$$

where ix is defined as the impact intensity for the "x" impact category; x is the total impact derived from the LCA analysis; cs is the compressive strength of the brick.

The economic analysis used the functional unit "1 m^3 of brick mixture". The functional unit "1 m^3 of brick mix" was employed for the life cycle cost analysis. The "per brick" functional unit was employed for the total cost analysis.

This study considered the "cradle-to-grave" life cycle of products. It consisted of four life cycle stages:

- Raw material extraction/production, which presents the production and preparation of different materials used in the later production stage; those materials included Na_2SiO_3, the two brown coal fly ashes, the extraction of aggregates, and the production of the PC. The Na_2SiO_3, NaOH, brown coal fly ashes, and aggregates were used in the fly ash bricks' production. The aggregates and PC were used to produce the PC concrete.

- The brick production stage represented the transportation of raw materials for the production of the bricks and the production process.
- Distribution and usage represented the transportation of the bricks and the brick wall construction process.
- End-of-life represented the transportation of demolished brick walls and the landfill.

The system boundary of the geopolymer and PC brick wall construction is presented in Figure 1.

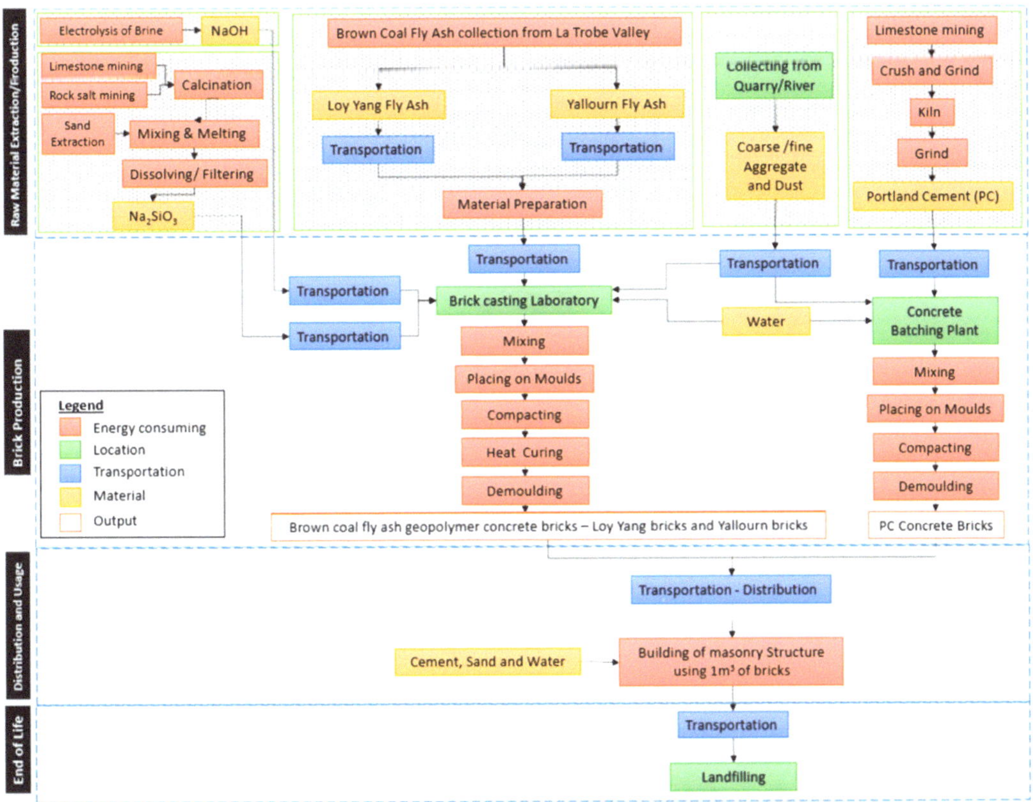

Figure 1. System boundary of the life cycle for the concrete brick wall.

2.3. Models and Testing Scenarios

The model used in this study was built in the SimaPro software. The method of using existing software has been adopted by researchers such as Farina et al. [29] and Zhang et al. [28].

As the selected impact assessment tool in this research, the ReCiPe Mid-Point (Europe H) method provided an exhaustive evaluation of the environmental implications. This method is capable of generating a total of eighteen diverse impact categories, out of which twelve were considered for the midpoint impact categories analysis, including climate change, ozone depletion, terrestrial acidification, human toxicity, photochemical oxidant formation, particulate matter formation, terrestrial ecotoxicity, freshwater ecotoxicity, marine ecotoxicity, water depletion, metal depletion, and fossil fuel depletion.

This study analyzed and compared the environmental impact data of brown coal fly ash brick with PC concrete blocks. Two brown coal brick mix designs and a PC mix designs were adopted [33]: (1) the Loy Yang brown coal fly ash (LYFA) mix achieved a 21.7 MPa compressive strength at 28 days, which corresponds to the application as fire bricks in Australia; (2) the Yallourn brown coal fly ash (YFA) mix had a lower compressive strength 6.8 MPa, and hence, can be used as a general-purpose brick [33]. Table 1 shows the mix proportions of the two types of bricks.

Table 1. Geopolymer and PC brick mix design [33].

Bricks	Mix Design (kg/m^3)								Water to Solid	28-Day Compressive Strength (MPa)
	PC	Brown Coal Ash	Sand	Dust	White Stone (7 mm)	Water	Activator Solution			
							Na_2SiO_3	NaOH		
PC	160	-	728	182	291	76.8	-	-	0.48	15.0
LYFA	-	160	728	182	291	10	208	12	0.52	21.7
YFA	-	152	689	172	276	0	271	17	0.58	6.8

3. Life Cycle Inventory Analysis

3.1. Life Cycle Phases

The LCA phases selected for each brick category are summarized in Table 2. These phases were selected based on the specific production conditions for the geopolymer and PC bricks. The raw material manufacturing stage consisted of the raw material extraction/production phase and the collection/drying process. The brick production phase consisted of both the mixing and heat-curing processes stated in Table 2.

Table 2. Detailed life cycle phases considered for the analysis.

Material	PC	Brown Coal Ash	Fine Aggregate	Coarse Aggregate	NaOH	Sodium Silicate	Geopolymer Brick	PC Brick
Raw material extraction and production	√	-	√	√	√	√	√	√
Collection and drying	-	√	-	-	-	-	-	-
Transportation	√	√	√	√	√	√	√	√
Mixing	-	-	-	-	-	-	√	√
Heat curing	-	-	-	-	-	-	√	-
Distribution	-	-	-	-	-	-	√	√
Usage	-	-	-	-	-	-	√	√
End of life	-	-	-	-	-	-	√	√

3.2. Raw Material Acquisition

The brown coal ash was obtained from two power plants in the La Trobe Valley, Victoria, Australia, namely Loy Yang (LYFA) and Yallourn (YFA) power plants. Although both power plants are close in proximity, the two ash types vary in composition and properties. The variations in the materials are attributed to differences between the two coal seams and storage regimes. The Loy Yang ash is relatively high in aluminosilicates, while the Yallourn ash is relatively low [34]. The mix design utilized identical specific fine and coarse aggregates. The fine aggregates employed were Chelvon sand and Hanson dust, while the coarse aggregates were Chelvon white stone (7 mm). A Grade-D sodium silicate solution and a sodium hydroxide (NaOH) 15 M solution were used as the alkali activator for the geopolymer bricks. General-purpose PC was adopted for the PC bricks' production for the comparative study.

3.3. Transportation Details

The transportation scenarios and distances for all the types of bricks are summarized in Table 3. The brick manufacturing was considered to be located in Melbourne, Australia. The transportation mode for all phases (including the transportation of raw materials and distribution of bricks) was considered as by road, with diesel heavy trucks. All the travel distance of the transportation of materials was based on real-life data.

Table 3. Transportation distances of this LCA study.

Transportation Stage	Distance (km)
LYFA to manufacturing plant	160
YFA to manufacturing plant	145
PC to manufacturing plant	50
Sodium silicate to manufacturing plant	38.5
NaOH to manufacturing plant	26.1
Chelvon sand to manufacturing plant	16.7
Chelvon dust to manufacturing plant	16.7
White stone to manufacturing plant	29.8
Distribution	50
Disposal (landfilling) distance	56

3.4. Energy Consumption

The Australian electricity grid mix [35] was used for all processes shown in the system boundary. The primary source of electricity was coal (61%), followed by natural gas (19%), oil (2%), hydropower (7%), wind (6%), solar (3%), bio-energy (2%), and other renewable energy.

4. Results

4.1. Comparative Analysis

A comparative analysis of the twelve environmental impact categories included in the ReCiPe midpoint methodology was undertaken for the two brown coal geopolymer bricks and compared with the PC bricks. The characterized impact intensities for the twelve midpoint categories are presented in Table 4. All results are presented as the unit of functional performance (compressive strength) for all categories. Figure 2 illustrates the percentage variation for the three brick types for all midpoint categories. The results showed that LYFA had a similar variation for climate change (1.97×10^1 kg CO_2 eq/m^3. MPa) during the "cradle-to-grave" phases compared to the PC bricks (1.94×10^1 kg CO_2 eq/m^3. MPa). However, slightly higher impacts for all other impact categories were observed for LYFA when compared with the PC bricks, except ozone depletion, water depletion, and metal depletion. When considering the "cradle-to-grave" approach, ozone depletion ($\sim$ 27%), water depletion ($\sim$ 30%) , and metal depletion ($\sim$ 47%) for LAFA showed reduced environmental impacts compared to the PC bricks. The LAFA geopolymer bricks showed higher impact values for terrestrial acidification ($\sim$ 67%), human toxicity ($\sim$ 40%), photochemical oxidant formation ($\sim$ 34%), particulate matter formation ($\sim$ 51%), terrestrial ecotoxicity ($\sim$ 24%), freshwater ecotoxicity ($\sim$ 94%), marine ecotoxicity ($\sim$ 92%), and fossil fuel depletion ($\sim$ 55%) compared to the PC bricks, as shown in Figure 2. The YFA bricks showed higher impacts ranging between 72% and 76% and 61% and 98% for all midpoint categories compared to the LYFA and PC bricks, respectively.

Table 4. Quantified environmental impact values for geopolymer and PC bricks.

Impact Category	Unit	Impact Intensity		
		LYFA	YFA	PC
Climate change	kg CO_2 eq/m^3. MPa	1.97×10^1	7.86×10^1	1.94×10^1
Ozone depletion	kg CFC-11 eq/m^3. MPa	4.53×10^{-7}	1.71×10^{-6}	6.19×10^{-7}
Terrestrial acidification	kg SO_2 eq/m^3. MPa	1.34×10^{-1}	5.58×10^{-1}	4.47×10^{-2}
Human toxicity	kg 1,4-DB eq/m^3. MPa	3.17×10^0	1.26×10^1	1.91×10^0
Photochemical oxidant formation	kg NMVOC/m^3. MPa	7.36×10^{-2}	2.97×10^{-1}	4.83×10^{-2}
Particulate matter formation	kg PM10 eq/m^3. MPa	3.40×10^{-2}	1.39×10^{-1}	1.68×10^{-2}
Terrestrial ecotoxicity	kg 1,4-DB eq/m^3. MPa	3.50×10^{-4}	1.41×10^{-3}	2.65×10^{-4}
Freshwater ecotoxicity	kg 1,4-DB eq/m^3. MPa	4.52×10^{-1}	1.92×10^0	2.91×10^{-2}
Marine ecotoxicity	kg 1,4-DB eq/m^3. MPa	3.91×10^{-1}	1.66×10^0	2.86×10^{-2}
Water depletion	m^3/m^3. MPa	1.46×10^{-1}	5.31×10^{-1}	2.08×10^{-1}
Metal depletion	kg Fe eq/m^3. MPa	1.16×10^{-1}	4.16×10^{-1}	2.17×10^{-1}
Fossil fuel depletion	kg oil eq/m^3. MPa	5.61×10^0	2.31×10^1	2.53×10^0

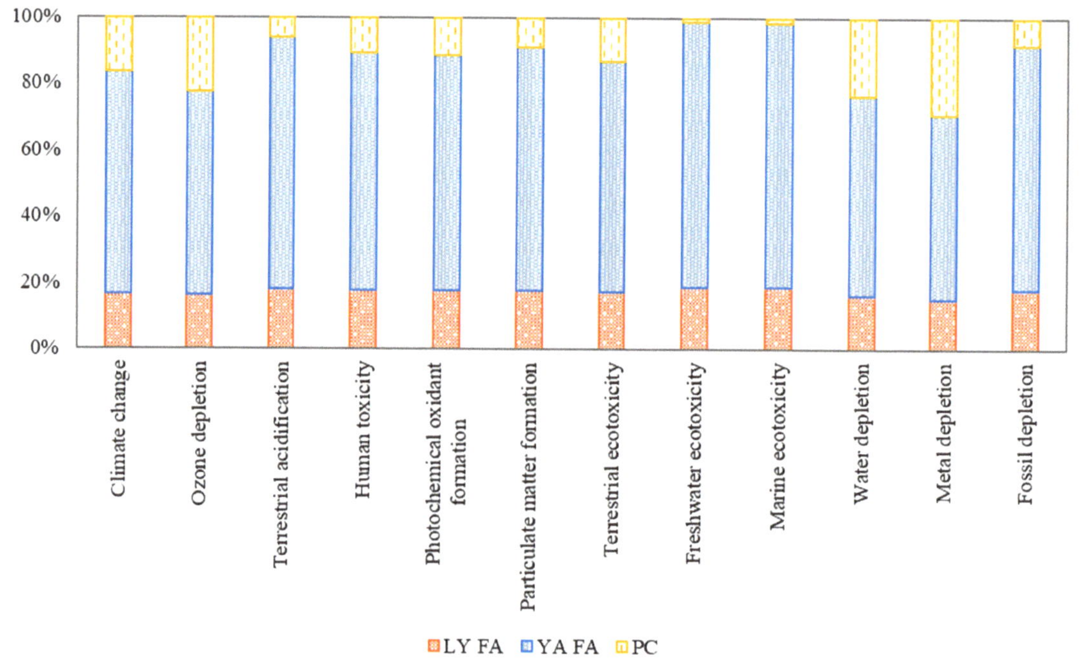

Figure 2. Comparative percentage midpoint characterization values for LYFA, YFA, and PC bricks.

4.2. Contribution Analysis

The impact categories' proportional impacts and intensities relevant to every life cycle stage of the brick production for the LYFA, YFA, and PC bricks are illustrated with regard to the twelve midpoint categories in Figure 3. The fly ash collection location (transportation distance), mix design, and brick compressive strength were the main differences between the LYFA and YFA bricks. Both the LYFA bricks and YFA bricks displayed a similar percentage variation for all midpoint categories, as shown in Figure 3. Climate change was the highest impact associated with the material manufacturing phase for all types of bricks. However, the total PC brick contributions were more than 80% in the climate change category, while both brown coal geopolymer bricks contributed approximately 62% to climate change in the stage of material manufacturing. Furthermore, fossil fuel depletion contributed approximately 22% and 9% for the total impact in the stage of material manufacturing for the geopolymer bricks and PC bricks, respectively. The transportation of raw materials, the phase of distribution usage, and the stage of end-of-life showed a similar variation of the percentage for all the bricks. Moreover, the brick manufacturing phase alone accounted for a higher share of climate change (~ 74%) , fossil fuel depletion (~ 19%), and human toxicity (~ 2%) for the PC bricks.

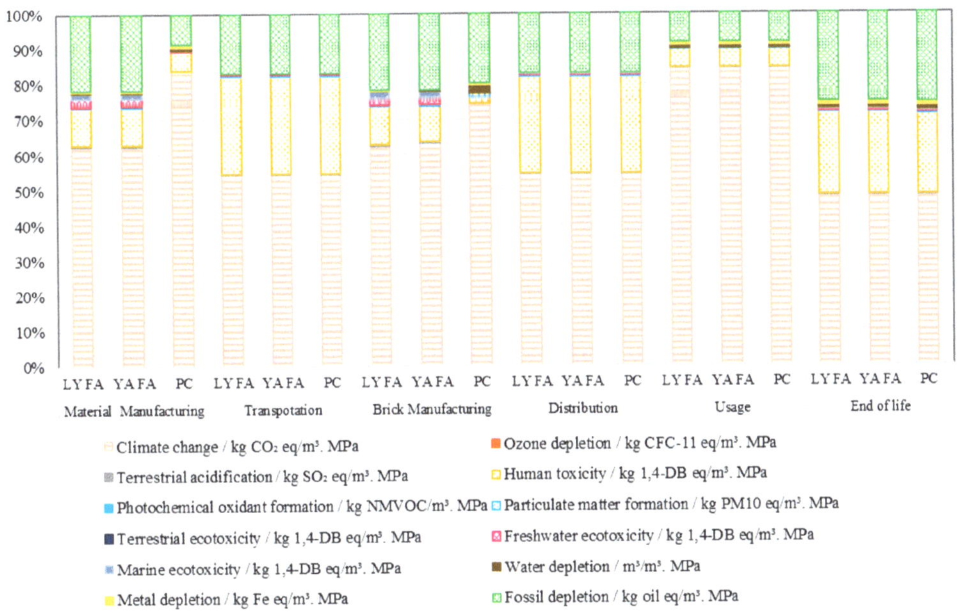

Figure 3. Comparison of the percentage of the environmental impacts of brown coal geopolymer bricks and PC bricks.

The impact intensities for the twelve selected environmental impact categories and key LCA phases for the geopolymer and PC bricks are illustrated in Figure 4. The results illustrated that the YFA bricks produced higher impact intensities for all impact categories based on six LCA phases, as verified by the total percentage of the impacts in the comparative analysis. However, the material manufacturing and usage phases showed higher values for both geopolymer and PC bricks in the climate change impact category. A higher percentage for the material manufacturing, usage, and end-of-life phases are reported for ozone depletion (Figure 4b), while the material manufacturing phase was accountable for the highest proportion for LYFA and YFA compared to PC, which were terrestrial acidification (Figure 4c), human toxicity (Figure 4d), photochemical oxidant formation (Figure 4e), particulate matter formation (Figure 4f), terrestrial ecotoxicity (Figure 4g), freshwater ecotoxicity (Figure 4h), and marine ecotoxicity (Figure 4i). Furthermore, material manufacturing alone was responsible for more than a 50% share of the whole process for all impact categories, except ozone depletion and metal depletion, in both geopolymer bricks.

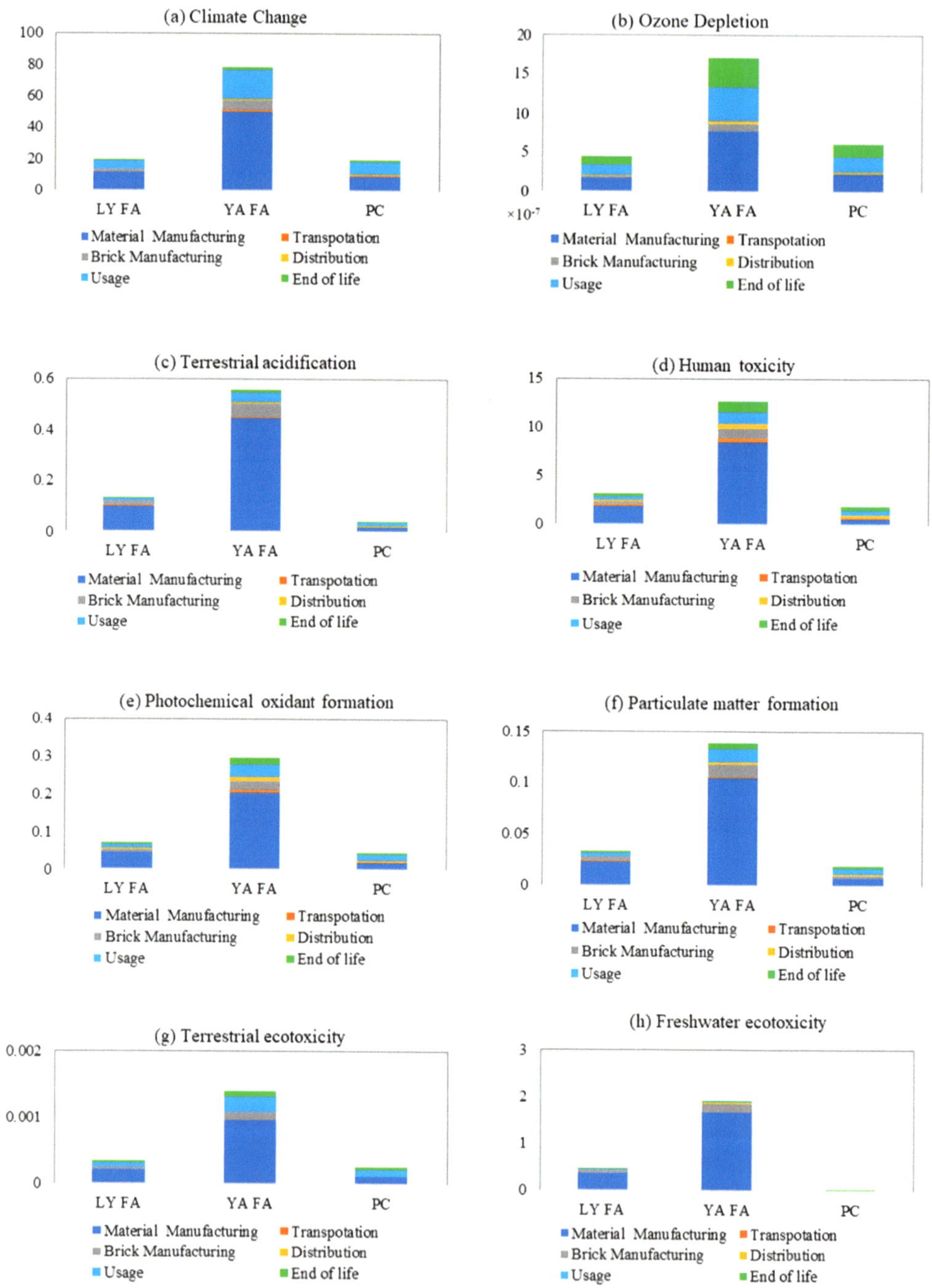

Figure 4. *Cont.*

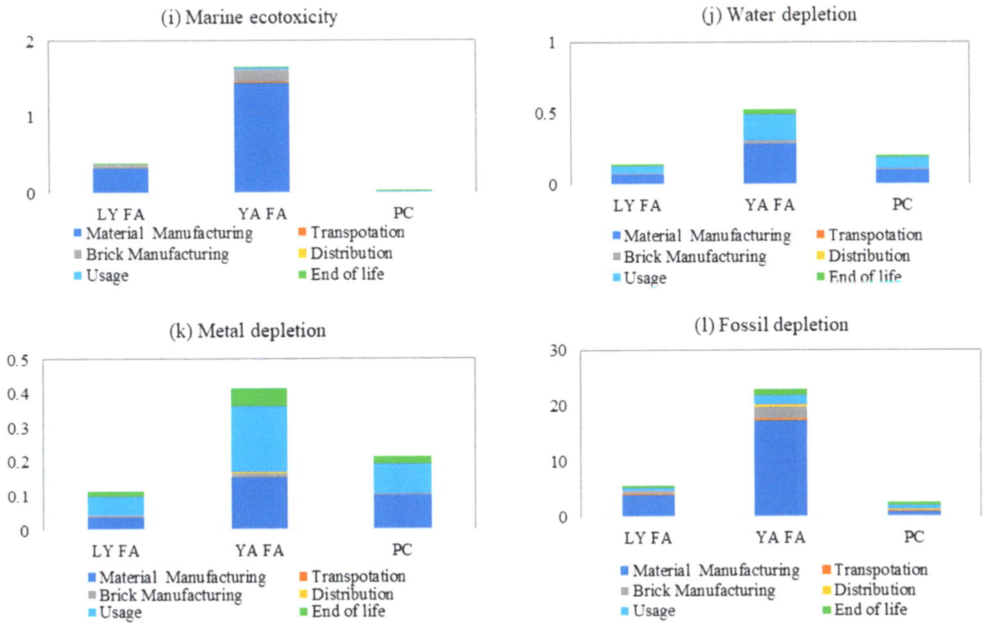

Figure 4. Comparison of the environmental impact intensities of brown coal geopolymer bricks and PC bricks; (**a**) climate change (kg CO_2 eq/m^3. MPa), (**b**) ozone depletion (kg CFC11 eq/m^3. MPa), (**c**) terrestrial acidification (kg SO_2 eq/m^3. MPa), (**d**) human toxicity (kg 1,4-DB eq/m^3. MPa), (**e**) photochemical oxidant formation (kg NMVOC/m^3. MPa), (**f**) particulate matter formation (kg PM_{10} eq/m^3. MPa), (**g**) terrestrial ecotoxicity (kg 1,4-DB eq/m^3. MPa), (**h**) freshwater ecotoxicity (kg 1,4-DB eq/m^3. MPa), (**i**) marine ecotoxicity (kg 1,4-DB eq/m^3. MPa), (**j**) water depletion (m^3/m^3. MPa), (**k**) metal depletion (kg Fe eq/m^3. MPa), and (**l**) fossil fuel depletion (kg oil eq/m^3. MPa).

The contribution analysis showed that the material manufacturing phase contributed from 34% to 84%, 32 to 87%, and 32% to 49% of all the impacts for the LYFA, YFA, and PC bricks, respectively. The transportation and distribution phase had the lowest proportion (i.e., 1% to 13%) for all impact categories for all brick types. The brick manufacturing phase contributed 3% to 13% for the LYFA geopolymer bricks, 3% to 10% for the YFA geopolymer bricks, and 1% to 15% for the PC bricks for all impact categories. The usage phase was the highest contributor to the metal depletion impact category for both the geopolymer and PC bricks. The LYFA, YFA, and PC bricks contributed 50%, 47%, and 40% for metal depletion in the usage phase, respectively. The end-of-life phase contributed 1% to 27% of the impact for all impact categories for the geopolymer and PC brick types.

The material manufacturing phase consisted of the material preparation, including the collection, drying, and processing of the raw materials for the geopolymer and PC bricks. Figure 5 shows the detailed comparison of the environmental impacts for each raw material used for the brick production at the material manufacturing stage. The results clearly identified that the alkaline activators were responsible for over 80% of the total impacts associated in the material manufacturing stage for both the LYFA and YFA bricks for all impact categories other than water and metal depletion. For the PC bricks, PC alone was responsible for the highest share among all the impacts during the stage of material manufacturing, other than the water depletion impact category.

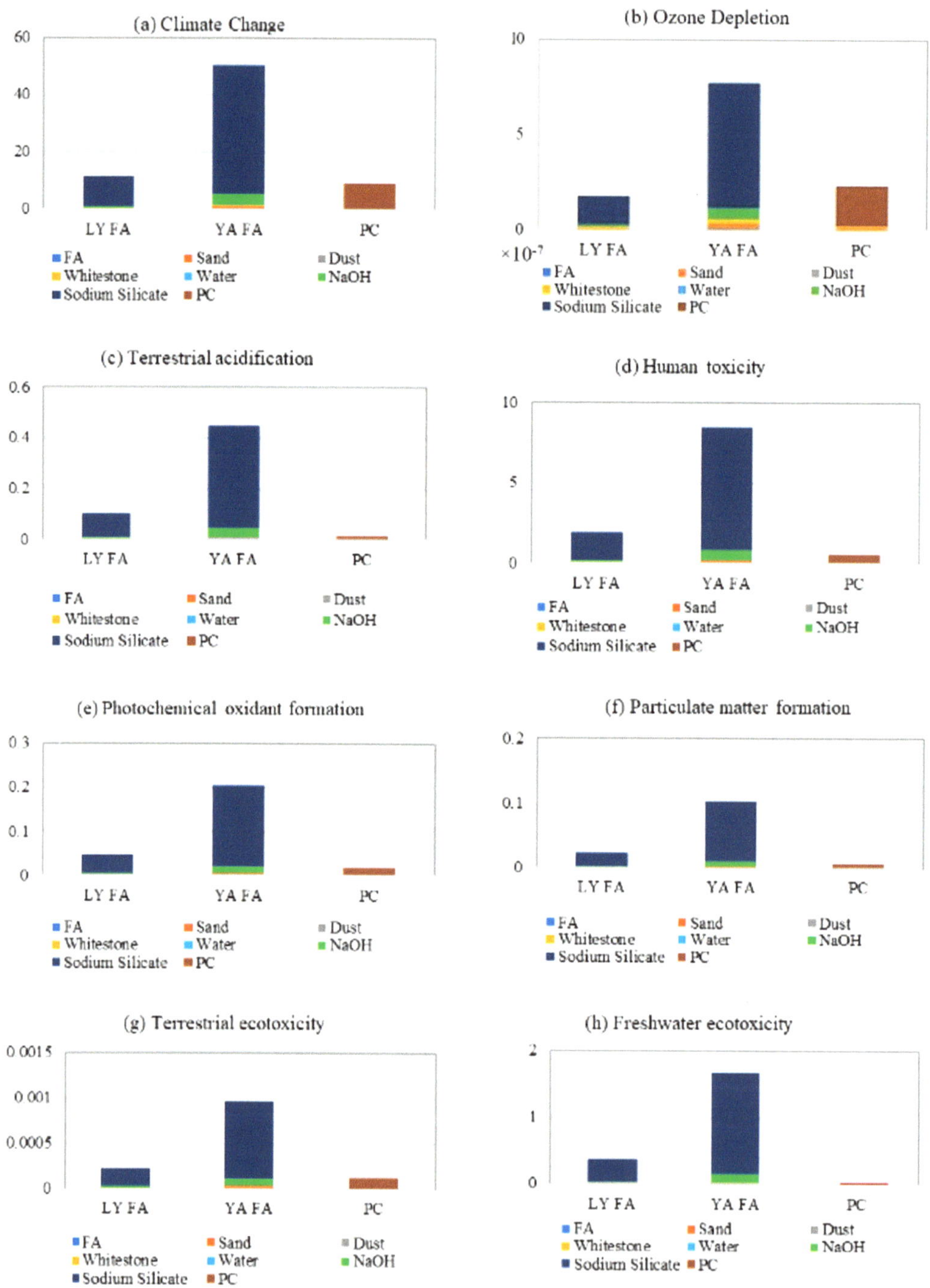

Figure 5. *Cont.*

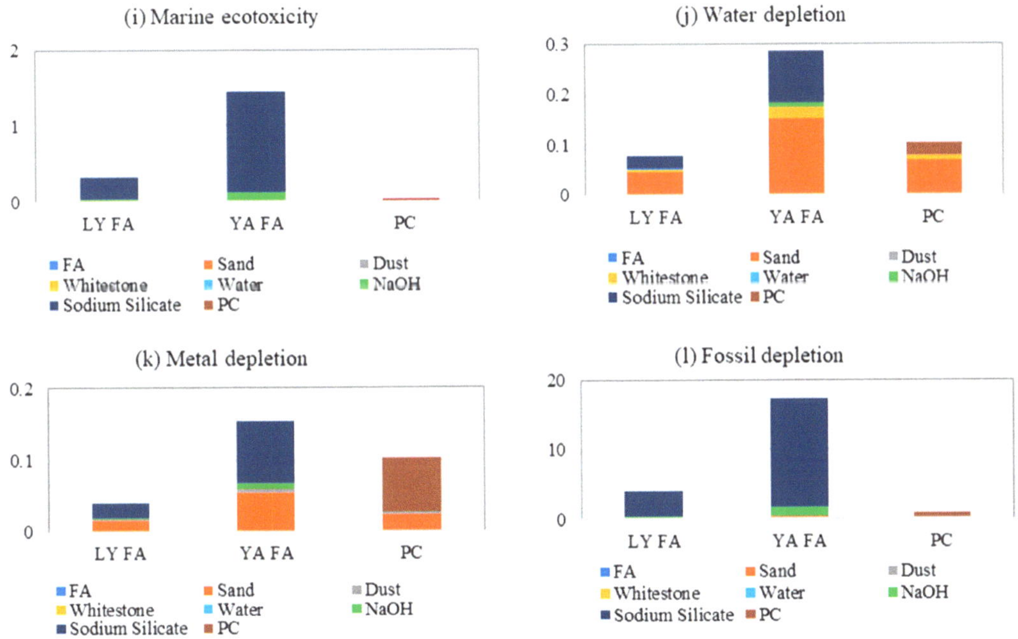

Figure 5. Comparison of the environmental impact intensities for raw material manufacturing stage; (**a**) climate change (kg CO_2 eq/m^3. MPa), (**b**) ozone depletion (kg CFC-11 eq/m^3. MPa), (**c**) terrestrial acidification (kg SO_2 eq/m^3. MPa), (**d**) human toxicity (kg 1,4-DB eq/m^3. MPa), (**e**) photochemical oxidant formation (kg NMVOC/m^3. MPa), (**f**) particulate matter formation (kg PM10 eq/m^3. MPa), (**g**) terrestrial ecotoxicity (kg 1,4-DB eq/m^3. MPa), (**h**) freshwater ecotoxicity (kg 1,4-DB eq/m^3. MPa), (**i**) marine ecotoxicity (kg 1,4-DB eq/m^3. MPa), (**j**) water depletion (m^3/m^3. MPa), (**k**) metal depletion (kg Fe eq/m^3. MPa), and (**l**) fossil fuel depletion (kg oil eq/m^3. MPa).

Table 5 summarizes the impact intensities for the brick manufacturing phase. According to the results, the higher impacts associated with the geopolymer brick production phase were due to heat curing in the brick production process. The mixing process was only required in the PC bricks' production phase, while heat curing was not required. When considering the mixing process, the energy consumption for a unit volume was the same for all the brick types. However, the impact intensity was lower for LYFA when compared to the PC brick mixing process, while the YFA brick manufacturing phase showed higher impact intensities for all categories due to lower mechanical performance. The brick manufacturing phase was directly correlated with energy consumption. Fossil fuel depletion and climate change were identified as the categories with the highest impact during the brick production stage.

Table 5. Detailed impact category intensities of the brick production phase.

Impact Category	Unit	Mixing			Heat Curing		
		LYFA	YFA	PC	LYFA	YFA	PC
Climate change	kg CO_2 eq/m^3. MPa	3.83×10^{-2}	4.18×10^{-1}	1.90×10^{-1}	1.70×10^{0}	5.76×10^{0}	0.00×10^{0}
Ozone depletion	kg CFC-11 eq/m^3. MPa	5.63×10^{-10}	1.01×10^{-8}	4.58×10^{-9}	2.50×10^{-8}	8.46×10^{-8}	0.00×10^{0}
Terrestrial acidification	kg SO_2 eq/m^3. MPa	3.44×10^{-4}	7.74×10^{-4}	3.51×10^{-4}	1.53×10^{-2}	5.17×10^{-2}	0.00×10^{0}
Human toxicity	kg 1,4-DB eq/m^3. MPa	6.51×10^{-3}	8.17×10^{-3}	3.71×10^{-3}	2.89×10^{-1}	9.78×10^{-1}	0.00×10^{0}

Table 5. *Cont.*

Impact Category	Unit	Mixing			Heat Curing		
		LYFA	YFA	PC	LYFA	YFA	PC
Photochemical oxidant formation	kg NMVOC/m^3. MPa	1.56×10^{-4}	1.61×10^{-3}	7.29×10^{-4}	6.93×10^{-3}	2.34×10^{-2}	0.00×10^{0}
Particulate matter formation	kg PM10 eq/m^3. MPa	7.99×10^{-5}	4.19×10^{-4}	2.85×10^{-3}	3.55×10^{-3}	1.20×10^{-2}	0.00×10^{0}
Terrestrial ecotoxicity	kg 1,4-DB eq/m^3. MPa	7.37×10^{-7}	4.99×10^{-6}	2.26×10^{-6}	3.27×10^{-5}	1.11×10^{-4}	0.00×10^{0}
Freshwater ecotoxicity	kg 1,4-DB eq/m^3. MPa	1.30×10^{-3}	2.12×10^{-4}	9.59×10^{-5}	5.78×10^{-2}	1.95×10^{-1}	0.00×10^{0}
Marine ecotoxicity	kg 1,4-DB eq/m^3. MPa	1.12×10^{-3}	2.35×10^{-4}	1.06×10^{-4}	4.99×10^{-2}	1.69×10^{-1}	0.00×10^{0}
Water depletion	m^3/m^3. MPa	8.77×10^{-5}	1.26×10^{-2}	5.70×10^{-3}	3.90×10^{-3}	1.32×10^{-2}	0.00×10^{0}
Metal depletion	kg Fe eq/m^3. MPa	7.36×10^{-5}	3.36×10^{-3}	1.52×10^{-3}	3.27×10^{-3}	1.11×10^{-2}	0.00×10^{0}
Fossil fuel depletion	kg oil eq/m^3. MPa	1.34×10^{-2}	1.10×10^{-1}	4.99×10^{-2}	5.94×10^{-1}	2.01×10^{0}	0.00×10^{0}

4.3. Benefit Analysis

The location in which the fly ash was obtained led to different environmental impacts of the brown coal fly ash in the bricks' production. Figure 6 illustrates the environmental benefits obtained by utilizing the brown coal ash from landfills. A substantial decrease in the category of human toxicity is noted, corresponding to LYFA being 3.18 kg 1,4-DB eq/m^3. MPa and YFA being 9.63 kg 1,4-DB eq/m^3. MPa. There were decreased impacts from freshwater ecotoxicity of 0.27 1,4-DB eq/m3 MPa and marine water ecotoxicity of 0.82 1,4-DB eq/m^3 MPa for LYFA and 2.56×10^{-1} 1,4-DB eq/m^3 MPa and 7.75×10^{-1} 1,4-DB eq/m^3 MPa for YFA, respectively. Climate change, acidification, photochemical oxidant formation, particulate matter formation, and fossil fuel depletion demonstrated benefits of less than 1% for LYFA and YFA. The highest benefit was obtained in the category of human toxicity, 100.34% and 76.19 % for LYFA and YFA, respectively. Furthermore, terrestrial, water, and marine ecotoxicity accounted for the second-largest benefits for both LYFA and YFA due to the avoidance of the storage of brown coal ash.

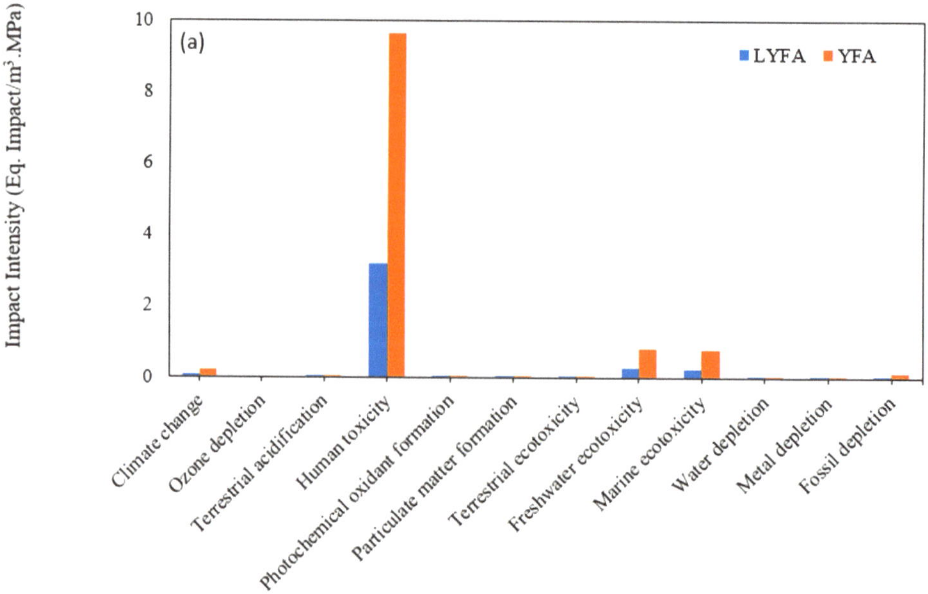

Figure 6. *Cont.*

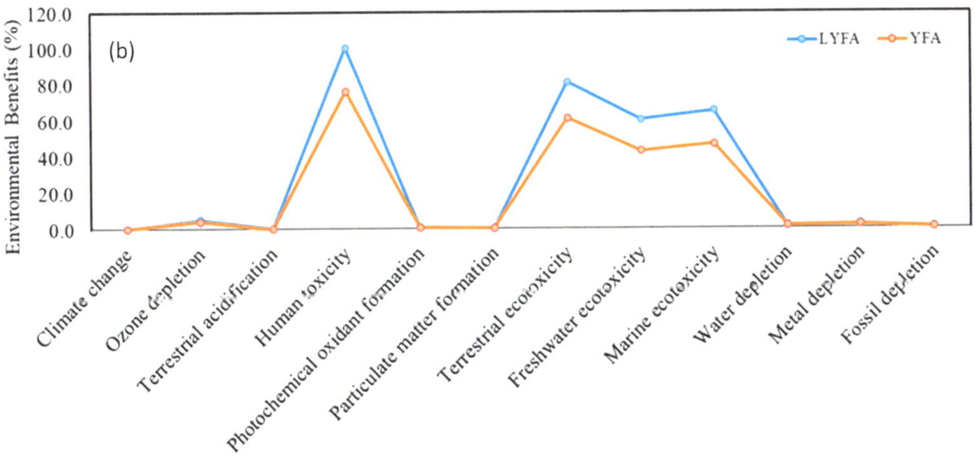

Figure 6. Utilization of waste brown coal ash's benefit for environmental (**a**) intensities and (**b**) variation of the percentage.

4.4. Cost Analysis

The life cycle cost analysis encompassed the extraction of the raw materials, the transportation of the raw materials, and the bricks' manufacturing (cradle-to-gate). The outputs are summarized in Table 6. Figure 7 compares the percentage distribution of the brown coal ash bricks and PC bricks for the raw material manufacturing and transportation phases. The unit costs for the extraction/production and transportation of raw materials were determined by accounting for current market values and sourcing from local Australian suppliers. PC had a cost of AUD 64, while brown coal ash can be freely obtained. When considering both LYFA and YFA geopolymer bricks, sodium silicate at the stage of material manufacture was the largest contributor to the cost, 51% and 57%, respectively. Sand had the highest cost attribution for the PC bricks, which constituted nearly 52% of the total cost.

Table 6. Cost analysis of the "cradle-to-gate" for geopolymer and PC bricks.

Phases	Brick	Cost (AUD) per 1 m^3								
		PC	Brown Coal Ash	Sand	Dust	White Stone (7 mm)	Activator Solution		Total Cost per 1 m^3	Cost per Brick
							Na_2SiO_3	NaOH		
Material Manufacturing	PC	64.00	0.00	400.40	100.10	144.77	0.00	0.00	709.28	1.33
	LYFA	0.00	0.00	400.40	100.10	144.77	1033.11	157.24	1835.63	3.45
	YFA	0.00	0.00	378.95	94.60	137.31	1346.03	222.76	2179.65	4.10
Material Transportation	PC	17.30	0.00	5.78	5.81	10.31	0.00	0.00	39.21	0.07
	LYFA	0.00	58.14	5.78	5.81	10.31	9.03	13.32	102.40	0.19
	YFA	0.00	50.18	5.78	5.81	10.31	9.03	13.32	94.44	0.18
Brick Manufacturing	PC								10.80	0.02
	LYFA				-				69.40	0.13
	YFA								69.40	0.13

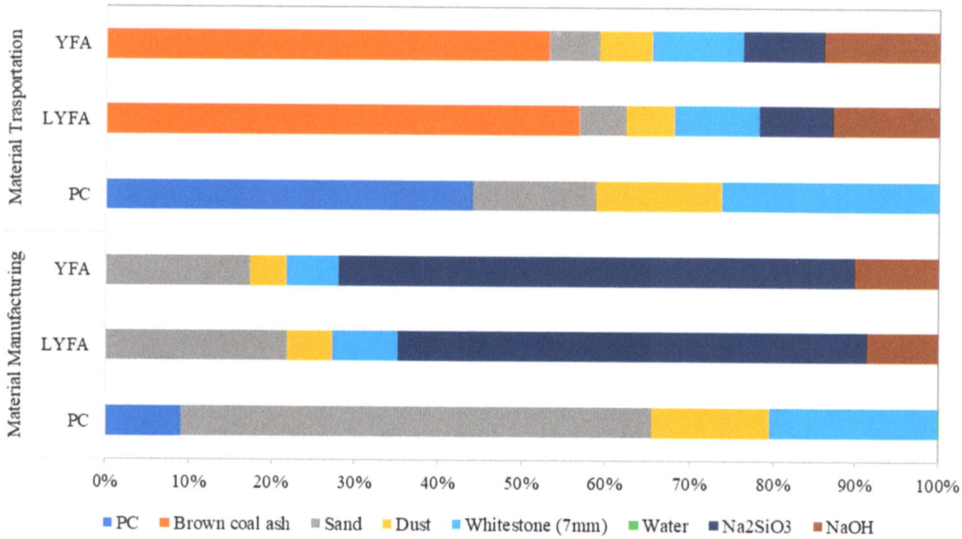

Figure 7. Percentage cost distribution for raw material production and transportation stages for LYFA, YFA, and PC bricks.

LYFA contributed the most to the material transport stage, followed by YFA and PC When compared to PC, LYFA and YFA had the largest cost of transportation (i.e., AUD 58.14 and AUD 50.18). In the production of the PC bricks, the transportation of PC cost AUD 17.3, making it the highest contributor in the transportation stage. Hence, when consideration was given to the overall transport costs, the geopolymer bricks were more costly compared to the PC bricks.

Only the electricity consumption during the brick manufacturing stage was regarded as a part of the brick manufacturing process. The high costs involved in the brick manufacturing stage of the brown coal bricks were attributed to the process of heat curing. The heat curing demanded a large amount of electricity and, therefore, resulted in higher costs. However, for the PC and brown coal geopolymer bricks, only about 1.4% and 3.4% of the total cost needed to be accounted for in the material manufacturing stage.

The PC concrete bricks' cost for cradle-to-gate manufacturing was AUD 1.43, while the LYFA and YFA geopolymer bricks cost AUD 3.78 and AUD 4.41, respectively. Both geopolymer bricks had a total brick cost increment of 162% (LYFA) and 167% (YFA) as compared to the PC bricks. Furthermore, the YFA bricks had a 14% higher total cost compared to the LYFA bricks.

5. Discussion

The comparative analysis of the two brown coal geopolymer bricks with traditional PC bricks highlighted that the YFA bricks had the highest associated impacts, followed by the LYFA bricks, both of which were higher than the traditional PC bricks. The main reason for the higher impacts associated with the YFA geopolymer bricks was the lower compressive strength. The LYFA and YFA bricks had a slight variation in the impact values due to the minor differences in the transportation phase of the bricks' production. Moreover, during the "cradle-to-grave" phases, LYFA demonstrated slightly higher climate change variation compared with the PC bricks. This was due to the higher impact in the manufacturing phase for the LYFA bricks, while both the manufacturing and usage phases contributed to the higher climate change impact for the PC bricks. Although a waste by-product was utilized for the geopolymer bricks' production, a higher impact was observed due to the alkali activators. From the detailed analysis of the LCA, it was clearly noted that the alkaline activators contributed significantly to the climate change impact category. Furthermore, LYFA and YFA not only had a greater impact on climate change, but also on all the other impact categories (except water depletion) during the material manufacturing stage due to the use of the alkaline activators (i.e., sodium silicate and sodium hydroxide). The sodium silicate production process was responsible for the higher impact of all other impact categories, mainly as a result of calcination in the manufacturing process. The manufacturing of sodium silicate involves dissolution, processing, and filtration, which entails a significant energy consumption (i.e., electricity and heat) and yields significant air and water emissions, as well as solid waste [36,37]. Additionally, the electrolysis of sodium chloride is an energy-intensive process in the manufacture of NaOH. This results in higher emissions and environmental impact due to electricity use, natural gas use, and waste disposal [38].

According to the detailed contribution analysis of the material manufacturing phase, PC contributed the greatest share for all impact categories (except water depletion). Clinker production during the PC manufacturing process is an energy-intensive activity that is responsible for the highest emissions and environmental impacts [39]. The brick production phase alone contributed a minor impact for all brick types. Hence, the higher impacts for the LYFA and YFA bricks than the PC bricks were primarily due to the heat-curing process, which consisted of additional energy consumption during the brick manufacturing process. In Australia, the national energy grid includes non-renewable energy, but is primarily coal combustion (almost 60% of the total energy grid). Hence, direct emissions derived from coal combustion have a primary effect on energy consumption during activator production and PC production.

The remaining categories, i.e., terrestrial acidification, human toxicity, photochemical oxidant formation, particulate matter formation, terrestrial ecotoxicity, freshwater ecotoxicity, marine ecotoxicity, and fossil fuel depletion, showed higher impacts for LYFA compared to the PC bricks, even with relatively higher compressive strength. This was again directly related to the higher energy consumption during the alkaline activator production. However, for the YFA bricks, all these impacts had significantly higher values due to their lower compressive strength.

PC production is the principal process responsible for metal depletion. Here, the term metal depletion focuses on the depletion of the resource, except for fossil fuels. This is mainly due to the consumption of natural resources such as limestone and clay and silica stone during clinker manufacturing through the pyrolysis process [40]. The resource depletion that occurred during the PC bricks' production was larger than the resource depletion that occurred for the LYFA geopolymer bricks' production.

Human toxicity was the second-highest impact for the transportation, distribution and end-of-life phases. This was due to the leaching of toxic material from diesel consumption in the transportation of the materials and other products (waste and bricks). The higher human toxicity in the geopolymer bricks was due to the release of toxic elements during the production of sodium silicate.

Water depletion is defined as water scarcity, which means a lack of sufficient available freshwater resources to meet the water demand. According to the results, sand (the extraction of the raw material) was the principal reason for the highest quantity of water depletion, making the material manufacturing stage the phase with the highest impact regarding the water depletion impact category. Generally, irrigation wells and groundwater sources are seriously threatened due to excessive sand extraction near rivers, which negatively affects groundwater recharge. In addition, falling groundwater levels are a major threat to water supplies, exacerbating the occurrence (frequency and periodicity) and severity of droughts, as tributaries of major rivers dry up when sand extraction meets a particular threshold [41]. This was the reason for the higher impact intensities associated with sand in all bricks, including the respective compressive strength within the selected mix design. However, the manufacturing of PC contributed to the water depletion minimally for the PC bricks, while the alkaline activators were the second-highest material resulting in water depletion for the geopolymer bricks. Hence, the impact on water depletion was not significantly affected by the type of brick (geopolymer or PC bricks).

The most-encouraging result was the reduction of ozone depletion by using the brown coal ash in the geopolymer bricks. This was primarily due to the elimination of the PC found in traditional PC bricks. Ozone depletion occurs when the anthropogenic emissions (chlorine and bromine atoms) come into contact with ozone in the stratosphere [42]. This is principally influenced by chlorinated or brominated hydrocarbons emitted during the production of fossil fuels, which was higher in the PC bricks than both geopolymer bricks.

The benefits analysis showed a performance improvement in sustainability due to the reduction of waste disposal (brown coal ash). This was because the waste was transformed into a useful and valuable product. A significant benefit with regard to the impact intensities was shown for human toxicity, fresh water ecotoxicity, and marine water ecotoxicity compared to the impact intensities for the LYFA and YFA bricks. Dumping of fly ash leads to contaminated water and soil with heavy metals and other toxic elements present in the ash itself. Coal ash includes toxic elements, i.e., arsenic (As), barium (Ba), boron (B), beryllium (Be), cadmium (Cd), chromium (Cr), cobalt (Co), lead (Pb), lithium (Li), manganese (Mn), mercury (Hg), molybdenum (Mo), radium (Ra), selenium (Se), thallium (Tl), and other hazardous chemicals [43]. This may be responsible for a range of health problems, affecting every major organ in the body. These effects include cancer, kidney disease, infertility, and compromising the nervous system, especially in children [43]. In addition, brown coal ash pollutes the soil and water sources surrounding coal-fired power stations. Vegetation growing in the vicinity has higher levels of heavy substances (i.e., selenium, zinc, nickel, copper, manganese, cadmium, and lead) from elements leaching from fly ash [44]. Researchers have also discovered that contaminants from coal ash, such as selenium and arsenic, accumulate to "very high concentrations" in fish and wildlife exposed to coal dump leachate or run-off and that these accumulated toxins could eventually lead to deformities or the death of the animal [45]. The use of brown coal ash to produce bricks, therefore, mitigates the environmental impacts of human, water, marine, and soil toxicity.

The life cycle cost analysis illustrated that the stage of raw material manufacturing was responsible for the higher cost associated with the brown coal geopolymer bricks as compared to the PC bricks. The alkaline activator was the main reason for this increased cost. The brick production costs were also higher for the brown coal bricks due to the cost of electricity consumption associated with the thermal curing process. However, in terms of cost, the transport and brick-making stages were of minor relevance compared to the raw material manufacturing stage.

Currently, coal-fired electricity production is recognized as un-environmentally friendly [46], and many countries are seeking "cleaner" energy, such as renewable energy including solar, hydro, wind, and bio. This will lead to a decrease in coal fly ash supply, which could result in the limited production of brown coal fly ash. However, the high penetration of renewable energy can increase the risk of power outages in the absence of an adequate protective measure [47,48]. Fossil energy is still a reliable energy source and accounted for 75% of the global net electricity generation in 2017 [49]. The evidence shows that coal is still the world's largest single source of electricity, set to still contribute 22% in 2040. In Southeast Asia, coal will provide 39% of electricity in 2040 [50]. In Australia, coal-fired electricity occupies 61% of the electricity production [35]. Since many rely on coal as a crucial electricity production source, the waste by-product, coal fly ash, still needs to be treated. Using coal fly ash in brick production can be a suitable method or this.

6. Conclusions and Future Research

Life cycle assessment was employed to carry out an environmental analysis for two brown coal ash geopolymer bricks. The following conclusions can be made based on the results:

- The Loy Yang FA (LYFA) bricks demonstrated slightly higher climate change impact intensities compared to the Portland cement (PC) bricks.
- The Yallourn FA (YFA) bricks showed higher environmental impact intensities for all midpoint categories when compared to both the LYFA and PC bricks due to the lower compressive strength.
- Fossil fuel depletion and climate change were identified as the highest impacted categories during the brick production stage.
- The combination of sodium silicate and sodium hydroxide was responsible for approximately 90% of the total impact for all categories except metal ($\sim$ 50%) and water depletion ($\sim$ 30%) for both brown coal geopolymer bricks.
- Terrestrial acidification, human toxicity, photochemical oxidant formation, particulate matter formation, terrestrial ecotoxicity, freshwater ecotoxicity, marine ecotoxicity, and fossil fuel depletion showed higher impacts for the LYFA bricks compared to the PC bricks.
- Significant environmental benefits in terms of human, freshwater, and marine water ecotoxicity can be obtained by utilizing brown coal ash for the brick manufacturing process.
- The most-significant benefits for the LYFA geopolymer bricks over the PC bricks were recorded for the ozone depletion, water depletion, and metal depletion (natural resources other than fossil fuels) categories due to the replacement of PC as a raw material.

The study's findings indicated that there is considerable potential for reducing the environmental impact of the brown coal geopolymer bricks, especially the LYFA bricks, compared to the PC bricks. The replacement of fossil fuels with renewable energy sources during heat curing and the optimization of the activator concentration, type, and ratio can significantly reduce emissions and energy consumption [46,48,51]. Additionally, proper precautions during chemical activator handling and usage can mitigate human toxicity risks. Future research should consider these factors to minimize the environmental impacts during brown coal geopolymer brick production.

Furthermore, energy consumption and material preparation are critical issues that require attention in brick production. Similar results were found in [52,53]. Moving material and brick production to locations where renewable energy sources are available can help control energy consumption. Furthermore, end-of-life management considerations can be included to calculate environmental impacts and benefits.

The study did not consider the durability or lifespan of the geopolymer and PC bricks. However, research has shown that geopolymers' durability can be superior to PC concrete [54–56]. An extended lifespan could enhance the environmental benefits and promote sustainability in future building construction applications.

Fly ash storage is a significant contributor to air, water, and soil pollution, which can be harmful to humans, biodiversity, and flora and fauna. Therefore, ecotoxicology impacts, including terrestrial, freshwater, and marine toxicity, should be considered to improve sustainability in construction. Leaching tests, chemical analyses, and toxicity tests can evaluate their ecotoxicity. Accounting for these impacts can help determine the potential environmental risks associated with the materials used in the construction sector.

Author Contributions: Conceptualization, D.W.L. and C.G.; methodology, S.F. and J.Z.; software, J.Z. and S.F.; validation, J.Z. and S.F.; formal analysis, J.Z.; resources, D.W.L. and C.G.; writing—original draft preparation, S.F., J.Z., D.W.L., G.Z. and C.G.; writing—review and editing, D.W.L., G.Z., J.Z. and M.S.; visualization, G.Z. and M.S.; supervision, D.W.L., G.Z., S.S. and C.G.; project administration, D.W.L., S.S., C.G. and G.Z. All authors have read and agreed to the published version of the manuscript.

Funding: This research received no external funding.

Institutional Review Board Statement: Not applicable.

Informed Consent Statement: Not applicable.

Data Availability Statement: Data is available on request.

Conflicts of Interest: The authors declare no conflict of interest.

References

1. Olhoff, A.; Christensen, J.M. Emissions Gap Report 2018. 2018. Available online: https://www.unep.org/resources/emissions gap-report-2018 (accessed on 1 November 2020).
2. Salas, D.A.; Ramirez, A.D.; Ulloa, N.; Baykara, H.; Boero, A.J. Life cycle assessment of geopolymer concrete. *Constr. Build. Mater* **2018**, *190*, 170–177. [CrossRef]
3. Sentman, S.D.; Del Percio, S.T.; Koerner, P. A Climate for Change: Green Building Policies, Programs, and Incentives. *J. Green Build* **2008**, *3*, 46–63. [CrossRef]
4. Reber, A.P.J.K.L. Taking the LEED: Determining the Appropriate Amount of Government Regulation in Green Building Projects *Ky. Law J.* **2009**, *98*, 573.
5. Furr, J.E. *Green Building and Sustainable Development: The Practical Legal Guide*; American Bar Association: Chicago, IL, USA, 2009
6. Weyant, C.; Athalye, V.; Ragavan, S.; Rajarathnam, U.; Lalchandani, D.; Maithel, S.; Baum, E.; Bond, T.C. Emissions from South Asian Brick Production. *Environ. Sci. Technol.* **2014**, *48*, 6477–6483. [CrossRef]
7. Bricks, T. Ever Wondered How Bricks Are Made? Think Brick Australia: St Leonards, Australia, 2020.
8. Hall, M.R.; Lindsay, R.; Krayenhoff, M. *Modern Earth Buildings: Materials, Engineering, Constructions and Applications*; Elsevier Science & Technology: Cambridge, UK, 2012; Volume 33.
9. Gartner, E. Industrially interesting approaches to "low-CO_2" cements. *Cem. Concr. Res.* **2004**, *34*, 1489–1498. [CrossRef]
10. Meyer, C. The greening of the concrete industry. *Cem. Concr. Compos.* **2009**, *31*, 601–605. [CrossRef]
11. Davidovits, J. Geopolymers: Inorganic polymeric new materials. *J. Therm. Anal. Calorim.* **1991**, *37*, 1633–1656. [CrossRef]
12. Duxson, P.; Fernández-Jiménez, A.; Provis, J.L.; Lukey, G.C.; Palomo, A.; van Deventer, J.S.J. Geopolymer technology: The current state of the art. *J. Mater. Sci.* **2007**, *42*, 2917–2933. [CrossRef]
13. *ISO14040*; Environmental Management-Life Cycle Assessment-Principles Framework. ISO: Geneva, Switzerland, 2006.
14. Celik, K.; Meral, C.; Gursel, A.P.; Mehta, P.K.; Horvath, A.; Monteiro, P.J.M. Mechanical properties, durability, and life-cycle assessment of self-consolidating concrete mixtures made with blended portland cements containing fly ash and limestone powder *Cem. Concr. Compos.* **2015**, *56*, 59–72. [CrossRef]
15. Ferone, C.; Liguori, B.; Capasso, I.; Colangelo, F.; Cioffi, R.; Cappelletto, E.; Di Maggio, R. Thermally treated clay sediments as geopolymer source material. *Appl. Clay Sci.* **2015**, *107*, 195–204. [CrossRef]
16. De Felice, F.; Petrillo, A.; Cooper, O. An integrated conceptual model to promote green policies. *Int. J. Innov. Sustain. Dev.* **2013**, *7*, 333. [CrossRef]
17. Britt, A.F.; Summerfield, D.; Senior, A.B.; Roberts, D.; Kay, P.; Hitchman, A.; Champion, D.C.; Huston, D.; Sexton, M.; Schofield, A.; et al. *Australia's Identified Mineral Resources 2016*; Geoscience Australia: Canberra, Australia, 2016.
18. Macphee, D.; Black, C.; Taylor, A. Cements incorporating brown coal fly ash from the latrobe valley region of Victoria, Australia. *Cem. Concr. Res.* **1993**, *23*, 507–517. [CrossRef]
19. Habert, G.; d'Espinose de Lacaillerie, J.B.; Roussel, N. An environmental evaluation of geopolymer based concrete production: Reviewing current research trends. *J. Clean. Prod.* **2011**, *19*, 1229–1238. [CrossRef]
20. Turner, L.K.; Collins, F.G. Carbon dioxide equivalent (CO_2-e) emissions: A comparison between geopolymer and OPC cement concrete. *Constr. Build. Mater.* **2013**, *43*, 125–130. [CrossRef]
21. Weil, M.; Dombrowski, K.; Buchwald, A. Life-cycle analysis of geopolymers. *Geopolymers* **2009**, 194–210. [CrossRef]
22. Yan, H.; Shen, Q.; Fan, L.C.; Wang, Y.; Zhang, L. Greenhouse gas emissions in building construction: A case study of One Peking in Hong Kong. *Build. Environ. Manag.-Life Cycle Assess.-Princ.* **2010**, *45*, 949–955. [CrossRef]
23. Youssef, N.; Rabenantoandro, A.Z.; Dakhli, Z.; Chehade, F.H.; Lafhaj, Z. Environmental evaluation of geopolymer bricks. *MATEC Web Conf.* **2019**, *281*, 03005. [CrossRef]
24. Petrillo, A.; Cioffi, R.; Ferone, C.; Colangelo, F.; Borrelli, C. Eco-sustainable Geopolymer Concrete Blocks Production Process. *Agric. Agric. Sci. Procedia* **2016**, *8*, 408–418. [CrossRef]
25. Abdulkareem, M.; Havukainen, J.; Nuortila-Jokinen, J.; Horttanainen, M. Environmental and economic perspective of waste-derived activators on alkali-activated mortars. *J. Clean. Prod.* **2020**, *280*, 124651. [CrossRef]
26. Xing, S.; Xu, Z.; Jun, G. Inventory analysis of LCA on steel- and concrete-construction office buildings. *Energy Build.* **2008**, *40*, 1188–1193. [CrossRef]
27. Lozano-Miralles, J.A.; Hermoso-Orzáez, M.J.; Gago-Calderón, A.; Brito, P. LCA Case Study to LED Outdoor Luminaries as a Circular Economy Solution to Local Scale. *Sustainability* **2019**, *12*, 190. [CrossRef]
28. Zhang, J.; Zhang, G.; Bhuiyan, M.A.; Agostino, J. Comparative life cycle assessment of kerbside collection methods for waste glass cullet for asphalt production. *J. Clean. Prod.* **2022**, *374*, 134055. [CrossRef]
29. Farina, A.; Zanetti, M.C.; Santagata, E.; Blengini, G.A. Life cycle assessment applied to bituminous mixtures containing recycled materials: Crumb rubber and reclaimed asphalt pavement. *Resour. Conserv. Recycl.* **2017**, *117*, 204–212. [CrossRef]

30. Ding, T.; Xiao, J. Estimation of building-related construction and demolition waste in Shanghai. *Waste Manag.* **2014**, *34*, 2327–2334. [CrossRef] [PubMed]
31. Bare, J.C.; Hofstetter, P.; Pennington, D.W.; De Haes, H.A.U. Midpoints versus endpoints: The sacrifices and benefits. *Int. J. Life Cycle Assess.* **2000**, *5*, 319–326. [CrossRef]
32. Damineli, B.L.; Kemeid, F.M.; Aguiar, P.S.; John, V.M. Measuring the eco-efficiency of cement use. *Cem. Concr. Compos.* **2010**, *32*, 555–562. [CrossRef]
33. Khodr, M. The Use of Brown Coal Fly Ash as a Replacement of Cement in Concrete Masonry Bricks. Ph.D. Thesis, RMIT University, Melbourne, Australia, 2020.
34. Khodr, M.; Law, D.; Gunasekara, C.; Setunge, S.; Brkljaca, R. Compressive strength and microstructure evolution of low calcium brown coal fly ash-based geopolymer. *J. Sustain. Cem. Mater.* **2020**, *9*, 17–34. [CrossRef]
35. AEU. Australian Energy Update 2020. Available online: https://www.energy.gov.au/publications/australian-energy-update-2020 (accessed on 1 November 2020).
36. Fawer, M.; Concannon, M.; Rieber, W. Life cycle inventories for the production of sodium silicates. *Int. J. Life Cycle Assess.* **1999**, *4*, 207–212. [CrossRef]
37. Gomes, K.C.; Carvalho, M.; Diniz, D.D.P.; Abrantes, R.D.C.C.; Branco, M.A.; de Carvalho, P.R.O., Jr. Carbon emissions associated with two types of foundations: CP-II Portland cement-based composite vs. geopolymer concrete. *Matéria* **2019**, *24*. [CrossRef]
38. Thannimalay, L.; Yusoff, S.; Zawawi, N. Life cycle assessment of sodium hydroxide. *Aust. J. Basic Appl. Sci.* **2013**, *7*, 421–431.
39. Chen, C.; Habert, G.; Bouzidi, Y.; Jullien, A. Environmental impact of cement production: Detail of the different processes and cement plant variability evaluation. *J. Clean. Prod.* **2010**, *18*, 478–485. [CrossRef]
40. Cucchiella, F.; D'adamo, I.; Gastaldi, M. Sustainable management of waste-to-energy facilities. *Renew. Sustain. Energy Rev.* **2014**, *33*, 719–728. [CrossRef]
41. GreenFacts. *The Mining of Sand, a Non-Renewable Resource*; Authority of the GreenFacts Scientific Board: Brussels, Belgium, 2021.
42. Goedkoop, M.; Heijungs, R.; Huijbregts, M.A.J.; De Schryver, A.; Struijs, J.; Van Zelm, R. ReCiPe 2008: A Life Cycle Impact Assessment Method Which Comprises Harmonised Category Indicators at the Midpoint the Endpoint Level. 2009, pp. 1–126. Available online: https://www.researchgate.net/publication/302559709_ReCiPE_2008_A_life_cycle_impact_assessment_method_which_comprises_harmonised_category_indicators_at_the_midpoint_and_the_endpoint_level (accessed on 1 November 2020).
43. Gottlieb, B.; Gilbert, S.G.; Evans, L.G. *Coal Ash: The Toxic Threat to Our Health and Environment*; Physicians for Social Responsibility: Bethesda, MD, USA, 2010.
44. Adak, S.; Adhikari, K.; Brahmachari, K. Effect of Fly Ash on Crop Coverage around coal-fired Thermal Power Plant in Rural India. *Int. J. Environ. Agric. Biotechnol.* **2016**, *1*, 238565. [CrossRef]
45. Rowe, C.L.; Hopkins, W.A.; Congdon, J.D. Ecotoxicological implications of aquatic disposal of coal combustion residues in the United States: A review. *Environ. Monit. Assess.* **2002**, *80*, 207–276. [CrossRef]
46. Asif, Z.; Chen, Z.; Wang, H.; Zhu, Y. Update on air pollution control strategies for coal-fired power plants. *Clean Technol. Environ. Policy* **2022**, *24*, 2329–2347. [CrossRef]
47. Liu, D.; Zhang, X.; Tse, C.K. Effects of High Level of Penetration of Renewable Energy Sources on Cascading Failure of Modern Power Systems. *IEEE J. Emerg. Sel. Top. Circuits Syst.* **2022**, *12*, 98–106. [CrossRef]
48. Yi, W.; Hill, D.J.; Song, Y. Impact of High Penetration of Renewable Resources on Power System Transient Stability. In Proceedings of the 2019 IEEE Power & Energy Society General Meeting (PESGM), Atlanta, GA, USA, 4–8 August 2019; pp. 1–5. [CrossRef]
49. Wang, Y.; Byers, E.; Parkinson, S.; Wanders, N.; Wada, Y.; Mao, J.; Bielicki, J.M. Vulnerability of existing and planned coal-fired power plants in Developing Asia to changes in climate and water resources. *Energy Environ. Sci.* **2019**, *12*, 3164–3181. [CrossRef]
50. Association, W.C. Coal & Electricity. 2020. Available online: https://www.worldcoal.org/coal-facts/coal-electricity/ (accessed on 10 April 2023).
51. Abdulkareem, M.; Havukainen, J.; Horttanainen, M. Environmental Assessment of Alkali-Activated Mortars Using Different Activators. In *SARDINIA*; CISA: Padova, Italy, 2019.
52. Marwa, D.; Heinonen, J.; El-Mahdy, D.; Hassan, D.M. A comparative study of life cycle carbon emissions and embodied energy between sun-dried bricks and fired clay bricks. *J. Clean. Prod.* **2020**, *275*, 122998.
53. Huarachi, D.A.R.; Gonçalves, G.; de Francisco, A.C.; Canteri, M.H.G.; Piekarski, C.M. Life cycle assessment of traditional and alternative bricks: A review. *Environ. Impact Assess. Rev.* **2020**, *80*, 106335. [CrossRef]
54. Bakharev, T. Resistance of geopolymer materials to acid attack. *Cem. Concr. Res.* **2005**, *35*, 658–670. [CrossRef]
55. Song, X.; Marosszekya, M.; Brungsb, M.; Munna, R. Durability of fly ash based geopolymer concrete against sulphuric acid attack. In Proceedings of the International Conference on Durability of Building Materials and Components, Lyon, France, 17–20 April 2005.
56. Thokchom, S.; Ghosh, P.; Ghosh, S. Effect of water absorption, porosity and sorptivity on durability of geopolymer mortars. *Mater. Sci.* **2009**, *4*, 28–32.

sustainability

MDPI

Review

Applications of Solar Panel Waste in Pavement Construction—An Overview

Malindu Sandanayake [1,*], Le Li [1], Junhai Zhao [2] and Paul Joseph [1]

[1] Institute for Sustainable Industries and Liveable Cities, Victoria University, Melbourne, VIC 3011, Australia
[2] School of Civil Engineering, Chang'an University, Xi'an 710061, China
* Correspondence: malindu.sandanayake@vu.edu.au

Abstract: Waste from used solar panels will be a worldwide problem in the near future mainly due to the strong uptake in solar energy and the necessity of disposing solar panel systems at the end–of–life stage, as these materials are hazardous. While new techniques and strategies are often investigated to manage the end–of–life of solar panels effectively, there is huge potential in recycling and reusing solar panel waste as components for alternate products. Numerous studies have been conducted on using alternate materials instead of conventional materials in pavement construction. The current study presents a detailed review and a discussion on using solar panel waste materials in pavement construction. The findings present opportunities to use different solar panel waste materials such as glass, aluminium (Al), silicon (Si), and polymer waste as potential replacement materials in various types of pavement construction. The study also presents the current progress and future focus on experimental developments in pavements with solar panel waste to benchmark short–term and long-term characteristics. Finally, the review discusses the impediments that restrict and the drivers that can facilitate the implementation of solar panel waste in pavement construction. The main findings from this review can be used as a quantitative foundation to facilitate decisions on using different solar panel waste materials in pavement construction applications. Furthermore, such findings will also be beneficial for policymakers and industry stakeholders to implement effective supply chain strategies for promoting solar panel waste as a potential pavement construction material.

Keywords: solar panel waste; waste utilization; recycling of materials; pavement construction

Citation: Sandanayake, M.; Li, L.; Zhao, J.; Joseph, P. Applications of Solar Panel Waste in Pavement Construction—An Overview. *Sustainability* **2022**, *14*, 14823. https://doi.org/10.3390/su142214823

Academic Editor: Ramchandra Pode

Received: 25 August 2022
Accepted: 1 November 2022
Published: 10 November 2022

1. Introduction

Rapid population growth and uptake of technologies due to industrialization have intensified energy consumption across the globe. Traditional methods such as coal–based electricity production are the leading form of electricity generation despite being known for the highest carbon emissions per kWh [1]. This is mainly due to the low cost of electricity production, simple conversion processes and easy access to raw materials. Owing to these enormous demands in energy consumption and the perceived environmental impacts, industries, and researchers have experimented with several alternative energy sources that are efficient and environmentally friendly. Solar power is such a form of renewable energy, and it offers several advantages including safety, conversion efficiencies, reliability, and minimised environmental impacts due to cleaner production technologies [2]. Despite the high initial costs, including installation, there is an enormous market potential for solar energy, i.e., using photovoltaic (PV) energy in most developed countries such as the United States, Japan, and Australia [2]. This is mainly due to indirect economic benefits through the observed life–cycle cost savings, government rebates, and increases in asset values. With the commitments to "affordable and clean energy" in the United Nations Sustainability Development Goals (SDGs), it is highly likely that solar power will be the predominant renewable energy type in the near future [3].

One governing concern of solar energy is the end–of–life (EOL) management of solar panels, which are recognised as hazardous waste. Solar panels have a relatively long life cycle of around 30 years and were not a major waste issue during the initial implementation and development phases. However, with the strong uptake and reach of the first EOL cycle since its first implementation, the accumulation of solar panel waste is becoming a serious concern. According to a recent publication, in 2047, Australia will accrue about 1 million tonnes of solar panel waste, which is equivalent to 19 Sydney Harbour Bridges, which further rationalizes the magnitude of the problem [4]. Moreover, due to government incentives for upgrades and replacements, these panels are often completely replaced after about 12–15 years of life cycle even with only minor damage to some panels [2]. This could lead to further acceleration in the waste accumulation of solar panels. The optimum solution would be to facilitate the complete or partial recycling of the panel through the recovery of materials, which can reduce the costs of solar panel production. However, currently, recycling is a small portion of solar panel EOL management and they often end up in huge piles of E–waste as landfill waste. Therefore, every attempt to divert any quantity of solar panel waste from landfills would be considered a benefit and could be used as a promotional response to the upsurge in solar energy use, highlighting the life–cycle benefits.

Construction is one of the leading energy–intensive industries mainly due to the excessive virgin material usage [5,6]. Therefore, both industries and academic researchers are continuously searching for ways to partially or completely replace virgin materials in construction materials [7,8]. Alternative pavement materials replacing virgin materials in both flexible and rigid pavements are widely researched across the world, both as an environmentally friendly and cost–effective practice [9–11]. Using composite waste materials from solar panel waste as a roadway subbase material can be an effective solution to the growing concern regarding solar panel waste. This study aims to provide a contemporary review of the potential applications of solar panels in pavement construction applications. The review also intends to present qualitative and semi–quantitative findings based on previous studies and benchmark future research directions. Furthermore, the findings of the study are also important to policy and industry decision makers in understanding the future opportunities to benchmark the sustainable uses of solar panels over their life cycles.

2. Research Methodology

This review study aims to undertake a review investigating the potential of using solar panel waste constituents in pavement construction applications, including concrete and asphalt pavements. A detailed review of the relevant literature precedents is conducted to understand the composition of solar panel systems and identify the potential waste materials that can be effectively used as raw construction materials for pavement. The review study then explores the possibilities of using different material constituents of a waste solar panel as a material replacement in concrete and asphalt pavement construction, focusing on mechanical properties. Subsequently, the review focuses on previous studies that have used various similar waste constituents in different types of pavement construction, with a view to obtaining an understanding of the potential behaviour and future research considerations. Finally, the study discusses barriers and key success factors for using solar panel waste in pavement construction with a focus on future research directions. These findings aim to inform both industry stakeholders and research communities on commercialisation aspects and research directions to improve the sustainability aspect of the product.

3. Composition and Material Properties of a Solar Panel

In this section, the material composition of solar panels is introduced. The environmental hazard of solar panel waste and the end–of–life (EOL) management of solar panel materials is also introduced. The section shows the benefits of recycling wasted solar panels in pavement construction to eliminate these environmental hazards.

A typical solar energy system consists of a solar panel, a solar controller, an inverter and a group of batteries, with the configuration shown in Figure 1a [2,12]. Effectively a solar panel (also known as a photovoltaic or a PV module) converts solar radiation into electrical energy. The solar controller regulates the voltage and current to prevent overcharging batteries. The battery group stores the energies, and the inverter converts the direct current into alternate current to use in the household [2]. A solar panel element is the most critical component of the solar energy system, and there are three main types of this component [13]. *Crystalline silicon (c–Si)* is the most common solar panel type used in the commercial market, which can be either monocrystalline or multi–crystalline The thin–film solar panel consists of amorphous silicon, cadmium telluride (CdTe), and copper indium gallium selenide (CIGS). Concentrator phonotactic solar panels can be dye–sensitized, organic, or hybrid panels. Dye–sensitized panels consist of cells with light–absorbing dye and a metal oxide semiconductor that carries the electric current. C–Si solar panels are extensively used in the market, with a share of over 95% of total solar panel usage. while thin–film and concentrator phonotactics account for around 4% and 0.3% respectively [2,14]. Due to current excessive usage and the potentially high possibility of waste collection in the future, only the application of waste c–Si panels is considered in this review. It is also essential to understand the composition of c–Si panels to investigate future recycling and reuse options for solar panel waste. Figure 1b illustrates the key components of a c–Si solar panel, which includes aluminium (Al) frames, glasses (tempered glass) polymer (encapsulant and back sheet foil), solar cells, and a junction box [15,16].

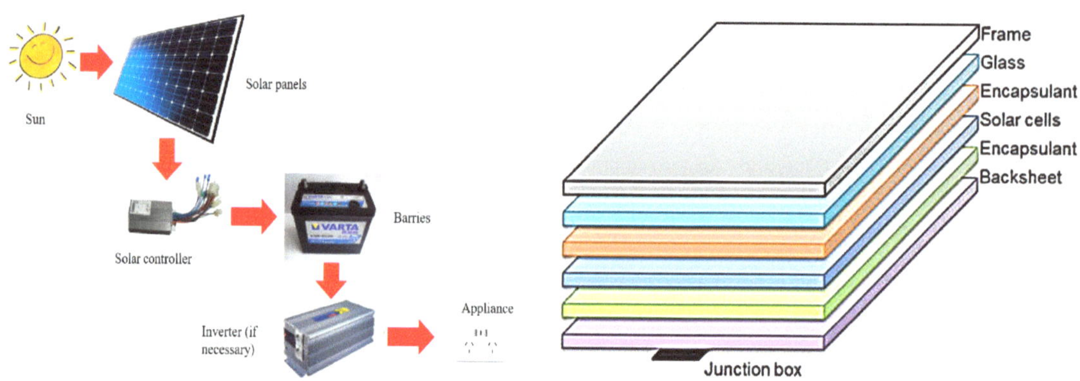

(**a**) (**b**)

Figure 1. Solar energy system configuration and composition of a c–Si panel [2,17]. (**a**) Configuration of solar energy system; (**b**) composition of c–Si panel.

Solar cells are the most critical part of the panels; they generate energy and are composed of a silicon (Si) wafer, a silver (Ag) electrode on the front side, and an Al electrode on the rear side. The cells are electrically interconnected (with tabbing) by copper (Cu) wires, creating a string of cells in a series (60 or 72 cells are the standard numbers that are generally used), which assemble into modules [2,17]. The weight percentage of the material in solar panels and their average market values are summarized in Table 1. The results indicate that solar panels are mainly made out of glass (74–76%), polymer (10%), aluminium (8–10%), and silicon (3–5%) [14,16]. These four materials are the dominant material type in solar panels; therefore, potential applications of these waste constituents in pavements would lead to sustainable practices [14].

Table 1. Material composition of c–Si panels.

No	Material	Weight (%)	Price, USD	Reference/s
1	Glass	74–76	0.10/kg	[14,16]
2	Polymer (Encapsulant and back sheet foil)	10	37/m^2 (encapsulant) 20/m^2 (back–sheet foil)	[18,19]
3	Al	8–10	2/kg	
4	Si	3–5	0.95/kg	
5	Cu	0.6–1	5.00/kg	[14,16]
6	Ag	0.06–0.1	574.23/kg	
7	Others (Sn, Pb, etc.)	<0.1	–	

The wider application of solar panels also leads to waste accumulation at the end–of–life (service life 25–30 years) [20,21]. For example, Paiano [21] predicted that the total waste generated by solar panels in 2050 (1,783,268 tons) could be 2125 times the waste generated in 2022 (839 tons) in Italy. Similarly, KEI [22] estimated that the accumulative solar panel waste could be up to 820,000 tons in Korea by 2040. This waste contains environmentally hazardous substances, making its management a challenging task. Specifically, crushed glass powders can cause the lung condition known as silicosis when inhaled [23]. Waste glass also deteriorates in the atmosphere, leading to calcium leaching [24]. Additionally, polymer fractions are a potential pollutant that causes cancer and neurological damage, and it can impair the development of reproductive systems [25]. Aluminium waste can damage the quality of ground and surface waters [26,27]. It can also cause loss of plasma– and hemolymph ions, leading to regulatory failure in gill–breathing animals such as fish and invertebrates [28,29]. Silicon (Si), copper (Cu), silver (Ag), tin (Sn), and lead (Pb) in the waste can also be toxic and harmful to the environment [30]. It is important to recycle these wastes considering their environmental impact and market values [31].

The end–of–life (EOL) management of solar panels is evolving, and it considers the harmful effects and market values of substances in the waste. The EOL management of solar panels is summarized in Figure 2. Solar panels, including their junction boxes and cables, are cleaned as a general step. Visual inspection is then carried out to detect any damage to the panels. Subsequently, three treatments can be carried out: mechanical treatment (also called physical treatment), chemical treatment, and thermal treatment [16]. Mechanical treatment is a physical separation, where crushing and seizing processes are applied to the PV panel modules [13,32]. Prior to this treatment, the frame, electrical cables, and junction box are removed. The remaining parts of the solar panels are crushed and refined to pieces of 4 to 5 mm in size using a hammer mill. During the refinement, glass and polymers are naturally separated from other large pieces due to the size of the mill cutting. The remainder goes through either a thermal treatment or a chemical treatment.

Thermal treatment is the heating and cooling process for separating and recovering valuable materials. The mechanically pre–treated panels are heated to 400–650 °C and cooled down afterwards [33,34]. Polymer components are burned/cracked during the heating process [13]. The treatment can further separate glass from solar cells, recovering glass in the remaining pieces. An overall glass recovery rate of 91% can be achieved by combing mechanical and thermal treatments [33].

Chemical treatment refers to the chemical etching and recovery until the targeted metals are recovered and the remainder from the mechanical and thermal treatments are subjected to chemical treatment. In this treatment, metals are dissolved using various reagents. For example, sodium hydroxide (NaOH) can be used for Si etching, methanesulfonic acid (CH_3SO_3H) and hydrogen peroxide (H_2O_2) can be used for Al etching, and nitric acid (HNO_3) can be used etching of Cu and Ag [16]. After chemical etching, a simple filtration process can be applied to leaching solutions to recover Si. Subsequently, a combination of filtration and heating processes can be applied to recover Al. Copper can then be recovered by adding hydroxy–5–nonyl acetophenone oxime and H_2SO_4 to the leaching solution and using the electro–winning method. Ag can later be recovered by applying hydrochloride acid (HCl), sodium hydroxide (NaOH), and hydrazine hydrate ($N_2H_4 \cdot H_2O$) to the solutions [35].

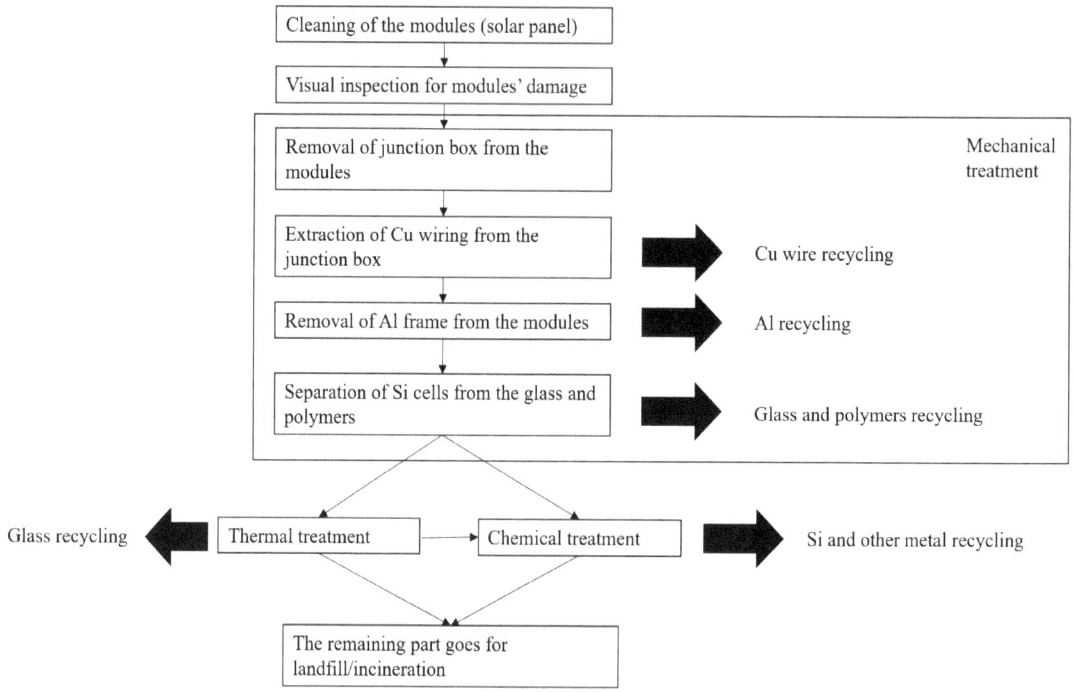

Figure 2. Waste treatment process of a typical solar panel [13].

Although different EOL management methods have been developed for solar panels, they still have negative impacts. Firstly, the uncovered fractions after the mechanical treatment, chemical treatment, and thermal treatment are sent to a landfill and, in some cases, partially incinerated (last step in Figure 3). These fractions still contain Cu, Ag, Sn, and Pb, as none of the current treatments can achieve a 100% recovery rate for these metals [13,35]. Secondly, the treatment procedures, especially the thermal and chemical treatments, are energy–intensive and create harmful impacts on the environment. Weckend et al. [14] mentioned that polymer decomposition in the thermal treatment produces toxic gases and results in high energy consumption. Chowdhury et al. [13] indicated that the silicon etching and rinsing procedure can release toxic gases such as nitrous oxide (NO_2) into the environment. In addition, chemical treatment can be hazardous to human health due to the use of acidic solutions. The remaining acidic solutions after chemical treatment can also be an issue.

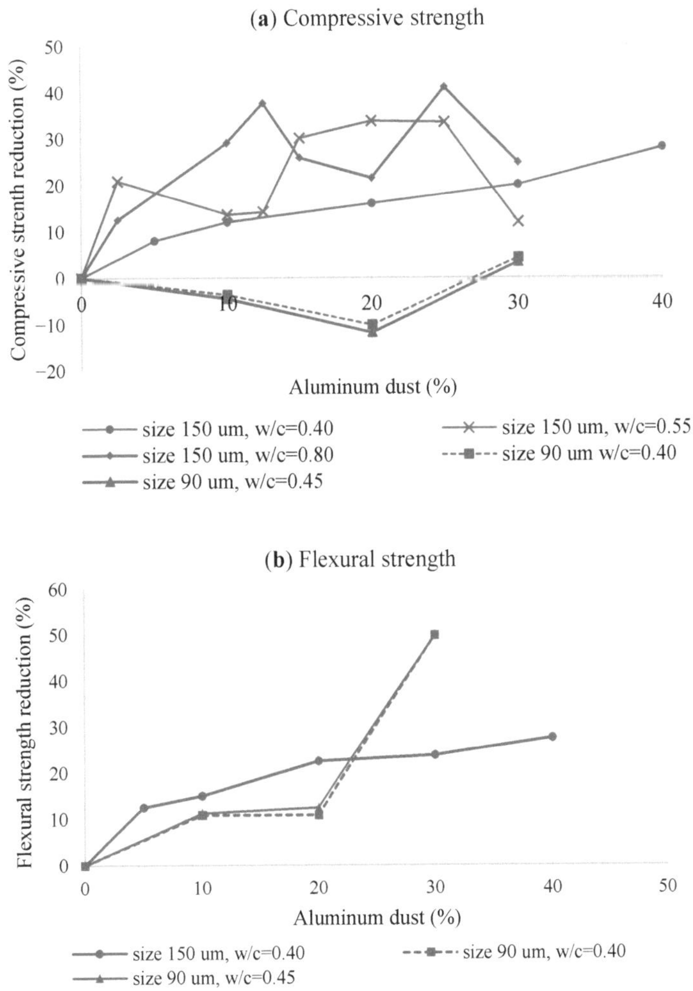

Figure 3. Concrete mechanical properties vs. Al dust levels [36–38]. Al dust content is expressed as the weight percentage of cement replaced by Al dust. (**a**) Compressive strength. (**b**) Flexural strength.

Thirdly, the glass and solar panels can deteriorate under actual working conditions. This can affect the quality of glass and metals recovered from the waste treatment. For example, Ardente et al. [39] raised concerns about low glass quality after recovery. To overcome the limitations cited above (i.e., Chowdhury et al. [13], Weckend et al. [14], and Huang et al. [35]), Imteaz et al. [40], Panditharadhya et al. [41], and Idrees et al. [42] suggested that some components of wasted solar panels (e.g., glass, aluminium, silicon) can be used in pavement construction applications following the mechanical treatment process. The potential feasibility of using c–Si panel waste in the two main types of pavements, i.e., rigid (concrete) and flexible (asphalt) pavements, was investigated in another study [43].

4. Concrete Pavement Applications

Glass, aluminium frames, polymer, and c–Si cells in a c–Si panel are the potential waste materials that can be used for surface, base, and subbase construction in a concrete pavement. Previous studies predominantly focused on investigating the fundamental properties of concrete pavements using glass waste, including compressive, flexure strength and durability. These studies also considered waste an aggregate replacement, a cement replacement, or a combination of both aggregate and cement replacement material in concrete pavements. In addition, other solar panel waste materials, such as aluminium and silicon, have also been researched as filler materials in concrete pavements.

4.1. Waste Glass as an Aggregate

The use of glass waste in concrete is not a novel research area, as initial studies were reported back in early 1963 by Schmidt and Saia [44]. Some studies attempted to investigate the mechanical properties of using waste glass as a natural aggregate replacement material in concrete. The results highlighted a degradation in compressive and flexure strength with the introduction of waste glass as coarse aggregate replacement material in concrete. This is mainly due to preventions in energy releasement during the hydration reaction, as glass aggregate cannot absorb water. The irregular shape of waste glasses can also affect the bond between aggregates and cement pastes in concrete. It is also worth investigating the angularity number of glass, which can quantify waste glass shape effects on the properties of concrete. However, this quantification has not been carried out so far. Polley et al. [45] and Zheng [46] further indicated that the alkali–silica reaction (ASR) initiated in waste glass particles creates alkali oxides in cement. This reaction can cause pressure accumulation inside the aggregate, leading to concrete expansion and strength degradation.

The shape, size, and type of glass, as well as the mix–design and curing time of concrete, are some of the key factors that affect the mechanical properties of concrete and a summarised representation of previous studies using glass waste in concrete pavements are highlighted in Table 2 [47–50]. As shown in Table 2, Topcu and Canbaz [47] reported a higher level of reduction in compressive strength for concrete containing waste glass compared with other studies, with a loss of 49% in compressive strength when 60% of crushed stone (coarse aggregate) was replaced with glass waste in the concrete. However, according to other studies, the loss in compressive strength is only 23.8% to 27.0%, respectively, when 100% of the crushed stone is replaced with waste glass [48,51]. The resultant comparisons between these two studies are reliable considering the similar type of cement (CEM II) and water–cement ratios in the concrete mix. This could mainly be due to the irregular shape of the waste glass used, which can improve the bond between aggregates and cement pastes [47].

Table 2. A summary of concrete strength changes with different levels of glass used as an aggregate.

Concrete Mix					Concrete Properties		
w/c *	S/A *	Cement Type *	Glass Content * (%)	Glass Resources	Compressive Strength Degradation	Flexure Strength Degradation	Reference/s
Coarse aggregate							
0.48	0.60	CEM I	10–100	Waste bottle	1.3% to 23.8%	–	Terro [48]
0.35	–	CEM I	10–30	–	−7.2% to −34.0%	−10.6% to −15.2%	Turgut and Yahlizade [52]
0.54	0.47	CEM II/B–M 32.5 R	15–60	Waste bottle	8% to 49%	−16% to 33%	Topcu and Canbaz [47]
0.55	0.49	CEM II A–L 42.5 R	5–20	–	0% to 2.5%	–	de Castro and de Brito [53]
0.50	–	CEM I	20–30	Window glass	−5.3% to −28.5%	10.8% to −21.7%	Keryou and Ibrahim [54]
0.55	–	CEM II A–L 42.5	5–20	–	6.5% to 10.5%	7.2% to 19.3%	Serpa et al. [55]
0.52	–	CEM II/A–L 42.5 N	12.5–100	Waste bottle	4.4% to 27.0%	–	Omoding et al. [51]

Table 2. *Cont.*

Concrete Mix					Concrete Properties		Reference/s
w/c *	S/A *	Cement Type *	Glass Content * (%)	Glass Resources	Compressive Strength Degradation	Flexure Strength Degradation	
Fine aggregate							
0.50	0.47	CEM I	30–70	Waste bottle	0.6% to 13.6%	3.2% to 18.1%	Park et al. [56]
0.49	0.75	CEM I	50	–	24.1%	18.1%	Shayan and Xu [57]
0.48	0.60	CEM I	10–100	Waste bottle	1.3% to 41.2%	–	Terro [48]
0.35	–	CEM I	10–30	–	−31.5% to −68.9%	−22.3% to −90.0%	Turgut and Yahlizade [52]
0.53	–	CEM I	10–20	Waste bottle and window	9.1% to −4.3%	−3.6% to 11.2%	Ismail and Al–Hashmi [58]
0.55–0.58	0.49	CEM II A–L 42.5 R	5–20	–	11.0% to 17.0%	–	de Castro and de Brito [53]
0.55–0.58	–	CEM II A–L 42.5	5–20	–	15.3% to 20.5%	20.9 to 28.1%	Serpa et al. [55]
Mix of coarse and fine aggregate							
0.48	0.60	CEM I	10–100	Waste bottle	7.6% to 68.4%	–	Terro [48]
0.47	–	CEM I	15–45	Waste bottle	1.5% to 8.5%	–	Kou and Poon [59]
0.55–0.58	0.49	CEM II A–L 42.5 R	5–20	–	7.0% to 17.0%	–	de Castro and de Brito [53]
0.55–0.57	–	CEM II A–L 42.5	5–20	–	13.7% to 26.7%	17.9 to 34.8%	Serpa et al. [55]

* w/c is the water–cement ratio; * S/A is the sand to aggregate ratio. * CEM I—Portland cement; CEM II/A–L—Portland–limestone cement; CEM II/B–M—Portland–composite cement. * Glass content is expressed as weight percentage of coarse/fine aggregate replaced by glass throughout this paper.

A smaller glass size can reduce the strength degradation level of concrete by causing pozzolanic reactions and filling the pores in concrete mixes. The presence of larger glass particles might further weaken the concrete structure because of their high friability [45]. Ismail and Al–Hashmi [58] suggested that the gradation of waste glass with a size smaller than 0.3 mm can cause significant pozzolanic reactions. Ismail and Al–Hashmi [58], Turgut and Yahlizade [52], and Du and Tan [60] also found an increase in compressive and flexure strength in concrete with increasing fine glass waste contents and improved pozzolanic reactions. However, Terro [48] and de Castro and de Brito [53] showed that the level of reduction in compressive strength is larger when fine aggregates are replaced in concrete compared with coarse aggregate. The conflicting results in these studies show the importance of conducting further studies on optimizing the replaced glass size and content in concrete to obtain sustainable concrete with high strength for pavement material.

The study by Keryou and Ibrahim [54] is the only one of its kind that found strength increments in concrete with larger glass sizes used as coarse aggregate (>4 mm). They claimed that this was because of the interlocking and friction increments among mixed particles in concrete due to the existence of the glass. However, Omoding et al. [51] indicated that ASR over–dominates the interlocking effect and degrades the mechanical properties of concrete accordingly. Therefore, further studies are necessary to investigate the interlocking effects of different glass particles. Moreover, the chemical composition of the glass type (e.g., toughened glass, soda–lime glass, laminated glass, etc.) can also affect the alkali–silica reaction (ASR) and their degradation effect on concrete strength. There are also conflicting findings on concrete strength degradation based on various glass types. Park et al. [56] indicated that green glass showed less ASR expansion than brown glass due to the sizeable Cr_2O_3 component; however, other studies have highlighted contradictory results, such as Dhir et al. [61]. Therefore, future studies are required to establish how the strength of the concrete is affected by the presence of glass waste of varying chemical compositions. The c–Si panel uses tempered glass, and there are limited studies on the effect of using waste tempered glass on concrete strength. Instead, most studies have focused on bottle glass waste in concrete (i.e., soda–lime glass, treated soda–lime glass, or borosilicate glass), as shown in Table 2.

For the concrete mix design, several studies indicated that a lower water–cement ratio can decrease the ASR between glass and cement, leading to smaller strength reductions in the concrete [48,49,52]. Furthermore, Du and Tan [62] showed that concrete containing a large portion of fly ash and slag cement contributes to pozzolanic reactions, potentially increasing strength in concrete containing glass. In addition, the level of strength increase with the extension of the concrete curing time due to the longer pozzolanic reaction time. Besides concrete strength analyses, some recent studies have also estimated the durability of concrete containing different levels of waste glass. The results illustrate an increase in concrete durability due to the addition of waste glass. de Castro and de Brito [53] indicated an enhancement in concrete chloride penetration resistance with the increasing proportion of glass components. The chloride penetration depth reduces by 20% when 10% of the course and fine aggregate is replaced by cement. Du and Tan [62] carried out a rapid chloride penetration test. The results indicate that the total charge passing the concrete was reduced by 66.7% when 30% of fine aggregate was replaced by glass, indicating a significant increase in the chloride penetration resistance.

4.2. Waste Glass as a Cement Replacement Material

Glass is generally converted into powders sized less than 100 μm, and then it is added as a partial cement replacement material in concrete. Some recent studies on using waste glass as a cement replacement in concrete are summarized in Table 3. Similar to the case of aggregate replacement, cement replacement in concrete using glass waste also demonstrates a reduction in the mechanical properties of concrete. Similar to the case of aggregate replacement, the reductions in mechanical properties can be affected by the size, glass and cement type, and mix design of the concrete. The strength reduction is significantly higher as compared with aggregate replacement, mainly due to weaker bonds between cement and aggregates with the introduction of glass particles in the place of cement [63]. The studies further highlighted that if glasses were used with a highly reactive pozzolana in the concrete mix, such as silica fume, the pozzolanic activity of the glass could be promoted. However, silica fume can also contribute to ASR [64]. Therefore, further experimental studies are required to justify improvements in concrete strength due to the addition of silica fume. Pozzolanic activity in glass–cement mixtures can also be promoted by performing heat treatment [65]. However, more experimental investigations are needed to assess the overall impact of heat treatment on the strength of concrete. Moreover, heat treatment can be energy–intensive, which could lead to additional costs, as well as environmental and practical handling implications [66].

Table 3. A summary of concrete strength changes with different levels of glass used as cement.

Concrete Mix			Concrete Properties			
w/c	Cement Type	Glass Content (%)	Glass Resources	Compressive Strength Degradation	Flexure Strength Degradation	Reference/s
0.75	CEM I	30	Fluorescent lamps	9.1% (38–75 μm glass) * to 31.8% (75–150 μm glass)	–	Shao et al. [63]
0.49	CEM I	20–30	–	21.2% (<10 μm glass)	–	Shayan and Xu [67]
0.49	CEM I	20	Glass beads	12.5% (30–100 μm glass)	–	Shi et al. [68]
0.42	CEM I	10	Window plate glass	6.7% (1–100 μm glass)	–	Schwarz et al. [69]
0.57	CEM I	30	Container (green)	31.9% (<40 μm glass)	–	Khmiri et al. [70]
0.45	CEM I	11–15	Container (green)	4.1% to 21.0% (18–80 μm glass)	5.4% to 47.8%	AL–Zubaid, Shabeeb [71]

Note: * The grain size indicates that over 90% of the glass particles are 35–75 μm, the same for all the grain size indications.

Adding waste glass to concrete can generally reduce its strength. This strength reduction level decreases with the decreasing water–cement ratio of the concrete, owing to the reasons mentioned in case 1 [63,70]. Additionally, Shao et al. [63] compared the strength reduction level for concrete cured after 28 and 90 days. They found that the reduction level lessens with curing time increments since it allows for more time for pozzolanic activity. The curing time increments can also form denser and less permeable concrete microstructures because of the filling effect of glass particles. However, the test results by Schwarz et al. [69] do not support this finding. Therefore, further investigations are required in this area.

4.3. Waste Glass Together as an Aggregate and Cement Replacement Material

Similar to the case of the solo replacement of aggregate or cement with glass, the combined replacement of cement and sand in concrete also results in a reduction in the compressive and flexural strength of the concrete. The reduction mechanism and the parameters that affect the reductions are similar to those given in previous cases. There are only a limited number of studies where the effect of using waste glass as cement and aggregate replacement in concrete is highlighted (see Table 4). However, these studies investigated only a maximum of 20% cement replacement and 50% aggregate replacement in concrete.

Table 4. A summary of concrete strength changes with different levels of glass used as cement and aggregate.

Concrete Mix				Concrete Properties		
w/c	Cement Type	Waste Glass Content (%)	Glass Resources	Compressive Strength Reduction	Flexure Strength Reduction	Reference/s
0.49	CEM I	20% for cement; 50% for coarse and fine aggregate	-	23.9%	-	Shayan and Xu [67]
0.38	CEM I	20% for cement; 50% for fine glass aggregate	-	19.2%	7.8%	Taha and Nounu [72]
0.38	CEM I	20% for cement; 50% for coarse aggregate	-	22.0% (14 days) 2.0% (56 days)	15.8% (14 days) 10.3% (56 days)	Wang and Huang [73]

4.4. Aluminium (Al) Waste in Concrete

In most of the previous studies, aluminium (Al) waste was crushed into a powder (i.e., aluminium dust) and added as a cement replacement material in concrete. These investigations highlight a general reduction in the mechanical properties of the modified materials. Mailar et al. [38] quantified this reduction by testing the compressive and flexure strengths of concrete with different Al dust contents. Concrete samples with two different water–cement ratios (0.40 and 0.45) were tested, with an average Al dust size of 90 μm. Similar tests investigated the mechanical properties of Al–waste–incorporated concrete with an average Al size of 150 μm and water–cement ratios of 0.40, 0.55, and 0.80 (see, for example, Elinwa and Mbadike [36] and Mbadike and Osadere [37]). The resulting compressive and flexural strengths are summarized in Figure 3. The observed test results clearly show that mechanical properties in concrete further reduce with an increase in Al dust content. This can be attributed to the fact that Al dust affects the bonding strength between aggregate and cement paste, thereby reducing the mechanical properties of the concrete [74]. In addition, Al dust can absorb water in the concrete mix, reducing water content and thus affecting the strength of the concrete. Furthermore, Al dust generates hydrogen gas when in contact with water, increasing pressure in the concrete and reducing its strength [36].

Figure 3 highlights a reduction in the compressive and flexural strength of concrete with different Al dust contents. The observed reduction in the mechanical properties of concrete strength is not linear for an increase in Al dust content [36]. Mailar et al. [38] even found an increase in the mechanical properties of concrete with Al dust proportions of 10% and 15% as cement replacement material. This phenomenon is further explained by Hay and Ostertag [75], as Al can be absorbed onto the amorphous silica surface via reactive aggregation in concrete, limiting the ASR expansion and leading to strength increments. Nonetheless, more studies are needed to investigate the effect of Al dust on concrete's mechanical properties comprehensively. Furthermore, Al dust size can affect mechanical properties, but limited studies have made attempts to quantify these changes [37]. In addition, there is a lack of microstructural analysis studies, which could find the mechanism behind the changes in the mechanical properties of Al–incorporated concrete. Despite claims by Mailar et al. [38] that an increase in curing time may improve the mechanical properties of concrete containing Al dust, the test conducted by Elinwa and Mbadike [36] contradict these results.

The durability characteristics of concrete, such as water penetration resistance, acid attack through water absorption, and acid resistance, have been enhanced with the addition of Al dust to the concrete mix [38]. The mass loss after being immersed in sulphuric acid with 5% weight for 30 days was reduced by 57.2% for concrete with a 30% replacement of cement with aluminium dross. This can be attributed to the fact that Al dust can fill the voids in concrete due to its small size, which reduces the pores of the concrete [38]. However, there is a lack of other durability studies on concrete with Al dust, including air permeability, chloride resistance, and sulphate attack resistance tests.

It should be noted here that one study has used Al dust as a partial fine aggregate replacement material in concrete [76]. In this study, 1%, 2%, and 5% of the sand were replaced by aluminium waste in the concrete, and the resulting mechanical strengths indicated a reduction of 3.6%, 18.7%, and 21% in the compressive strength, respectively. In addition, there was an increment in concrete durability based on the water absorption test (66.3% reduction in the water absorption rate at a 5% sand replacement). The decrease in bond strength of the concrete aggregate and the reduction in concrete porosity led to strength reduction and durability increments. Inspired by fibre–reinforced concrete, Muwashee et al. [77] added Al strips to the concrete mix during production, and 22% and 238% increases in compressive and flexural strengths were observed by adding 2.5% Al strips to concrete by volume. This was mainly because Al strips can delay the formation of cracks and make the concrete matrix stronger.

4.5. Polymer Waste in Concrete

Polymer waste in a c–Si panel mainly consists of encapsulant and back–sheet foil. For encapsulant, Dulsang et al. [78] and Khan et al. [79] replaced the cement with different levels of waste encapsulant in concrete manufacturing. They tested the 28–day compressive strength, with the results summarized in Figure 4. The water–cement ratio was 0.40 in both studies. Dulsang et al. [78] found a 68.8% reduction in the compressive strength with waste encapsulant content increasing from 3 to 10%. However, Khan et al. [79] found that compressive strength increases with encapsulant content increments. Waste encapsulant had an average size of 4.5 mm in Dulsang et al. [78] and 0.41 mm in Khan et al. [79]. Large encapsulant size can create internal voids in concretes and affect the bond between the plastic aggregate and the cement paste. This can be avoided with small encapsulant waste sizes [79]. Dulsang et al. [78] also mentioned that encapsulant is a water–reducing polymer. A small encapsulant size may increase its water–reducing effect, enhancing the bond between cement hydrates and the inert aggregates.

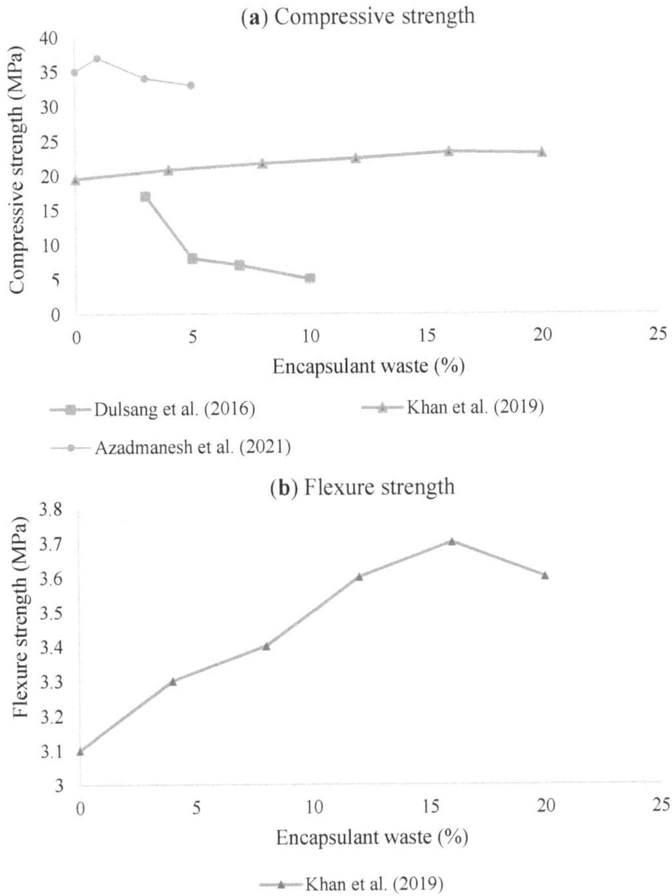

Figure 4. Concrete mechanical properties vs. encapsulant waste levels [78–80]. Encapsulant waste content is expressed as the weight percentage of cement replaced by encapsulant waste. (**a**) Compressive strength. (**b**) Flexure strength.

Apart from studies on regular concrete, Azadmanesh, Hashemi [80] added different levels of encapsulant to Engineered Cementitious Composites (ECC) and tested the 28–day compressive strength afterwards. The encapsulant size ranged from 1 to 7 μm. It could be seen that there was a limited reduction in compressive strength with an encapsulant content of 5%. The study also indicated that a small encapsulant size can prevent strength reduction in concrete. However, it did not show that a small encapsulant size contributes to its water–reducing effect. Otherwise, compressive strength should be increased with encapsulant content increments. Further studies are, therefore, needed.

Only Khan et al. [79] quantified changes in the 28–day flexure strength among these studies. Flexure strength increases by 17% when encapsulant waste reaches 20% due to the water–reducing effect of the encapsulant. More studies are also needed in this field.

For durability, Dulsang et al. [78] showed that concrete's sorptivity coefficient can be reduced by over 92.1% when encapsulant content reaches 10%. They also found that concrete weight loss reduces from 15% to 5% when encapsulant waste increases from 3% to 10%. Increased tortuosity due to encapsulant waste leads to sorptivity reduction and acid–resistance enhancement.

To the best of the authors' knowledge, no studies have been conducted to test the strength and durability of concrete with back–sheet foil (i.e., polyvinyl fluoride (PVF)).

4.6. Silicon Waste in Concrete

There are few studies that have investigated the behaviour of using recycled silicon solar cell waste in concrete. Ren et al. [81] replaced 5% to 20% of cement with silicon carbide (SiC) particles (silicon solar cell contains SiC) in concrete manufacturing. The concrete had a water–to–cement ratio of 0.40. The 28–day compressive strength was tested, and the results are shown in Figure 5. Similar tests have been carried out by Małek et al. [82] and Idrees et al. [42]. Particle sizes of SiC were 50 μm, 5.5 mm, and 27 μm in these three studies. There was an increase in the compressive strength of concrete with an increase in SiC content overall, despite when the SiC content was 5% and 10% in Ren et al. [81].

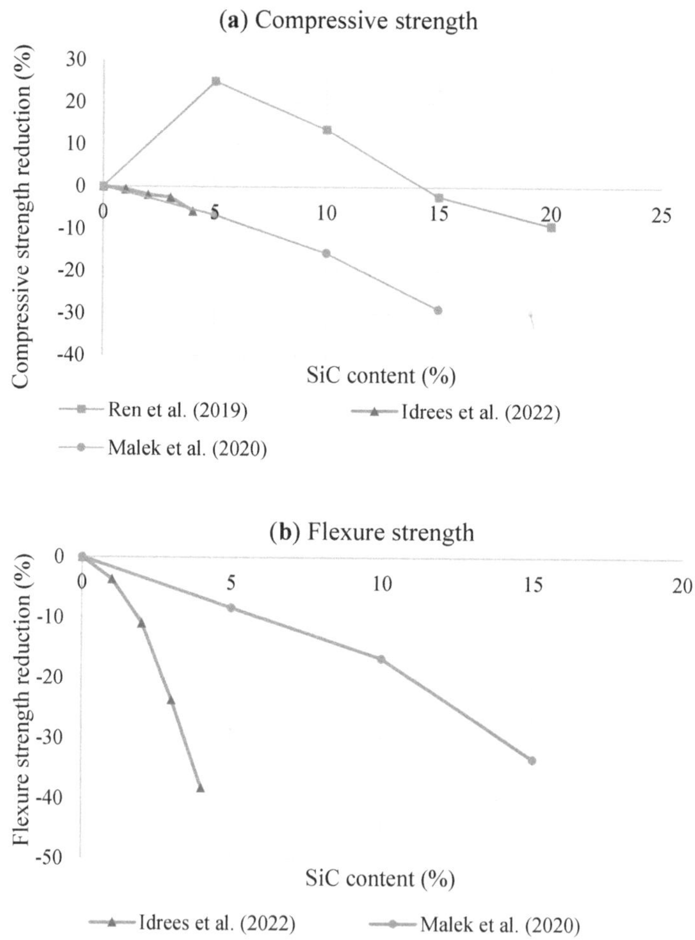

Figure 5. Concrete mechanical properties vs. SiC levels [42,81,82]. SiC content is expressed as the weight percentage of cement replaced by SiC particles. (**a**) Compressive strength. (**b**) Flexure strength.

Based on Ren et al. [81], there is less cement in concrete to produce hydrates when SiC particles partially replace cement. This leads to strength reduction at SiC levels of 5% and 10%. However, SiC is highly abrasive and can act as a reinforcing filter in concrete mixes with significant SiC content. This reinforcing effect dominates the strength reduction effect and eventually leads to an increase in concrete strength. In addition, large SiC particle content can lead to capillary suction, vapour diffusion, and capillary condensation, which can transport water from SiC waste to cement paste, promotes the hydration of the cement in the paste, and increases concrete strength.

Małek et al. [82] and Idrees et al. [42] indicated that the highly abrasive nature of SiC leads to an increase in concrete strength, even with a small amount of SiC content. Małek et al. [82] and Idrees et al. [42] also found an increase in flexure strength due to the highly abrasive nature of SiC in concrete mix and the promotion of hydration in the cement (Figure 5b). More studies are, therefore, needed to clarify the effect of SiC on concrete properties.

Besides concrete, Jiang et al. [83] used SiC particles extracted from silicon solar cell waste in CEM I mortars and observed an increase in compressive and flexural strength. However, Fernández et al. [84] found that an increased proportion of silicon waste can reduce the compressive and tensile strength of concrete with calcium aluminate cement (CAC), as the bond between the cement paste and aggregate can be reduced due to the existence of silicon waste. These contradictory finds indicate that more studies are needed to investigate cement properties containing SiC particles. In addition, based on a study on concrete with waste glass [48,49,52], it is likely that the differences between cement types could also lead to variations in concrete strength, but this needs further studies.

Ren et al. [81] tested chloride resistance with a rapid chloride permeability test for durability. The total charging recorded in 6 h reduced by 85.7% in the rapid chloride permeability test with 20% SiC, indicating a significant increase in chloride resistance. Ren et al. [81] also found water absorption was reduced by 10.7% when SiC content increased to 20%. Adding SiC increases concrete's durability due to the reduction in its porosity. However, Idrees et al. [42] found a reduction in chloride resistance for concrete containing SiC particles. This is because Idrees et al. [42] used SiC particles with s large size (5.5 mm compared with 50 μm in Ren et al. [81]). The large size of SiC particles increased concrete porosity instead. Comparing these two studies indicates that size control over SiC particles is essential before it is added to concrete.

5. Asphalt Pavement Applications

Asphalt is a mix of sand, gravel, broken stones, soft materials, and bituminous binder (asphalt binder) that can be used in the wearing surface and base construction of pavements. Only a handful of studies have been carried out to estimate the properties of asphalt mixtures made with c–Si panel waste components. These studies primarily focused on using different levels of waste glass in asphalt mixtures in pavements and seldom considered other waste constituents in a solar panel [85–90].

It should be noted here that 4.75 mm was the maximum size of the glass aggregates reported in these studies, and 1–2% hydrated lime was added to the asphalt mixtures to improve the cohesion between stone and glass aggregates with bitumen coatings. The observed optimum asphalt content in the glass–asphalt mixtures and the resulting stability (fatigue resistance) in those studies are summarized in Figures 5 and 6. According to the summarised results from the studies in Figure 6, no significant reduction in the optimal binder content of the composite asphalt mixture was observed with an increase in the waste glass percentage of up to 20% as a result of using slaked lime.

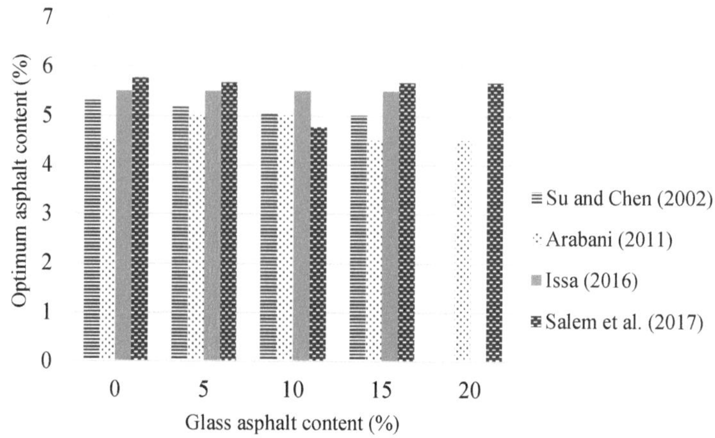

Figure 6. Optimum asphalt content vs. glass asphalt mixtures [86–89].

The results in Figure 6 highlight that stability is slightly reduced with the increase in waste glass in the asphalt mixture. However, until the waste glass content in the asphalt mixture reached 20%, no significant reduction in stability was observed. Adhesion loss between the binder and glass, skid resistance loss, reduced stripping resistance, and increased ravelling potential were identified as the main reasons for this reduction in stability [85]. The presence of broken glass in the mixture has also been reported to contribute to reductions in stability. However, these issues can be corrected by adding lime to the mixture [86]. Marandi and Ghasemi [91] indicated that adding rubber polymers can also eliminate degradation in the stability of an asphalt mixture with a glass content of up to 5%.

Arabani [87], Issa [88], and Salem et al. [89] also summarized the properties of asphalt mixtures with the additions of waste glass based on the Marshall test, as summarized in Table 5. Table 5 shows the changes in flow, voids in the mineral aggregates in asphalt concrete with and without waste glass, the void percentage filled with bitumen in asphalt concrete with and without waste glass, and so on. It can be seen from Table 5 that there was no significant reduction in these properties with the increase in the waste glass percentage in the asphalt–concrete mixture. However, further studies are needed to find a more reliable estimation of these property changes due to conflicts in the current studies. For example, Arabani [87] found an increase in flow with glass content increments. However, Issa [88] found a decrement in the flow instead. In addition, Arabani [87] tested the stiffness modulus of asphalt–concrete mixtures with waste glass content, shown in Figure 7. It can be seen from Figure 7 that adding waste glass to asphalt mixtures increases their stiffness due to the interlocking effect of glass between mixed particles.

Table 5. Properties of asphalt–concrete mixtures with waste glass content.

Glass Content	Bitumen Content (%)	Flow (mm)	Unit Weight g cm^{-3}	Air Void (%)	Voids in Mineral Aggregates (%)	Voids Filled with Asphalt (%)	Reference
0	4.5	2.31	2.337	4.74	13.60	65.13	Arabani [87]
5	4.5	2.26	2.323	5.01	13.95	64.08	
10	4.5	2.42	2.305	5.33	14.45	63.11	
15	4.5	2.63	2.331	5.03	13.31	62.22	
20	4.5	2.63	2.314	5.4	13.78	60.81	

Table 5. *Cont.*

Glass Content	Bitumen Content (%)	Flow (mm)	Unit Weight g cm^{-3}	Air Void (%)	Voids in Mineral Aggregates (%)	Voids Filled with Asphalt (%)	Reference
0	–	2.93	2.40	4.74	–	–	Issa [88]
5	–	2.80	2.25	4.53	–	–	
10	–	2.87	2.13	4.30	–	–	
15	–	2.73	2.10	4.16	–	–	
0		4.32	2.213	4.2	16.5	73.5	Salem et al. [89]
5		4.45	2.248	2.8	15.35	81.0	
10		4.06	2.225	4.4	16.35	72.5	
15		4.57	2.24	3.5	16	77.0	
20		4.11	2.247	2.5	15.2	83.5	

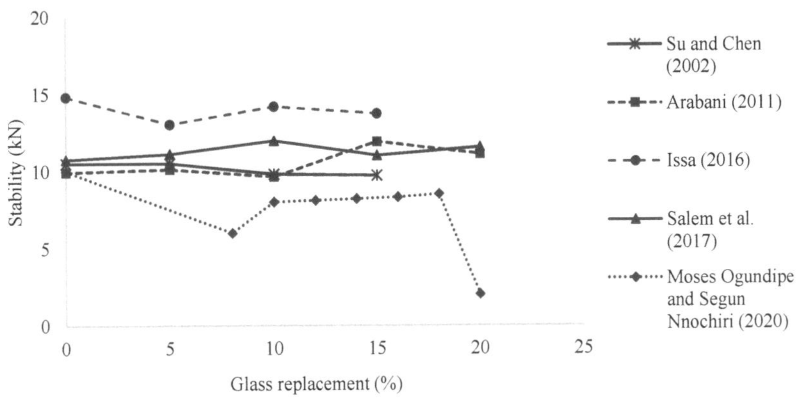

Figure 7. Stability value of glass asphalt mixtures with different levels of glass content [86–90].

In addition to stability, Arabani [87] and Su and Chen [86] found that asphalt mixtures with recycled glass can increase the skid resistance and night visibility of asphalt pavements, eventually leading to improved driving conditions at night. Lachance–Tremblay et al. [92] found no noticeable degradation in compaction ability, rutting resistance, thermal cracking resistance, or asphalt mixture stiffness with 25% waste glass content. Shafabakhsh and Sajed [93] also found that adding waste glass can increase the stiffness modulus, dynamic properties, and resistance of asphalt mixtures against deformation and rutting. Glass can increase the interlocking effect between aggregates, helping asphalt maintain its workability, which includes properties such as stability, skid resistance, night visibility, compaction ability, rutting resistance, thermal cracking resistance, stiffness modulus, dynamic properties, deformation resistance, and rutting resistance. The Federal Highway Administration [94] and Wu et al. [95], through their findings, showed that asphalt mixtures with glass contents of up to 15% and 25% can be used for wearing surfaces and base construction, respectively, in pavement construction. Moses Ogundipe and Segun Nnochiri [90] tested the stability of asphalt with glass sizes of up to 25 mm. They found a significant degradation in mechanical properties, as a large glass size can affect the bond between glass particles and asphalt. The collective interpretation of these studies highlights that larger than 4.75 mm glass pieces can reduce stability in the asphalt mixture. Therefore, it is essential to maintain the crushed glass size at 4.75 mm to avoid the workability degradation of asphalt pavement.

However, asphalt grades and mixing design can also affect workability. The optimum asphalt content and workability in an asphalt–glass mixture can significantly change with different asphalt grades and mixing temperatures [90]. However, there is a severe lack of precedents in the literature relating to this aspect. In addition, most studies have considered incorporating waste glass in hot mix asphalt (HMA) mixtures, while other categories, such as stone mastic asphalt (SMA), have not been considered. Further studies are, therefore needed to compare properties, such as the workability of SMA asphalt mixtures with different glass percentages [94] (Figure 8).

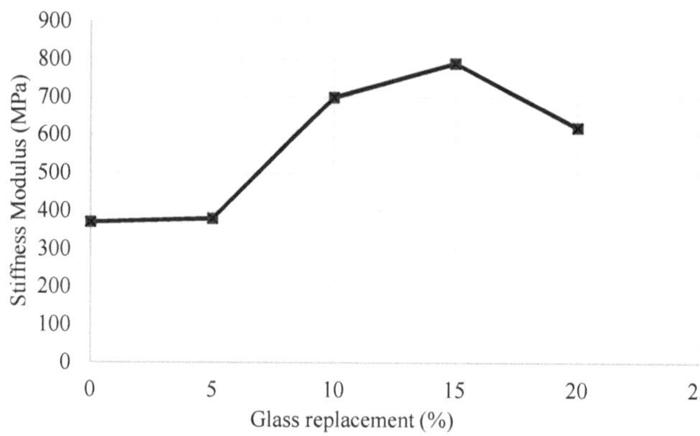

Figure 8. Stiffness modulus of glass–asphalt mixtures with different levels of glass content.

6. Impediments and Drivers of Using Solar Panel Waste as a Pavement Material

Despite numerous research initiatives to promote the use of waste materials in building and construction materials, several limitations hinder the product's translation into a market–ready product. Understanding the problem and parallel research to mitigate or minimize the effects of these barriers is a contemporary requirement to improve the marketability of the product. Despite the satisfactory mechanical properties in using the material either for structural or non–structural applications, the practical implementation of the product is limited due to workability and handling issues. Previous studies highlighted that a reduction in workability as a result of introducing glass powder and other articles is minimized by the introduction of superplasticizer [96]. In addition, the pumpability of concrete can also be affected in circumstances where large concrete pumps are utilised for mass concrete construction activities. This can also result in additional costs and the need for technical skills for either the introduction of supplementary materials or alternative processes. Despite the perceived environmental benefits, contractors and decision–making stakeholders in the construction industry are often hesitant to invest in alternative materials due to low profit margins and a lack of returns. Presently, few to no standards or policies are available that define systematic procedures for using the different waste constituents of a solar panel system in construction materials in order to promote sustainability. A lack of government incentives is another major issue that needs to be addressed to encourage the increased uptake of green materials. Moreover, the majority of countries treat solar panels as general E–waste and lack specific guidelines for safe EOL management and disposal [97]. This is considered a major barrier, and the availability of a guideline would encourage stakeholders to explore and implement innovative methods of using various waste constituents in pavement applications. Supply chain management issues with handling waste materials are a key problem that could demote the potential of using solar panel waste in pavement construction. Often, these solar panel systems are disposed in large quantities, and converting them into useful concrete and asphalt pavement raw materials requires significant supply chain phases that could be energy–intensive and have high costs. Special storage and transportation processes may be required for handling these materials due to the hazardous nature of solar panels [98]. At present, there are no systematic procedures for the systematic handling of solar panel waste. However, if research can be translated into commercial products and applications, there will be increased job opportunities in the supply chain for converting solar panel waste into raw construction material. Due to the potential availabilities of waste across all countries and regions, in the future, local conversion plants can be set up to promote sustainable development.

Potential environmental impacts due to the presence of hazardous and toxic materials, such as lead, lithium, and cadmium, can curtail the use of solar panel waste as a pavement material [97]. Therefore, further research should be focused on investigating leachate impacts due to dumping solar panel waste on pavements, and comprehensive life cycle assessment (LCA) studies should be used to benchmark the environmental benefits [99]. This can also enhance the commercial promotion of the product through environmental and sustainable labelling. At the moment, there is high potential demand for renewable energies across the world, which will lead to an abundance of solar panel waste materials in the future. The availability and abundance of waste materials due to an enhanced degree of end–of–life disposal is a driving factor that can also promote the possibilities of re–engineering them as pavement construction material. In addition to EOL solar panel waste, high amounts of premature solar panel waste, due to poor handling during transportation, installation, and operations, can also magnify the availability of solar panel waste. This can be also considered a driving factor to use it as a pavement material. Moreover, there are multiple potential applications for solar panel waste materials in pavement construction, which can encourage stakeholders to accelerate the market promotion of the product, thus obtaining required certifications and approvals. The circular economy status of a product needs to achieve life–cycle benefits, including material substitution, sustainable design, benchmark environmental impacts through LCA, and end–of–life management [100]. Pertinent policies and regulations should be developed to systematically define the circular–economy–based reverse supply chain processes for both the products (solar panels and pavements) to improve market intake. Similar to government stimulus for solar energy implementations, incentives should be introduced to develop effective EOL management techniques for solar panel waste.

7. Conclusions and Future Research

The heavy uptake of solar panels in many countries is predicted to cause a huge waste problem in the near future as the initial end–of–life periods are approaching since their introduction. Previous studies made significant attempts to improve the production efficiency of solar panels. Some studies also made attempts to recover, recycle, and reuse the waste constituents of solar panels as an end–of–life (EOL) management strategy. Despite many attempts to identify potential EOL solutions, there is still huge potential in reusing solar panel waste due to the predicted massive disposal of solar panels. Construction, on the other hand, is known to be one of the most energy–intensive industries due to excessive virgin material usage. Pavement construction is a similar energy–intensive construction type, and over the past few decades, several studies focused on researching alternate raw materials to replace virgin materials. Therefore, addressing these issues together could pave the way for rapid sustainable development solutions. The current study presented a detailed review of the potential of using different solar panel waste materials as a raw construction materials in both flexible (asphalt) and rigid (concrete) pavement types.

Adding waste constituents from solar panels is likely to affect the mechanical properties of concrete pavement. Most of the studies found that adding the key waste constituents of c–Si panels, such as Al, polymer, and silicon, to concrete can reduce the compressive and flexure strength of concrete. Our review's findings highlight that a reduction in the compressive and flexural strength is not substantial with a glass content of 10% in concrete as a partial replacement material for cement or aggregate. This reduction is often influenced by the glass type, size, and concrete mix design and, therefore, more studies are required to justify the relationship between glass content and the mechanical properties of concrete pavements. Compared with glass, studies have seldom investigated mechanical properties after the addition of Al, polymer, and silicon to concrete pavements. Current findings illustrate conflicting results and, therefore, additional studies are needed to find the optimum particle size and content for these waste elements to maximize their compressive and flexural strengths. Apart from compressive and flexure strengths, abrasion resistance stiffness, and fatigue cracking resistance are critical properties for pavement construction. To the best of the authors' knowledge, thus far, no studies have been carried out to study the properties of concrete using waste c–Si panels. Additionally, future studies are required to investigate the composite behaviour of the interaction effects between glass, Al, polymer, silicon, and other c–Si panel components on the compressive and flexure strength, abrasion resistance, stiffness, and fatigue cracking resistance of concrete. The durability of concrete pavements is enhanced with the addition of waste materials from solar panels such as glass, Al, polymer, and silicon. The majority of durability analyses have focused on water absorption and acid resistance tests, while a handful of studies have learned that the chloride penetration resistance of concrete can be improved by adding glass particles. However, limited studies have focused on finding the chloride penetration resistance of concrete composites with Al, polymer, or silicon. Moreover, there are also limited studies on the freezing and thawing durability of concrete filled with waste c–Si panels, which can be a future study focus. Previous studies emphasized that asphalt pavements generally have no noticeable degradation in the workability of asphalt at a glass content level of 25%. However, the effect of adding other c–Si panel components on asphalt workability must be studied. More studies can focus on the durability of asphalt pavement with c–Si panel waste.

The composite behaviour of using multiple solar panel waste materials as pavement material should also be investigated in future studies to upsurge the potential use of waste materials. Future research can also be focused on investigating the use of solar panel waste in low–stress applications, such as walking paths, driveways, and landscape blocks. The commercial implementation of these applications would be relatively simple as compared with structural applications, due to low structural standard requirements. A c–Si solar panel system includes several material elements that have different residual values. Therefore, it is important to develop a systematic framework that can identify the potential recyclable and recoverable elements of a solar panel and divert the residual waste to investigate the potential applications in pavement construction. Subsequently, based on the supply, research can focus on prioritizing the waste constituents of a solar panel system to replace virgin materials in different pavement types. However, the translation of these research findings to market products is often a daunting task due to practical and legislative implementation requirements. Therefore, further initiatives should be facilitated at the government and organizational levels to strengthen funding support with the intent of driving research commercialization. The results of this review demonstrate the need and potential of using solar panel waste in pavement construction applications. The findings of the study may enable interested stakeholders to understand current trends and future research regarding the use of solar panel waste in pavement construction.

Author Contributions: Conceptualization, M.S. and L.L.; methodology, L.L.; validation, M.S., L.L., J.Z. and P.J.; formal analysis, L.L.; investigation, M.S. and L.L.; resources, M.S. and L.L.; data curation, M.S. and L.L.; writing—original draft preparation, L.L.; writing—review and editing, M.S., L.L., J.Z. and P.J.; visualization, M.S. and L.L.; supervision, M.S.; project administration, M.S. All authors have read and agreed to the published version of the manuscript.

Funding: This research received no external funding.

Informed Consent Statement: Not applicable.

Data Availability Statement: Not applicable.

Conflicts of Interest: The authors declare no conflict of interest.

References

1. Varun; Prakash, R.; Bhat, I.K. Energy, economics and environmental impacts of renewable energy systems. *Renew. Sustain. Energy Rev.* **2009**, *13*, 2716–2721. [CrossRef]
2. Xu, Y.; Li, J.H.; Tan, Q.Y.; Peters, A.L.; Yang, C.R. Global status of recycling waste solar panels: A review. *Waste Manag.* **2018**, *75*, 450–458. [CrossRef] [PubMed]
3. Zappa, W.; Van Den Broek, M. Analysing the potential of integrating wind and solar power in Europe using spatial optimisation under various scenarios. *Renew. Sustain. Energy Rev.* **2018**, *94*, 1192–1216. [CrossRef]
4. Mahmoudi, S.; Huda, N.; Behnia, M. Photovoltaic waste assessment: Forecasting and screening of emerging waste in Australia. *Resour. Conserv. Recycl.* **2019**, *146*, 192–205. [CrossRef]
5. Sandanayake, M.; Lokuge, W.; Zhang, G.; Setunge, S.; Thushar, Q. Greenhouse gas emissions during timber and concrete building construction—A scenario based comparative case study. *Sustain. Cities Soc.* **2018**, *38*, 91–97. [CrossRef]
6. Sandanayake, M.; Zhang, G.; Setunge, S. A comparative method of air emission impact assessment for building construction activities. *Environ. Impact Assess. Rev.* **2018**, *68*, 1–9. [CrossRef]
7. Kuruvachalil, L.; Sandanayake, M.; Kumanayake, R.; Radhakrishna. Carbon Emission and Cost Analysis of Using Hybrid Fibre White Topping Overlays & mdash;A Road Rehabilitation Feasibility Study. *Future Transp.* **2022**, *2*, 263–280.
8. Sandanayake, M.; Law, D.; Sargent, P. A new framework for assessing the environmental impacts of circular economy friendly soil waste–based geopolymer cements. *Build. Environ.* **2022**, *210*, 108702. [CrossRef]
9. Maghool, F.; Arulrajah, A.; Ghorbani, B.; Horpibulsuk, S. Strength and permanent deformation properties of demolition wastes, glass, and plastics stabilized with foamed bitumen for pavement bases. *Constr. Build. Mater.* **2022**, *320*, 126108. [CrossRef]
10. Senanayake, M.; Arulrajah, A.; Maghool, F.; Horpibulsuk, S. Evaluation of rutting resistance and geotechnical properties of cement stabilized recycled glass, brick and concrete triple blends. *Transp. Geotech.* **2022**, *34*, 100755. [CrossRef]
11. Luo, W.; Sandanayake, M.; Zhang, G.; Tan, Y. Construction Cost and Carbon Emission Assessment of a Highway Construction—A Case towards Sustainable Transportation. *Sustainability* **2021**, *13*, 7854. [CrossRef]
12. Sherwani, A.; Usmani, J. Life cycle assessment of solar PV based electricity generation systems: A review. *Renew. Sustain. Energy Rev.* **2010**, *14*, 540–544. [CrossRef]
13. Chowdhury, M.S.; Rahman, K.S.; Chowdhury, T.; Nuthammachot, N.; Techato, K.; Akhtaruzzaman, M.; Tiong, S.K.; Sopian, K.; Amin, N. An overview of solar photovoltaic panels' end–of–life material recycling. *Energy Strategy Rev.* **2020**, *27*, 100431. [CrossRef]
14. Weckend, S.; Wade, A.; Heath, G.A. *End of Life Management: Solar Photovoltaic Panels*; National Renewable Energy Lab (NREL): Golden, CO, USA, 2016.
15. Jung, B.; Park, J.; Seo, D.; Park, N. Sustainable system for raw-metal recovery from crystalline silicon solar panels: From noble–metal extraction to lead removal. *ACS Sustain. Chem. Eng.* **2016**, *4*, 4079–4083. [CrossRef]
16. Tao, M.; Fthenakis, V.; Ebin, B.; Steenari, B.M.; Butler, E.; Sinha, P.; Corkish, R.; Wambach, K.; Simon, E.S. Major challenges and opportunities in silicon solar module recycling. *Prog. Photovolt. Res. Appl.* **2020**, *28*, 1077–1088. [CrossRef]
17. Lunardi, M.M.; Alvarez–Gaitan, J.P.; Bilbao, J.I.; Corkish, R. A Review of Recycling Processes for Photovoltaic Modules. In *Solar Panels and Photovoltaic Materials*; Intech Open: London, UK, 2018; pp. 9–27.
18. Chang, N.L.; Ho–Baillie, A.W.Y.; Vak, D.; Gao, M.; Green, M.A.; Egan, R.J. Manufacturing cost and market potential analysis of demonstrated roll–to–roll perovskite photovoltaic cell processes. *Sol. Energy Mater. Sol. Cells* **2018**, *174*, 314–324. [CrossRef]
19. James, T.; Goodrich, A.; Woodhouse, M.; Margolis, R.; Ong, S. *Building-Integrated Photovoltaics (BIPV) in the Residential Sector: An Analysis of Installed Rooftop System Prices*; National Renewable Energy Lab (NREL): Golden, CO, USA, 2011.
20. Ganescu, V.; Pascu, A. A study on prevalent factors behind efficiency deterioration of a typical low power solar panel. In Proceedings of the International Symposium on Microelectronics: International Microelectronics Assembly and Packaging Society, Orlando, FL, USA, 30 September–3 October 2013; pp. 384–388.
21. Paiano, A. Photovoltaic waste assessment in Italy. *Renew. Sustain. Energy Rev.* **2015**, *41*, 99–112. [CrossRef]
22. KEI. *Management Status and Improvement Plans of Waste Solar Panels*; Korea Environment Institute: Seoul, Korea, 2018.
23. Rai, A.K.; Singh, G.; Tiwari, A.K. Comparative study of soil stabilization with glass powder, plastic and e–waste: A review. *Mater. Today Proc.* **2020**, *32*, 771–776.
24. Gueli, A.M.; Pasquale, S.; Tanasi, D.; Hassam, S.; Lemasson, Q.; Moignard, B.; Pacheco, C.; Pichon, L.; Stella, G.; Politi, G. Weathering and deterioration of archeological glasses from late Roman Sicily. *Int. J. Appl. Glass Sci.* **2020**, *11*, 215–225. [CrossRef]
25. Rajmohan, K.V.S.; Ramya, C.; Viswanathan, M.R.; Varjani, S. Plastic pollutants: Effective waste management for pollution control and abatement. *Curr. Opin. Environ. Sci. Health* **2019**, *12*, 72–84. [CrossRef]
26. Shinzato, M.C.; Hypolito, R. Effect of disposal of aluminum recycling waste in soil and water bodies. *Environ. Earth Sci.* **2016**, *75*, 1–10. [CrossRef]
27. Borggaard, O.K.; Gimsing, A.L. Fate of glyphosate in soil and the possibility of leaching to ground and surface waters: A review. *J. Pest Manag. Sci. Former. Pestic. Sci.* **2008**, *64*, 441–456. [CrossRef]
28. Jaishankar, M.; Tseten, T.; Anbalagan, N.; Mathew, B.B.; Beeregowda, K.N. Toxicity, mechanism and health effects of some heavy metals. *Interdiscip. Toxicol.* **2014**, *7*, 60–72. [CrossRef]
29. Rosseland, B.O.; Eldhuset, T.D.; Staurnes, M. Environmental effects of aluminium. *Environ. Geochem. Health* **1990**, *12*, 17–27. [CrossRef] [PubMed]
30. Pandey, G.; Madhuri, S. Heavy metals causing toxicity in animals and fishes. *Res. J. Anim. Vet. Fish. Sci.* **2014**, *2*, 17–23.

31. Sevigné-Itoiz, E.; Gasol, C.M.; Rieradevall, J.; Gabarrell, X. Environmental consequences of recycling aluminum old scrap in a global market. *J. Resour. Conserv.* **2014**, *89*, 94–103. [CrossRef]
32. Elgharbawy, A.S.A.A. Review on corrosion in solar panels. *Int. J. Smart Grid* **2018**, *2*, 218–220.
33. Zhang, J.; Lv, F.; Ma, L.Y.; Yang, L.J. The status and trends of crystalline silicon PV module recycling treatment methods in Europe and China. *Adv. Mater. Res.* **2013**, *724–725*, 200–204. [CrossRef]
34. Pagnanelli, F.; Moscardini, E.; Granata, G.; Atia, T.A.; Altimari, P.; Havlik, T.; Toro, L. Physical and chemical treatment of end of life panels: An integrated automatic approach viable for different photovoltaic technologies. *Waste Manag.* **2017**, *59*, 422–431. [CrossRef]
35. Huang, B.; Zhao, J.; Chai, J.; Xue, B.; Zhao, F.; Wang, X. Environmental influence assessment of China's multi-crystalline silicon (multi-Si) photovoltaic modules considering recycling process. *Solar Energy* **2017**, *143*, 132–141. [CrossRef]
36. Elinwa, A.U.; Mbadike, E. The use of aluminum waste for concrete production. *J. Asian Archit. Build. Eng.* **2011**, *10*, 217–220. [CrossRef]
37. Mbadike, E.M.; Osadere, N. Effect of incorporation of aluminium waste in concrete matrix using different mix ratio and water cement ratio. *Biotechnol. Adv.* **2014**, *2*, 58–74. [CrossRef]
38. Mailar, G.; Sreedhara, B.; Manu, D.; Hiremath, P.; Jayakesh, K. Investigation of concrete produced using recycled aluminium dross for hot weather concreting conditions. *Resour. Effic. Technol.* **2016**, *2*, 68–80. [CrossRef]
39. Ardente, F.; Latunussa, C.E.; Blengini, G.A. Resource efficient recovery of critical and precious metals from waste silicon PV panel recycling. *Waste Manag.* **2019**, *91*, 156–167. [CrossRef] [PubMed]
40. Imteaz, M.A.; Ali, M.Y.; Arulrajah, A. Possible environmental impacts of recycled glass used as a pavement base material. *Waste Manag. Res.* **2012**, *30*, 917–921. [CrossRef]
41. Panditharadhya, B.J.; Sampath, V.; Mulangi, R.H.; Shankar, A.U.R. Mechanical properties of pavement quality concrete with secondary aluminium dross as partial replacement for ordinary portland cement. *IOP Conf. Ser. Mater. Sci. Eng.* **2018**, *431*, 032011. [CrossRef]
42. Idrees, M.; Chaudhary, H.; Akbar, A.; Mohamed, A.; Fathi, D. Effect of Silicon Carbide and Tungsten Carbide on Concrete Composite. *Materials* **2022**, *15*, 2061. [CrossRef]
43. Jain, S.; Joshi, Y.; Goliya, S. Design of rigid and flexible pavements by various methods & their cost analysis of each method. *Int. J. Eng. Res. Appl.* **2013**, *3*, 119–123.
44. Schmidt, A.; Saia, W. Alkali–aggregate reaction tests on glass used for exposed aggregate wall panel work. *ACI Mater. J.* **1963**, *60*, 1235–1236.
45. Polley, C.; Cramer, S.M.; Cruz, R.V.d.l. Potential for using waste glass in Portland cement concrete. *J. Mater. Civ. Eng.* **1998**, *10*, 210–219. [CrossRef]
46. Zheng, K. Pozzolanic reaction of glass powder and its role in controlling alkali–silica reaction. *J. Cem. Concr. Compos.* **2016**, *67*, 30–38. [CrossRef]
47. Topcu, I.B.; Canbaz, M. Properties of concrete containing waste glass. *Cem. Concr. Res.* **2004**, *34*, 267–274. [CrossRef]
48. Terro, M.J. Properties of concrete made with recycled crushed glass at elevated temperatures. *Build. Environ.* **2006**, *41*, 633–639. [CrossRef]
49. Sopov, V.; Korkh, O.; Izbash, M.Y. A study of the alkali–silica reaction in recycled glass concrete. *IOP Conf. Ser. Mater. Sci. Eng.* **2020**, *907*, 012062. [CrossRef]
50. Yuan, Z.; Jia, Y.; Materials, B. Mechanical properties and microstructure of glass fiber and polypropylene fiber reinforced concrete: An experimental study. *J. Constr.* **2021**, *266*, 121048. [CrossRef]
51. Omoding, N.; Cunningham, L.S.; Lane-Serff, G.F. Effect of using recycled waste glass coarse aggregates on the hydrodynamic abrasion resistance of concrete. *Constr. Build. Mater.* **2021**, *268*, 121177. [CrossRef]
52. Turgut, P.; Yahlizade, E. Research into concrete blocks with waste glass. *Int. J. Civ. Environ. Eng.* **2009**, *3*, 186–192.
53. de Castro, S.; de Brito, J. Evaluation of the durability of concrete made with crushed glass aggregates. *J. Clean. Prod.* **2013**, *41*, 7–14. [CrossRef]
54. Keryou, A.B.; Ibrahim, G.J. Effect of using windows waste glass as coarse aggregate on some properties of concrete. *Eng. Technol. J.* **2014**, *32*.
55. Serpa, D.; De Brito, J.; Pontes, J. Concrete made with recycled glass aggregates: Mechanical performance. *ACI Mater. J.* **2015**, *112*, 29–39. [CrossRef]
56. Park, S.B.; Lee, B.C.; Kim, J.H. Studies on mechanical properties of concrete containing waste glass aggregate. *Cem. Concr. Res.* **2004**, *34*, 2181–2189. [CrossRef]
57. Shayan, A.; Xu, A. Stainless steel reinforcement to increase durability in concrete structures. In Proceedings of the Australian Small Bridges Conference, Sydney, NSW, Australia, 12–13 October 2005.
58. Ismail, Z.Z.; Al-Hashmi, E.A. Recycling of waste glass as a partial replacement for fine aggregate in concrete. *Waste Manag.* **2009**, *29*, 655–659. [CrossRef] [PubMed]
59. Kou, S.C.; Poon, C.S. Properties of self-compacting concrete prepared with recycled glass aggregate. *Cem. Concr. Compos.* **2009**, *31*, 107–113. [CrossRef]
60. Du, H.; Tan, K.H. Effect of particle size on alkali–silica reaction in recycled glass mortars. *Constr. Build. Mater.* **2014**, *66*, 275–285. [CrossRef]

61. Dhir, R.; Dyer, T.; Tang, M. Expansion due to alkali–silica reaction (ASR) of glass cullet used in concrete. In Proceedings of the 2003 International Symposium "Advances in Waste Management and Recycling", Dundee, UK, 9–11 September 2003; pp. 751–760.
62. Du, H.; Tan, K.H. Waste glass powder as cement replacement in concrete. *J. Adv. Concr. Technol.* **2014**, *12*, 468–477. [CrossRef]
63. Shao, Y.; Lefort, T.; Moras, S.; Rodriguez, D. Studies on concrete containing ground waste glass. *Cem. Concr. Res.* **2000**, *30*, 91–100. [CrossRef]
64. Maas, A.J.; Ideker, J.H.; Juenger, M.C. Alkali silica reactivity of agglomerated silica fume. *J. Cem. Concr. Res.* **2007**, *37*, 166–174. [CrossRef]
65. Pan, Z.; Tao, Z.; Murphy, T.; Wuhrer, R. High temperature performance of mortars containing fine glass powders. *J. Clean. Prod.* **2017**, *162*, 16–26. [CrossRef]
66. Doni, A.; Dughiero, F. Electrothermal heating process applied to c-Si PV recycling. In Proceedings of the 2012 38th IEEE Photovoltaic Specialists Conference, Austin, TX, USA, 3–8 June 2012; pp. 757–762.
67. Shayan, A.; Xu, A. Value-added utilisation of waste glass in concrete. *Cem. Concr. Res.* **2004**, *34*, 81–89. [CrossRef]
68. Shi, C.; Wu, Y.; Riefler, C.; Wang, H. Characteristics and pozzolanic reactivity of glass powders. *Cem. Concr. Res.* **2005**, *35*, 987–993. [CrossRef]
69. Schwarz, N.; Cam, H.; Neithalath, N. Influence of a fine glass powder on the durability characteristics of concrete and its comparison to fly ash. *Cem. Concr. Compos.* **2008**, *30*, 486–496. [CrossRef]
70. Khmiri, A.; Samet, B.; Chaabouni, M. A cross mixture design to optimise the formulation of a ground waste glass blended cement. *Constr. Build. Mater.* **2012**, *28*, 680–686. [CrossRef]
71. AL–Zubaid, A.B.; Shabeeb, K.M.; Ali, A.I. Study the effect of recycled glass on the mechanical properties of green concrete. *Energy Procedia* **2017**, *119*, 680–692. [CrossRef]
72. Taha, B.; Nounu, G. Properties of concrete contains mixed colour waste recycled glass as sand and cement replacement. *Constr. Build. Mater.* **2008**, *22*, 713–720. [CrossRef]
73. Wang, H.-Y.; Huang, W.-L. Durability of self-consolidating concrete using waste LCD glass. *Constr. Build. Mater.* **2010**, *24*, 1008–1013. [CrossRef]
74. Paktiawal, A.; Alam, M. An experimental study on effect of aluminum composite panel waste on performance of cement concrete. *Ain Shams Eng. J.* **2021**, *12*, 83–98. [CrossRef]
75. Hay, R.; Ostertag, C.P. On utilization and mechanisms of waste aluminium in mitigating alkali-silica reaction (ASR) in concrete. *J. Clean. Prod.* **2019**, *212*, 864–879. [CrossRef]
76. Rahim, N.L.; Ibrahim, N.M.; Salehuddin, S.; Che Amat, R.; Mohammed, S.A.; Hibadullah, C.R. *The Utilization of Aluminum Waste as Sand Replacement in Concrete*; Trans Tech Publications: Bäch, Switzerland, 2014; Volume 594.
77. Muwashee, R.S.; Al-Jameel, H.A.; Jabal, Q.A. Investigating the behavior of concrete and mortar reinforced with Aluminum waste strips. *Int. J. Eng. Technol.* **2018**, *7*, 211–213. [CrossRef]
78. Dulsang, N.; Kasemsiri, P.; Posi, P.; Hiziroglu, S.; Chindaprasirt, P. Characterization of an environment friendly lightweight concrete containing ethyl vinyl acetate waste. *Mater. Des.* **2016**, *96*, 350–356. [CrossRef]
79. Khan, K.A.; Ahmad, I.; Alam, M. Effect of ethylene vinyl acetate (EVA) on the setting time of cement at different temperatures as well as on the mechanical strength of concrete. *Arab. J. Sci. Eng. Technol. J.* **2019**, *44*, 4075–4084. [CrossRef]
80. Azadmanesh, H.; Hashemi, S.A.H.; Ghasemi, S.H. The effect of styrene-butadiene rubber and ethylene vinyl acetate polymers on the mechanical properties of Engineered Cementitious Composites. *Compos. Commun.* **2021**, *24*, 100656. [CrossRef]
81. Ren, Q.; Jiang, Z.; Li, H.; Zhu, X.; Chen, Q. Fresh and hardened properties of self-compacting concrete using silicon carbide waste as a viscosity–modifying agent. *Constr. Build. Mater.* **2019**, *200*, 324–332. [CrossRef]
82. Małek, M.; Jackowski, M.; Życiński, W.; Łasica, W.; Owczarek, M.J.M.T. Influence of silicone carbide additions on the mechanical properties of concrete. *Mater. Technol.* **2020**, *54*, 595–599.
83. Jiang, Z.; Ren, Q.; Li, H.; Chen, Q. Silicon carbide waste as a source of mixture materials for cement mortar. *Front. Environ. Sci. Eng.* **2017**, *11*, 1–8. [CrossRef]
84. Fernández, L.J.; Ferrer, R.; Aponte, D.; Fernandez, P. Recycling silicon solar cell waste in cement–based systems. *Sol. Energy Mater. Sol. Cells* **2011**, *95*, 1701–1706. [CrossRef]
85. Kandhal, P.S. *Waste Materials in Hot Mix Asphalt—An Overview*; Report No: NCAT Report No. 92–6; National Center for Asphalt Technology: Auburn, AL, USA, 1992.
86. Su, N.; Chen, J. Engineering properties of asphalt concrete made with recycled glass. *Resour. Conserv. Recycl.* **2002**, *35*, 259–274. [CrossRef]
87. Arabani, M. Effect of glass cullet on the improvement of the dynamic behaviour of asphalt concrete. *Constr. Build. Mater.* **2011**, *25*, 1181–1185. [CrossRef]
88. Issa, Y. Effect of adding crushed glass to asphalt mix. *Arch. Civ. Eng.* **2016**, *62*, 35–44. [CrossRef]
89. Salem, Z.T.A.; Khedawi, T.S.; Baker, M.B.; Abendeh, R. Effect of waste glass on properties of asphalt concrete mixtures. *Jordan J. Civ. Eng.* **2017**, *11*, 117–131.
90. Moses Ogundipe, O.; Segun Nnochiri, E. Evaluation of the effects of waste glass in asphalt concrete using the Marshall test. *Eng. Rev.* **2020**, *40*, 24–33. [CrossRef]
91. Marandi, S.M.; Ghasemi, M. Laboratory studies of the effect of recycled glass powder additive on the properties of polymer modified asphalt binders. *Int. J. Eng.* **2013**, *26*, 1183–1190.

92. Lachance-Tremblay, É.; Vaillancourt, M.; Perraton, D. Evaluation of the impact of recycled glass on asphalt mixture performances. *Road Mater. Pavement Des.* **2016**, *17*, 600–618. [CrossRef]
93. Shafabakhsh, G.; Sajed, Y. Investigation of dynamic behavior of hot mix asphalt containing waste materials; case study: Glass cullet. *Case Stud. Constr. Mater.* **2014**, *1*, 96–103. [CrossRef]
94. Federal Highway Administration, U.S.D.O.T. User Guidelines for Waste and Byproduct Materials in Pavement Construction Waste Glass. Available online: http://www.fhwa.dot.gov/publications/research/infrastructure/pavements/97148/068.cfm (accessed on 5 July 2022).
95. Wu, S.; Yang, W.; Xue, Y. Preparation and properties of glass-asphalt concrete. *Wuham China Key Lab. Silic. Mater. Sci. Eng. Minist. Educ. Wuham Univ. Technol.* **2004**. Available online: https://www.researchgate.net/publication/237644653_Preparation_and_Properties_of_Glass-asphalt_Concrete (accessed on 5 July 2022).
96. Rahma, A.; El Naber, N.; Issa Ismail, S. Effect of glass powder on the compression strength and the workability of concrete. *Cogent Eng.* **2017**, *4*, 1373415. [CrossRef]
97. Jain, S.; Sharma, T.; Gupta, A.K. End–of–life management of solar PV waste in India: Situation analysis and proposed policy framework. *Renew. Sustain. Energy Rev.* **2022**, *153*, 111774. [CrossRef]
98. Jiang, W. Logistic Challenges of Large-Scale Solar Projects. Available online: https://www.solarfeeds.com/mag/logistic--challenges--of--large--scale--solar--projects/ (accessed on 20 January 2022).
99. Sharma, H.B.; Vanapalli, K.R.; Barnwal, V.K.; Dubey, B.; Bhattacharya, J. Evaluation of heavy metal leaching under simulated disposal conditions and formulation of strategies for handling solar panel waste. *Sci. Total Environ.* **2021**, *780*, 146645. [CrossRef]
100. Gautam, A.; Shankar, R.; Vrat, P. End–of–life solar photovoltaic e–waste assessment in India: A step towards a circular economy. *Sustain. Prod. Consum.* **2021**, *26*, 65–77. [CrossRef]

MDPI AG
Grosspeteranlage 5
4052 Basel
Switzerland
Tel.: +41 61 683 77 34

Sustainability Editorial Office
E-mail: sustainability@mdpi.com
www.mdpi.com/journal/sustainability

www.ingramcontent.com/pod-product-compliance
Lightning Source LLC
Chambersburg PA
CBHW041406210926
51850CB00013B/3193
* 9 7 8 3 7 2 5 8 3 8 3 6 3 *